Advanced Analysis and Design of Steel Frames

Advanced Analysis and Design of Steel Frames

Guo-Qiang Li
Tongji University, PR China

Jin-Jun Li
Shanghai Inspire Consultants Ltd, PR China

John Wiley & Sons, Ltd

Other Wiley Editorial Offices

John Wiley & Sons Inc., 111 River Street, Hoboken, NJ 07030, USA

Jossey-Bass, 989 Market Street, San Francisco, CA 94103-1741, USA

Wiley-VCH Verlag GmbH, Boschstr. 12, D-69469 Weinheim, Germany

John Wiley & Sons Australia Ltd, 42 McDougall Street, Milton, Queensland 4064, Australia

John Wiley & Sons (Asia) Pte Ltd, 2 Clementi Loop #02-01, Jin Xing Distripark, Singapore 129809

John Wiley & Sons Canada Ltd, 6045 Freemont Blvd, Mississauga, ONT, Canada L5R 4J3

Wiley also publishes its books in a variety of electronic formats. Some content that appears in print may not be
available in electronic books.

Anniversary Logo Design: Richard J. Pacifico

British Library Cataloguing in Publication Data

A catalogue record for this book is available from the British Library

ISBN 978-0-470-03061-5 (HB)

Typeset in 10/12 pt Times by Thomson Digital
Printed and bound in Great Britain by Antony Rowe Ltd, Chippenham, Wiltshire
This book is printed on acid-free paper responsibly manufactured from sustainable forestry
in which at least two trees are planted for each one used for paper production.

Contents

Preface

With advantages in high strength, good ductility and fast fabrication and erection, steel frames are widely used for industrial, commercial and residential buildings. Currently, the common procedures for the structural design of steel frames worldwide are: (1) to conduct linearly elastic structural analysis to determine the resultants of structural members under various actions; and (2) to check the resultants against the limit states of structural members specified in the codes, based on the reliability theory for the limit state of structural members. However, drawbacks of the current approach exist in the following two aspects. Firstly, the normal elastic analysis of steel frames takes account of only typical flexural, shear and axial deformations of frame components, and cannot consider effects such as shear deformation of joint-panels, flexibility of beam-to-column connections, brace buckling and nonprismatic sections (tapered members). Also, material and geometric nonlinearities and imperfection (residual stress and initial geometric imperfection) cannot be involved in linearly elastic analysis. Secondly, the structural members of a frame is generally in an elasto-plastic state when they approach limit states, whereas the member resultants used in limit state check are taken from the linearly elastic analysis of the frame. The incompatibility of the member resultants obtained in structural analysis for limit state check and those in real limit state results in uncertain member reliability.

To overcome the drawback mentioned above, the concept of Advanced Design has been proposed. Second-order inelastic analysis is used in Advanced Design of steel frames to determine the structural ultimate capacities, which considers all the effects significant for structural nonlinear behavior and is termed as *advanced analysis*.

A large amount of achievements have been made in the past two decades on advanced analysis of steel frames. However, in the view of structural design, the reliability evaluation of structural systems should be incorporated into advanced analysis to make the steel frames designed have certain system reliability. Such structural design with definite system reliability is termed as *advanced design*. Unfortunately, little progress was reported in this area. In this book, a concept of reliability-based advanced design is developed and proposed for steel frames.

The first author of this book began to study the theory of structural reliability design in 1982 when he was in Chongqing Institute of Architecture and Engineering for his Master degree and began to study the theory of advanced analysis for steel frames in 1985 when he was in Tongji University for his PhD degree. The main contents of this book are actually the summarization of our research achievements in structural reliability design and advanced analysis of steel frames for over 20 years, including the contribution from Ms. Yushu Liu and Ms. Xing Zhao, who are the former PhD students of the first author.

Two parts are included in this book. Part One is advanced analysis for beam (prismatic beam, tapered beam and composite beam), column, joint-panel, connection, brace, and shear beam elements in steel frames, and methods for stability analysis, nonlinear analysis and seismic analysis of steel frames. Part Two is reliability-based advanced design for steel portal frames and multi-storey frames.

We are grateful for the advice from Prof. Jihua Li and Prof. Zuyan Shen who supervised the first author's Master and PhD degree study and guide him to an attractive field in structural engineering.

We wish also to thank the following persons in helping to prepare the manuscript of the book, including typing and checking the text and drawing all the figures. They are Ms. Yamei He, Mr. Wenlong Shi, Mr. Peijun Wang, Mr. Baolin Hu, Mr. Wubo Li, Mr. Dazhu Hu, Mr. Hui Gao, Mr. Chaozhen Chen and Mr. Yang Zhang. Without their assistance, the timely outcome of the book would have been difficult.

We also want to thank our families for their continuous support and understanding in our research presented in this book for many years at Tongji University.

Finally, we want to thank the National Natural Science Foundation of China for continuous support on the research, many results of which are reported in this book.

Guo-Qiang Li and Jin-Jun Li
September 2006

Symbols

Unless the additional specification appears in the text, the physical or mathematical definitions of the symbols in this book are as follows:

1 Variable

A	Sectional area
A_f	Flange area of H- or box-shape sections
A_w	Web area of H- or box-shape sections
D	Displacement
D_c	Sway stiffness of frame columns (also termed to as the D-value of frame columns)
E	Elastic tensile modulus
F	Load, force
f	Force
f_y	Yielding stress
G	Elastic shear modulus
H	Structural height
H_i	Height of the ith story
h_c	Sectional height of column
h_g	Sectional height of beam
I	Inertial moment
i	Linear stiffness $\left(i = \dfrac{I}{l}\right)$
K	Stiffness
k	Stiffness, element of stiffness matrix
l	Length of beam, column or brace
M	Moment
M_p	Ultimate yielding moment (plastic moment)
M_{pN}	Ultimate yielding moment accounting effect of axial force
M_s	Initial yielding moment
M_{sN}	Initial yielding moment accounting effect of axial force
M_γ	Shear moment
$M_{\gamma p}$	Shear yielding moment
N	Axial force (positive in tension and negative in compression)
$N_{c\gamma}$	Buckling load in axial compression
N_E	Euler load
N_p	Yielding load in axial compression
P	Load
Q	Shear force

q	Uniformly distributed load, hardening factor
q_0	Amplitude of distributed triangle load
R	Recovery force parameter
t	Time, duration
t_f	Flange thickness of H- or box-shape sections
t_p	Thickness of joint panel
t_w	Web thickness of H- or box-shape sections
u	Horizontal displacement of frame floor
u_g	Horizontal ground movement
V	Shear force between floors
W	Sectional modulus
w	Vertical displacement of nodes
y	Deflection
Γ	Yielding function
γ	Shear strain, shear deformation of joint panel
δ	Displacement
Δ	Story drift
η	Effective length factor
θ	Rotation
λ	Slenderness
μ	Shear shape factor of sections
μ_p	Mean value of failure probability
ζ	Damp ratio
σ	Normal stress
σ_s	Yielding stress
σ_p	Deviation of failure probability
ε	Normal strain
ε_y	Yielding strain
τ	Shear stress
τ_s	Shear yielding stress
φ	Rotation of beam-to-column connection (relative rotation between adjacent beam and column)
Φ	Curvature parameter
ϕ	Curvature
ϕ_p	The elastic curvature corresponding to M_p $\phi_p = \left\| \dfrac{M_p}{EI} \right\|$
ϕ_{pN}	The elastic curvature corresponding to M_{pN} $\phi_{pN} = \left\| \dfrac{M_{pN}}{EI} \right\|$
$\{\phi\}$	Vector of vibration modes
χ_p	Plastic shape factor of sections
ω	Circular frequency
ρ_σ	Correlation factor of material yielding strength
ρ_w	Correlation factor of member section modulus

2 Superscript
T	Transfer of vector or matrix

3 Subscript
b	Brace
c	Column
e	Elastic
G	Geometrical nonlinearity
g	Beam
H	Horizontal
k	Number of frame floor

o	Reference point of frame floor
p	Plastic, elasto-plastic, ultimate yielding
s	Yielding, initial yielding
t	In tension
u	Unloading
u, v, w	Global coordinate axis
x, y, z	Local coordinate axis
γ	Joint panel
1, 2	Elemental ends without joint panel
i, j	Elemental ends with joint panel

4 Arithmetic operator

d	Differential, incremental
Δ	Incremental
$\sum$	Summation

Part One

Advanced Analysis
of Steel Frames

1 Introduction

1.1 TYPE OF STEEL FRAMES

Steel frames have been widely used in single-storey, low-rise industrial buildings (Figure 1.1(a)), power plants (Figure 1.1(b)), ore mines (Figure 1.1(c)), oil and gas offshore platforms (Figure 1.1(d)) and multi-storey, high-rise buildings (Figure 1.1(e)). The discussions contained in this book will be mainly on, but not limited to, the steel frames used in buildings. According to the elevation view, steel frames used in low-rise and high-rise buildings can be categorized into (1) pure frame (Figure 1.2), (2) concentrically braced frame (Figure 1.3), (3) eccentrically braced frame (Figure 1.4) and (4) frame tube (Figure 1.5).

A pure frame has good ductility with not so good sway stiffness for multi-storey buildings. Strengthened with braces to pure frame, the sway stiffness of a concentrically braced frame is much improved. However, its capacity against lateral loading will be easily reduced if braces in compression are buckled, which is unfavourable under conditions such as earthquakes. An eccentrically braced frame is a compromise in sway stiffness and capacity between the pure frame and the concentrically braced frame. Buckling of braces in compression can be prevented by introducing shear yielding of an eccentric shear beam, which provides good energy-consuming performance to the eccentrically braced frame (Li, 2004). A frame tube is actually a frame group with very close columns, where because of small span and relatively large stiffness of steel beams, columns in the peripheral bend as a thin-walled tube to resist sway loads. Because it has good sway stiffness and load capacity, the frame tube is generally used in high-rise buildings (Council on Tall Buildings, 1979).

1.2 TYPE OF COMPONENTS FOR STEEL FRAMES

For convenience of fabrication, the prismatic components with uniform section (Figure 1.6(a)) are usually used for steel frames. However, to reduce steel consumption, tapered beams and columns (Figure 1.6(b)) are normally employed for steel portal frames (Figure 1.7) to keep relatively uniform strength to resist the dominant vertical loads (Li, 2001). In multi-storey steel buildings, the cast in-site concrete is widely used for floor slabs (Figure 1.8). To utilize the capacity of concrete slabs, a composite beam can be designed, and with headed shear studs, the composite action between concrete slabs and steel beams can be obtained (Nethercot, 2003), as shown in Figure 1.9.

Figure 1.1 Application of steel frames: (a) single-storey industrial building; (b) power plant; (c) ore miners tower; (d) oil and gas offshore platform; (e) high-rise building

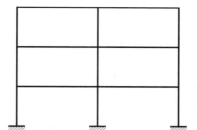

Figure 1.2 Pure frame

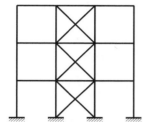 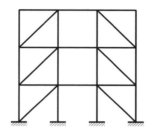 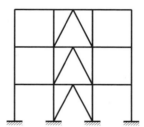

Figure 1.3 Concentrically braced frames

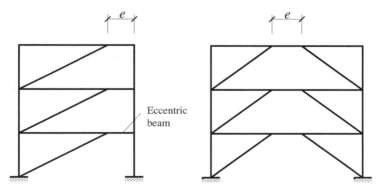

Figure 1.4 Eccentrically braced frames

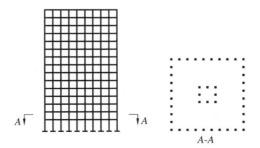

Figure 1.5 Frame-tube structures

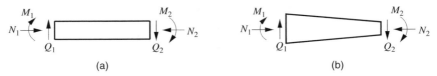

Figure 1.6 (a) Prismatic and (b) tapered members in steel frames

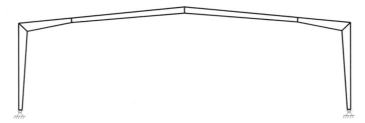

Figure 1.7 Steel portal frame with tapered members

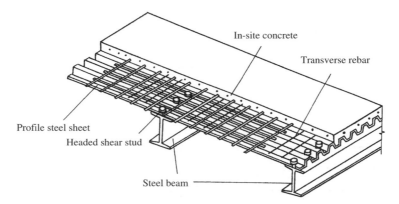

Figure 1.8 Floor system in multi-storey, high-rise steel buildings

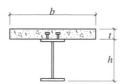

Figure 1.9 Section of a steel–concrete composite beam

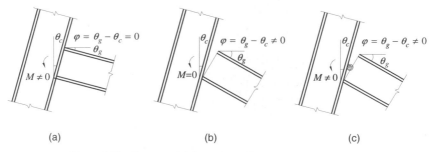

Figure 1.10 Forces and deformations of beam–column connections

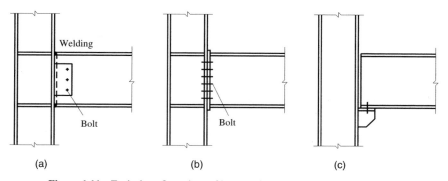

Figure 1.11 Typical configurations of beam–column connections in steel frames

1.3 TYPE OF BEAM–COLUMN CONNECTIONS

According to the moment–curvature characteristic, connections of beam to column in steel frames can be categorized into (Chen, Goto and Liew, 1996)

(1) rigid connection, where no relative rotation occurs between adjacent beams and columns and bending moment can be transferred fully from a beam to the neighbouring column (Figure 1.10(a));

(2) pinned connection, where relative rotation occurs and bending moment cannot be transferred at all (Figure 1.10(b));

(3) semi-rigid connection, where relative rotation occurs and bending moment can be transferred partially (Figure 1.10(c)).

Some typical beam-to-column connection configurations are illustrated in Figure 1.11(a)–(c) for rigid, pinned and semi-rigid connections, respectively. Semi-rigid connections are often engineering options in the application of steel frames.

1.4 DEFORMATION OF JOINT PANEL

Joint panel is the connection zone of beam and column members in steel frames, as shown in Figure 1.12. Subjected to reaction forces of the beam and column ends adjacent to a joint panel, three possible deformations can occur in the joint panel (Figure 1.13): (1) stretch/contract, (2) bending and (3) shear deformations.

Due to restraint of adjacent beams, stretch/contract and bending deformations of the joint panel are very small and can be ignored. Shear deformation is therefore dominant for the joint panel and an experimental deformation of the joint panel is shown in Figure 1.14 (Li and Shen, 1998).

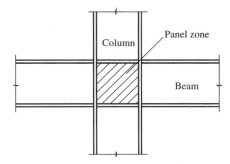

Figure 1.12 Joint panel in steel frames

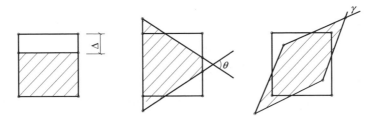

Figure 1.13 Deformations of the joint panel

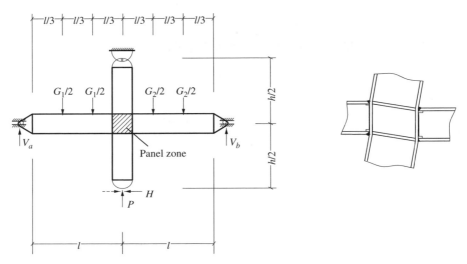

Figure 1.14 Joint panel in the experiment

1.5 ANALYSIS TASKS AND METHOD FOR STEEL FRAME DESIGN

The analysis tasks for the steel frame design include (1) linearly elastic frame analysis to determine resultant forces and deformation of frame members, (2) elastic stability analysis of the frame under vertical loads, (3) nonlinear frame analysis to determine the load-bearing capacity, and (4) elastic and elasto-plastic seismic frame analysis (Liu and Ge, 2005). The first analysis task is the most common practice in structural design, and the latter three analyses will be discussed in this book.

Traditional structure analysis methods such as the force method, displacement method and moment distribution method can be used in linearly elastic analysis of steel frames. However, for the frames with many storeys and bays when nonlinear analysis is performed, traditional analysis methods are not applicable. With the development of computer hardware and software, the matrix analysis method based on finite elements has been widely employed in structural engineering. For the analysis of steel frames, the procedures of the matrix analysis method based on finite component elements are (Bath, 1996)

(1) *Discretizing frame*. Generally, the whole beam, column or brace component can be represented with one element. In elasto-plastic analysis, subdivision is necessary if the plastic deformation occurs within two ends of beams. If effects of the joint panel are necessarily considered in the analysis, the joint panel should be represented with an independent element.

(2) *Establishing elemental stiffness equations*. Elemental stiffness equation is the relationship between nodal forces and deformations of the element, which can be expressed with the matrix and vector

equation as $[k]\{\delta\} = \{f\}$, where $[k]$ is the elemental stiffness matrix, and $\{\delta\}$ and $\{f\}$ are nodal deformation and force vectors. In linearly elastic analysis, $[k]$ is a constant matrix, whereas in nonlinear analysis it relates to the history of elemental force and deformation.

(3) *Assembling the global stiffness equation.* Elemental stiffness equations can be assembled into a global stiffness equation through incidence between the local node number of the elements and the global node number of the frame for the analysis, and with nodal force equilibrium.

(4) *Calculating nodal deformation.* With consideration of boundary conditions, nodal deformation in global coordinates can be solved from the global stiffness equation.

(5) *Determining elemental resultant.* Nodal deformation in global coordinates can be transformed to that in local coordinates, namely elemental deformation. Then the elemental resultant can be calculated using the elemental stiffness equation with the given elemental deformation.

It can be found from the above procedures that the key step in finite element analysis of steel frames is the development of the elemental stiffness equation because other steps are standard and commonplace in the finite element method.

1.6   DEFINITION OF ELEMENTS IN STEEL FRAMES

The following elements are defined in this book for the analysis of steel frames (Li and Shen, 1998):

(1) *Beam element.* A beam element is often subjected to uniaxial bending moment and minor axial force with negligible axial deformation. Generally, beam members in steel frames can be represented with the beam element due to restraints of floor slabs or floor braces. In addition, column members in steel frames can also be represented with the beam element if the axial deformation can be ignored. It should be noted that, although axial deformation is excluded, effects of axial force on bending stiffness can be involved in the beam element.

(2) *Column element.* A column element is usually subjected to uniaxial or biaxial bending moment and significant axial force. Column members in steel frames can be represented with the column element, and the beam in steel frames can also be represented with the column element if effects of axial deformation are considered. In addition, braces, for example eccentric braces, can be treated as column elements if buckling is precluded.

(3) *Brace element.* A brace element is subjected to no more than axial force. Brace members in steel frames are dominated by axial forces and can be represented with the brace element.

(4) *Shear beam element.* It is a special beam element where shear deformation and shear yielding failure are dominant. An eccentric beam in eccentrically braced frames should be represented with the shear beam element.

(5) *Joint-panel element.* It is a special element to represent the shear deformation of the joint panel in the beam–column connection zone.

2 Elastic Stiffness Equation of Prismatic Beam Element

2.1 GENERAL FORM OF EQUATION

Beam element is one of the basic element types in finite element analysis of frame systems, the elastic stiffness of which in commonplace can be found in many textbooks on structural analysis (McGuire, Gallagher and Ziemian, 1999; Norris, Wilbur and Utku, 1976). However, the effects of shear deformation and axial force on the stiffness of beam elements were seldom considered simultaneously in previous investigations (Tranberg, Meek and Swannell, 1976).

For steel-framed systems, simultaneous effects of shear deformation and axial force on the behaviour of beam elements cannot be ignored in certain cases (Li and Shen, 1995). This section describes the derivation of elastic stiffness equations from the differential equilibrium equation, for the beam elements including the above two effects.

2.1.1 Beam Element in Tension

The nodal forces and displacements of beam elements in tension are illustrated in Figure 2.1. Under the simultaneous action of moment, shear force and axial tension force, element deflection y consists of the portion induced by bending deformation y_M and that by shear deformation y_Q, i.e.

$$y = y_M + y_Q. \tag{2.1}$$

The curvature of the element caused by bending is

$$y_M'' = -\frac{M}{EI}, \tag{2.2}$$

where E is the elastic modulus, I is the moment of inertia of the cross section and M is the cross-sectional moment given by

$$M = M_1 - Q_1 z - Ny. \tag{2.3}$$

The work done by shear in the differential element is (see Figure 2.2; Timoshenko and Gere, 1961)

$$dW_Q = \frac{1}{2} Q \, dy_Q. \tag{2.4a}$$

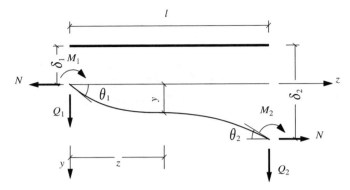

Figure 2.1   Forces and deformations of a beam element

and the shear strain energy is

$$dU_Q = \frac{\mu Q^2}{2GA}\, dz, \tag{2.4b}$$

where G is the elastic shear modulus, A is the area of the cross section, Q is the shear force of the cross section and μ is the shear shape factor of the cross section, considering effects of uneven distribution of shear deformation over the cross section, as shown in Figure 2.3.

By energy theory, $dW_Q = dU_Q$, one can derive from Equation (2.4) that

$$y'_Q = \frac{\mu Q}{GA} = \frac{\mu}{GA}\frac{dM}{dz}. \tag{2.5}$$

Substituting Equation (2.3) into Equation (2.5) yields

$$y'_Q = \frac{\mu}{GA}(-Q_1 - Ny'), \tag{2.6}$$

and differentiating Equation (2.6) once gives

$$y''_Q = -\frac{\mu N}{GA}y''. \tag{2.7}$$

Combining Equations (2.2) and (2.7), one has

$$y'' = -\frac{M_1 - Q_1 z - Ny}{EI} - \frac{\mu N}{GA}y''. \tag{2.8}$$

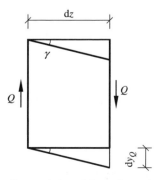

Figure 2.2   Shear and shear deformation on beam sections

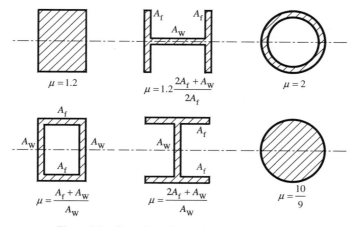

Figure 2.3   Shear shape factor of various beam sections

Let

$$\eta = 1 + \frac{\mu N}{GA},$$
(2.9)

$$\alpha^2 = \frac{N}{\eta EI},$$
(2.10)

and then rewrite Equation (2.8) as

$$y'' - \alpha^2 y = -\frac{M_1 - Q_1 z}{\eta EI}.$$
(2.11)

The solution of Equation (2.11) is

$$y = a \cosh \alpha z + b \sinh \alpha z + \frac{M_1 - Q_1 z}{N},$$
(2.12)

where a and b are the unknown coefficients depending on boundary conditions. Boundary conditions of the beam element are

- for $z = 0$:

$$y = 0,$$
(2.13a)

$$y' = y'_M + y'_Q = \theta_1 + \frac{\mu}{GA}(-Q_1 - Ny') \rightarrow y' = \frac{1}{\eta}\left(\theta_1 - \frac{\mu Q_1}{GA}\right);$$
(2.13b)

- for $z = l$:

$$y = \delta_2 - \delta_1,$$
(2.13c)

$$y' = y'_M + y'_Q = \theta_2 + \frac{\mu}{GA}(-Q_1 - Ny') \rightarrow y' = \frac{1}{\eta}\left(\theta_2 - \frac{\mu Q_1}{GA}\right).$$
(2.13d)

Substituting Equation (2.12) into Equations (2.13a)–(2.13d) yields four simultaneous linear equations in terms of a, b, θ_1 and θ_2, i.e.

$$a + \frac{M_1}{N} = 0, \tag{2.14a}$$

$$b\alpha - \frac{Q_1}{N} = \frac{1}{\eta}\left(\theta_1 - \frac{\mu Q_1}{GA}\right), \tag{2.14b}$$

$$a \cosh \alpha l + b \sinh \alpha l + \frac{M_1 - Q_1 l}{N} = \delta_2 - \delta_1, \tag{2.14c}$$

$$a\alpha \sinh \alpha l + b\alpha \cosh \alpha l - \frac{Q_1}{N} = \frac{1}{\eta}\left(\theta_2 - \frac{\mu Q_1}{GA}\right). \tag{2.14d}$$

Then, a and b are obtained from Equations (2.14a) and (2.14c) as

$$a = -\frac{M_1}{N}, \tag{2.15a}$$

$$b = \frac{1}{\sinh \alpha l}\left[\frac{M_1}{N}(\cosh \alpha l - 1) + \frac{Q_1 l}{N} + (\delta_2 - \delta_1)\right]. \tag{2.15b}$$

Substituting Equations (2.15a) and (2.15b) into Equations (2.14b) and (2.14d) yields

$$\theta_1 = \frac{\eta \alpha l}{\sinh \alpha l}\frac{M_1}{Nl}(\cosh \alpha l - 1) + \left(\frac{\eta \alpha l}{\sinh \alpha l} - 1\right)\frac{Q_1}{N} + \frac{\eta \alpha l}{\sinh \alpha l}\frac{\delta_2 - \delta_1}{l}, \tag{2.16a}$$

$$\theta_2 = \eta \alpha l \frac{M_1}{Nl}\frac{1 - \cosh \alpha l}{\sinh \alpha l} + \left(\frac{\eta \alpha l \cosh \alpha l}{\sinh \alpha l} - 1\right)\frac{Q_1}{N} + \frac{\eta \alpha l \cosh \alpha l}{\sinh \alpha l}\frac{\delta_2 - \delta_1}{l}. \tag{2.16b}$$

The equilibrium of elemental moments can be written as

$$M_1 + M_2 - Q_1 l - N(\delta_2 - \delta_1) = 0,$$

by which one obtains

$$Q_1 = \frac{M_1 + M_2}{l} - N\frac{\delta_2 - \delta}{l}. \tag{2.17}$$

From Equation (2.10), axial force can be expressed as

$$N = \alpha^2 \eta EI = \eta EI \frac{(\alpha l)^2}{l^2}. \tag{2.18}$$

Substituting Equation (2.17) into Equations (2.16a) and (2.16b) yields

$$\theta_1 - \frac{\delta_2 - \delta_1}{l} = -\frac{\sinh \alpha l - \eta \alpha l \cosh \alpha l}{\eta(\alpha l)^2 \sinh \alpha l}\frac{M_1 l}{EI} - \frac{\sinh \alpha l - \eta \alpha l}{\eta(\alpha l)^2 \sinh \alpha l}\frac{M_2 l}{EI}, \tag{2.19a}$$

$$\theta_2 - \frac{\delta_2 - \delta_1}{l} = -\frac{\sinh \alpha l - \eta \alpha l}{\eta(\alpha l)^2 \sinh \alpha l}\frac{M_1 l}{EI} - \frac{\sinh \alpha l - \eta \alpha l \cosh \alpha l}{\eta(\alpha l)^2 \sinh \alpha l}\frac{M_2 l}{EI}. \tag{2.19b}$$

From the above two equations, one obtains

$$M_1 = \frac{EI}{l}\left(4\psi_3 \theta_1 + 2\psi_4 \theta_2 - 6\psi_2 \frac{\delta_2 - \delta_1}{l}\right), \tag{2.20a}$$

$$M_2 = \frac{EI}{l}\left(2\psi_4 \theta_1 + 4\psi_3 \theta_2 - 6\psi_2 \frac{\delta_2 - \delta_1}{l}\right). \tag{2.20b}$$

Substituting Equations (2.20a), (2.20b) and (2.18) into Equation (2.17) yields

$$Q_1 = -Q_2 = \frac{EI}{l}\left(6\psi_2\frac{\theta_1}{l} + 6\psi_2\frac{\theta_2}{l} - 12\psi_1\frac{\delta_2 - \delta}{l^2}\right), \tag{2.20c}$$

where

$$\psi_1 = \frac{1}{12\psi_t}\eta^2(\alpha l)^3 \sinh \alpha l, \tag{2.21a}$$

$$\psi_2 = \frac{1}{6\psi_t}\eta(\alpha l)^2(\cosh \alpha l - 1), \tag{2.21b}$$

$$\psi_3 = \frac{1}{4\psi_t}\alpha l(\eta\alpha l \cosh \alpha l - \sinh \alpha l), \tag{2.21c}$$

$$\psi_4 = \frac{1}{2\psi_t}\alpha l(\sinh \alpha l - \eta\alpha l) \tag{2.21d}$$

and

$$\psi_t = 2 - 2\cosh \alpha l + \eta\alpha l \sinh \alpha l. \tag{2.21e}$$

Equations (2.20a)–(2.20c) are factually the elastic stiffness equations for beam elements considering effects of shear deformation and axial force simultaneously, which can be expressed in matrix form as

$$\frac{EI}{l}\begin{bmatrix} \frac{12}{l^2}\psi_1 & \frac{6}{l}\psi_2 & -\frac{12}{l^2}\psi_1 & \frac{6}{l}\psi_2 \\ \frac{6}{l}\psi_2 & 4\psi_3 & -\frac{6}{l}\psi_2 & 2\psi_4 \\ -\frac{12}{l^2}\psi & -\frac{6}{l}\psi_2 & \frac{12}{l^2}\psi & -\frac{6}{l}\psi_2 \\ \frac{6}{l}\psi_2 & 2\psi_4 & -\frac{6}{l}\psi_2 & 4\psi_3 \end{bmatrix}\begin{Bmatrix} \delta_1 \\ \theta_1 \\ \delta_2 \\ \theta_2 \end{Bmatrix} = \begin{Bmatrix} Q_1 \\ M_1 \\ Q_2 \\ M_2 \end{Bmatrix} \tag{2.22a}$$

or

$$[k_{ge}]\{\delta_g\} = \{f_g\}, \tag{2.22b}$$

where

$$\{\delta_g\} = \{\delta_1,\ \theta_1,\ \delta_2,\ \theta_2\}^{\mathrm{T}},$$
$$\{f_g\} = \{Q_1,\ M_1,\ Q_2,\ M_2\}^{\mathrm{T}},$$
$$[k_{ge}] = \frac{EI}{l}\begin{bmatrix} \frac{12}{l^2}\psi_1 & \frac{6}{l}\psi_2 & -\frac{12}{l^2}\psi_1 & \frac{6}{l}\psi_2 \\ \frac{6}{l}\psi_2 & 4\psi_3 & -\frac{6}{l}\psi_2 & 2\psi_4 \\ -\frac{12}{l^2}\psi & -\frac{6}{l}\psi_2 & \frac{12}{l^2}\psi & -\frac{6}{l}\psi_2 \\ \frac{6}{l}\psi_2 & 2\psi_4 & -\frac{6}{l}\psi_2 & 4\psi_3 \end{bmatrix} \tag{2.23}$$

and $[k_{ge}]$ is the elastic stiffness matrix of beam elements in tension.

2.1.2 Beam Element in Compression

The nodal forces and displacements of beam elements in compression are the same as those shown in Figure 2.1, but $N < 0$. The differential equilibrium equation can be established in the similar way as

$$y'' + \alpha^2 y = -\frac{M_1 - Q_1 z}{\eta EI}, \tag{2.24}$$

where η is defined by Equation (2.9) as well, but

$$\alpha^2 = -\frac{N}{\eta EI}. \tag{2.25}$$

The solution of Equation (2.24) becomes

$$y = a \cos \alpha z + b \sin \alpha z + \frac{M_1 - Q_1 z}{N}.$$

With the same boundary conditions as adopted in Equation (2.13) and the similar derivation in Section 2.1.1 for beam elements in tension, the elastic stiffness equation of beam elements in compression can be developed and has the same form as that in Equation (2.22), but

$$\psi_1 = \frac{1}{12\psi_c} \eta^2 (\alpha l)^3 \sin \alpha l, \tag{2.26a}$$

$$\psi_2 = \frac{1}{6\psi_c} \eta (\alpha l)^2 (1 - \cos \alpha l), \tag{2.26b}$$

$$\psi_3 = \frac{1}{4\psi_c} \alpha l (\sin \alpha l - \eta \alpha l \cos \alpha l), \tag{2.26c}$$

$$\psi_4 = \frac{1}{2\psi_c} \alpha l (\eta \alpha l - \sin \alpha l) \tag{2.26d}$$

and

$$\psi_c = 2 - 2 \cos \alpha l - \eta \alpha l \sin \alpha l. \tag{2.26e}$$

2.1.3 Series Expansion of Stiffness Equations

It can be found from Equations (2.21e) and (2.26e) that if $\alpha \to 0$ due to $N \to 0$, then ψ_t or $\psi_c \to 0$. Numerical instability may thus occur when axial forces of beam elements are zero or very small in the frame analysis using Equation (2.21) or (2.26) directly. Employment of a series expansion of $\sinh \alpha \, l$, $\cosh \alpha \, l$, $\sin \alpha \, l$ and $\cos \alpha \, l$ functions in the definitions of $\psi_1 - \psi_4$ can avoid numerical instability due to zero or small axial force. No matter the beam element is in tension or compression, $\psi_1 - \psi_4$ can be expressed uniformly in series expansions as

$$\psi_1 = \frac{1}{12\psi} [1 + \beta(\alpha l)^2]^2 \left\{ 1 + \sum_{n=1}^{\infty} \frac{1}{(2n+1)!} [(\alpha l)^2]^n \right\}, \tag{2.27a}$$

$$\psi_2 = \frac{1}{6\psi} [1 + \beta(\alpha l)^2]^2 \left\{ \frac{1}{2} + \sum_{n=1}^{\infty} \frac{1}{(2n+2)!} [(\alpha l)^2]^n \right\}, \tag{2.27b}$$

$$\psi_3 = \frac{1}{4\psi} \left\{ \frac{1}{3} + \sum_{n=1}^{\infty} \frac{2(n+1)}{(2n+3)!} [(\alpha l)^2]^n + \beta \left(1 + \sum_{n=1}^{\infty} \frac{1}{(2n)!} [(\alpha l)^2]^n \right) \right\}, \tag{2.27c}$$

$$\psi_4 = \frac{1}{2\psi} \left\{ \frac{1}{6} + \sum_{n=1}^{\infty} \frac{1}{(2n+3)!} [(\alpha l)^2]^n - \beta \right\} \tag{2.27d}$$

Table 2.1   The relative errors resulting from term number of series

m	10	12	14	16	18
Upper bound of R	3.49×10^{-4}	1.48×10^{-6}	3.47×10^{-9}	4.81×10^{-12}	4.20×10^{-15}

and

$$\psi = \frac{1}{12} + \sum_{n=1}^{\infty} \frac{2(n+1)}{(2n+4)!} [(\alpha l)^2]^n + \beta \left\{ 1 + \sum_{n=1}^{\infty} \frac{1}{(2n+1)!} [(\alpha l)^2]^n \right\}, \tag{2.27e}$$

where

$$\beta = \frac{\mu EI}{GAl^2}, \tag{2.28}$$

$$(\alpha l)^2 = \frac{Nl^2}{\eta EI}. \tag{2.29}$$

It should be noted that N is positive for axial tension force and negative for compression force.

If $N \to 0$, then $\psi \to 1/12 + \beta$ from Equation (2.27e) and numerical difficulty is avoided in the calculation of $\psi_1 - \psi_4$. In practical calculation, the first m terms of the series expansion can be used, where the truncation error can be estimated from Equation (2.27) as

$$R \leq \sum_{n=m+1}^{\infty} \frac{[(\alpha l)^2]^n}{(2n)!} \leq \frac{(\alpha l)^{2(m+1)}}{(2m+2)!} \left[1 + \sum_{i=1}^{\infty} \frac{(\alpha l)^{2i}}{(2m+3)^{2i}} \right] = \frac{(\alpha l)^{2(m+1)}}{(2m+2)!} \left[\frac{(2m+3)^2}{(2m+3)^2 - (\alpha l)^2} \right]. \tag{2.30}$$

The upper boundary value of (αl) is limited by the tensile yielding load for elements in tension and by the squash load or the Euler load for elements in compression. If the Euler load of the beam element is less than the squash load for the element in compression, the upper boundary value of (αl) is equal to 2π and the relationship between m and the truncation error is tabulated in Table 2.1. It can be concluded from Table 2.1 that a good accuracy can be obtained when $m = 10$.

2.1.4 Beam Element with Initial Geometric Imperfection

The initial imperfection of a steel member is random and of arbitrary shape in reality (McNamee and Lu, 1972). In design codes or researches, initial geometric imperfection is typically assumed to be in a half-sine curve, as shown in Figure 2.4. Therefore, the imperfection function along the element length z is adopted as

$$y_0 = y_{0m} \sin \frac{\pi z}{l}, \tag{2.31}$$

where y_{0m} is the maximum imperfection at mid-span of the member.

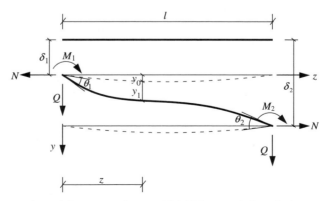

Figure 2.4   A beam element with initial geometric imperfection

As for the beam element in tension and with initial geometric imperfection defined in Equation (2.31), Equations (2.2), (2.3) and (2.7) become

$$y''_{1M} = -\frac{M}{EI},$$ (2.32)

$$M = M_1 - Q_1 z - N(y_1 + y_0),$$ (2.33)

$$y''_{1Q} = -\frac{\mu N}{GA}(y''_1 + y''_0).$$ (2.34)

The governing equation to determine the curvature of the deflection of the element can then be obtained as

$$\begin{aligned} y''_1 &= y''_{1M} + y''_{1Q} \\ &= -\frac{M_1 - Q_1 z - N(y_0 + y_1)}{EI} - \frac{\mu N}{GA}(y''_0 + y''_1). \end{aligned}$$ (2.35)

Substituting Equation (2.31) into Equation (2.35) gives

$$y''_1 - \alpha^2 y_1 = -\frac{M_1 - Q_1 z}{\eta EI} + \gamma \sin\frac{\pi z}{l},$$ (2.36)

where

$$\gamma = \left(\frac{1}{EI} + \frac{\mu}{GA}\frac{\pi^2}{l^2}\right)\frac{N}{\eta}y_{0m}.$$ (2.37)

The solution of Equation (2.36) is

$$y_1 = a\cosh\alpha z + b\sinh\alpha z + \frac{M_1 - Q_1 z}{N} - \frac{\gamma}{\left(\frac{\pi}{l}\right)^2 + \alpha^2}\sin\frac{\pi z}{l},$$ (2.38)

where a and b share the definitions expressed in Equations (2.15a) and (2.15b). Boundary conditions in Equations (2.13b) and (2.13d) become

$$y'_1 = \theta_1 + \frac{\mu}{GA}[-Q_1 - N(y'_1 + y'_0)] \rightarrow y'_1 = \frac{1}{\eta}\left[\theta_1 - \frac{\mu Q_1}{GA} + \frac{\pi}{l}y_{0m}\right] - \frac{\pi}{l}y_{0m},$$ (2.39a)

$$y'_1 = \theta_2 + \frac{\mu}{GA}[-Q_1 - N(y'_1 + y'_0)] \rightarrow y'_1 = \frac{1}{\eta}\left[\theta_2 - \frac{\mu Q_1}{GA} - \frac{\pi}{l}y_{0m}\right] + \frac{\pi}{l}y_{0m}.$$ (2.39b)

Following the same procedures as described in Section 2.1.1, one can develop the elastic stiffness equation for the beam element in tension and with initial geometric imperfection as

$$M_1 = \frac{EI}{l}\left(4\psi_3\theta_1 + 2\psi_4\theta_2 - 6\psi_2\frac{\delta_2 - \delta_1}{l} + \psi_5\frac{y_{0m}}{l}\right),$$ (2.40a)

$$M_2 = \frac{EI}{l}\left(2\psi_4\theta_1 + 4\psi_3\theta_2 - 6\psi_2\frac{\delta_2 - \delta_1}{l} - \psi_5\frac{y_{0m}}{l}\right),$$ (2.40b)

$$Q_1 = -Q_2 = \frac{EI}{l}\left(6\psi_2\frac{\theta_1}{l} + 6\psi_2\frac{\theta_2}{l} - 12\psi_1\frac{\delta_2 - \delta}{l^2}\right),$$ (2.40c)

where $\psi_1 - \psi_4$ and ψ_t are defined in Equation (2.21) and

$$\psi_5 = \frac{1}{\psi_t l}\frac{\pi/l}{(\pi/l)^2 + \alpha^2}\left[\frac{\mu\alpha^2}{GA}EI - 1\right]\eta(\alpha l)^3[2\sinh\alpha l - \eta\alpha l\cosh\alpha l - \eta\alpha l].$$ (2.41)

For beam elements in compression and with initial geometric imperfection, a similar derivation can be given and the stiffness equation has the same form as that in Equation (2.40) but replacing $\psi_1 - \psi_4$ with those in

Equations (2.26a)–(2.26d), replacing ψ_t with ψ_c in Equation (2.26e) and replacing ψ_5 in Equation (2.41) with the following expression:

$$\psi_5 = \frac{1}{\psi_c l}\frac{\pi/l}{\alpha^2 - (\pi/l)^2}\left[\frac{\mu\alpha^2}{GA}EI + 1\right]\eta(\alpha l)^3[2\sinh\alpha l - \eta\alpha l\cosh\alpha l - \eta\alpha l]. \tag{2.42}$$

2.2 SPECIAL FORMS OF ELEMENTAL EQUATIONS

The general form of the elastic stiffness equation for the beam element considering effects of shear deformation, axial force and initial geometric imperfection is given in Equation (2.40). It can be found that Equation (2.40) may simply regress to Equation (2.20) when $y_0 = 0$, i.e. initial geometric imperfection is neglected. In this section, special forms of elemental stiffness equations neglecting shear deformation or/ and axial force are presented based on Equation (2.20).

2.2.1 Neglecting Effect of Shear Deformation

If the effect of shear deformation is neglected, i.e. $GA \to \infty$, then $\eta = 1$ from Equation (2.9). Hence, $\psi_1 - \psi_4$ in Equation (2.22) become

- for the element in tension:

$$\psi_1 = \frac{1}{12\psi_t}(\alpha l)^3\sinh\alpha l, \tag{2.43a}$$

$$\psi_2 = \frac{1}{6\psi_t}(\alpha l)^2(\cosh\alpha l - 1), \tag{2.43b}$$

$$\psi_3 = \frac{1}{4\psi_t}\alpha l(\alpha l\cosh\alpha l - \sinh\alpha l), \tag{2.43c}$$

$$\psi_4 = \frac{1}{2\psi_t}\alpha l(\sinh\alpha l - \alpha l), \tag{2.43d}$$

in which

$$\psi_t = 2 - 2\cosh\alpha l + \alpha l\sinh\alpha l; \tag{2.43e}$$

- for the element in compression:

$$\psi_1 = \frac{1}{12\psi_c}(\alpha l)^3\sin\alpha l, \tag{2.44a}$$

$$\psi_2 = \frac{1}{6\psi_c}(\alpha l)^2(1 - \cos\alpha l), \tag{2.44b}$$

$$\psi_3 = \frac{1}{4\psi_c}\alpha l(\sin\alpha l - \alpha l\cos\alpha l), \tag{2.44c}$$

$$\psi_4 = \frac{1}{2\psi_c}\alpha l(\alpha l - \sin\alpha l), \tag{2.44d}$$

in which

$$\psi_c = 2 - 2\cos\alpha l - \alpha l\sin\alpha l. \tag{2.44e}$$

From Equation (2.28), $\beta = 0$ when the effect of shear deformation is neglected ($GA \to \infty$). The series expansions of $\psi_1 - \psi_4$ in Equation (2.27) become

$$\psi_1 = \frac{1}{12\psi}\left\{1 + \sum_{n=1}^{\infty}\frac{1}{(2n+1)!}[(\alpha l)^2]^n\right\}, \tag{2.45a}$$

$$\psi_2 = \frac{1}{6\psi}\left\{\frac{1}{2} + \sum_{n=1}^{\infty}\frac{1}{(2n+2)!}[(\alpha l)^2]^n\right\}, \tag{2.45b}$$

$$\psi_3 = \frac{1}{4\psi}\left\{\frac{1}{3} + \sum_{n=1}^{\infty}\frac{2(n+1)}{(2n+3)!}[(\alpha l)^2]^n\right\}, \tag{2.45c}$$

$$\psi_4 = \frac{1}{2\psi}\left\{\frac{1}{6} + \sum_{n=1}^{\infty}\frac{1}{(2n+3)!}[(\alpha l)^2]^n\right\}, \tag{2.45d}$$

in which

$$\psi = \frac{1}{12} + \sum_{n=1}^{\infty}\frac{2(n+1)}{(2n+4)!}[(\alpha l)^2]^n. \tag{2.45e}$$

If only the first term in series expansions of $(\alpha l)^2$ is adopted, $1/12\psi$ can be expressed as

$$\frac{1}{12\psi} \approx 1 - \frac{(\alpha l)^2}{15},$$

and then $\psi_1 - \psi_4$ in Equation (2.45) become

$$\psi_1 \approx \left[1 + \frac{(\alpha l)^2}{6}\right]\left[1 - \frac{(\alpha l)^2}{15}\right] \approx 1 + \frac{(\alpha l)^2}{10}, \tag{2.46a}$$

$$\psi_2 \approx \left[1 + \frac{(\alpha l)^2}{12}\right]\left[1 - \frac{(\alpha l)^2}{15}\right] \approx 1 + \frac{(\alpha l)^2}{60}, \tag{2.46b}$$

$$\psi_3 \approx \left[1 + \frac{(\alpha l)^2}{10}\right]\left[1 - \frac{(\alpha l)^2}{15}\right] \approx 1 + \frac{(\alpha l)^2}{30}, \tag{2.46c}$$

$$\psi_4 \approx \left[1 + \frac{(\alpha l)^2}{20}\right]\left[1 - \frac{(\alpha l)^2}{15}\right] \approx 1 - \frac{(\alpha l)^2}{60}. \tag{2.46d}$$

Substituting Equation (2.46) into Equation (2.23) and considering Equation (2.29) can yield the stiffness matrix of beam elements neglecting effects of shear deformation as

$$[k_{ge}] = [k_{g0}] + [k_{gG}], \tag{2.47}$$

Table 2.2   The relative error R_1 versus $(\alpha l)^2$

$(\alpha l)^2$	1	1.5	2	2.5	3
R_1	0.85%	1.93%	3.48%	5.49%	7.99%

where

$$
[k_{g0}] = \frac{EI}{l}
\begin{bmatrix}
\dfrac{12}{l^2} & \dfrac{6}{l} & -\dfrac{12}{l^2} & \dfrac{6}{l} \\[6pt]
\dfrac{6}{l} & 4 & -\dfrac{6}{l} & 2 \\[6pt]
-\dfrac{12}{l^2} & -\dfrac{6}{l} & \dfrac{12}{l^2} & -\dfrac{6}{l} \\[6pt]
\dfrac{6}{l} & 2 & -\dfrac{6}{l} & 4
\end{bmatrix},
\tag{2.48}
$$

$$
[k_{gG}] = N
\begin{bmatrix}
\dfrac{6}{5l} & \dfrac{1}{10} & -\dfrac{6}{5l} & \dfrac{1}{10} \\[6pt]
\dfrac{1}{10} & \dfrac{2l}{15} & -\dfrac{6}{l} & -\dfrac{l}{30} \\[6pt]
-\dfrac{6}{5l} & -\dfrac{6}{l} & \dfrac{6}{5l} & -\dfrac{1}{10} \\[6pt]
\dfrac{1}{10} & -\dfrac{l}{30} & -\dfrac{1}{10} & \dfrac{2l}{15}
\end{bmatrix}.
\tag{2.49}
$$

$[k_{g0}]$ is the elastic stiffness matrix of beam elements neglecting effects of both shear deformation and axial forces and $[k_{gG}]$ is the geometric stiffness matrix of beam elements due to axial force.

The effect of axial force on the stiffness of beam elements is reflected directly in Equation (2.47), where tension axial force $(N > 0)$ can increase the bending stiffness of beam elements whereas compression axial force $(N < 0)$ can reduce it. It should be noted that Equation (2.47) is the linear approximation of the stiffness matrix of beam elements considering effects of axial force. The maximum relative error R_1 in the elements of the matrix may be estimated with

$$
R_1 \le \sum_{n=2}^{\infty} \frac{[(\alpha l)^2]^n}{(2n+1)!} \le \frac{(\alpha l)^4}{5!}\left[1 + \sum_{i=1}^{\infty} \frac{(\alpha l)^{2i}}{7^{2i}}\right] = \frac{(\alpha l)^4}{120}\frac{49}{49 - (\alpha l)^2}.
\tag{2.50}
$$

The relationship between R_1 and $(\alpha l)^2$ is listed in Table 2.2. It can be found from Equation (2.50) that when $(\alpha l)^2 > 3$, relatively large error can be produced by Equation (2.47) in the calculation of the stiffness matrix of beam elements.

2.2.2   Neglecting Effect of Axial Force

When the effect of axial force is neglected, one may let $N = 0$ and then $(\alpha l)^2 = Nl^2/\eta EI = 0$. Expressions of $\psi_1 - \psi_4$ may be simplified from Equation (2.27) as

$$
\psi_1 = \frac{1}{1+r},
\tag{2.51a}
$$

$$
\psi_2 = \frac{1}{1+r},
\tag{2.51b}
$$

$$
\psi_3 = \frac{1 + r/4}{1+r},
\tag{2.51c}
$$

$$
\psi_4 = \frac{1 - r/2}{1+r},
\tag{2.51d}
$$

where

$$r = 12\beta = \frac{12\mu EI}{GAl^2}. \tag{2.51e}$$

Substituting Equation (2.51) into Equation (2.23), one can derive the stiffness matrix of beam elements including only the effect of shear deformation:

$$[k_{ge}] = \frac{EI}{(1+r)l} \begin{bmatrix} \dfrac{12}{l^2} & \dfrac{6}{l} & -\dfrac{12}{l^2} & \dfrac{6}{l} \\ \dfrac{6}{l} & 4+r & -\dfrac{6}{l} & 2-r \\ -\dfrac{12}{l^2} & -\dfrac{6}{l} & \dfrac{12}{l^2} & -\dfrac{6}{l} \\ \dfrac{6}{l} & 2-r & -\dfrac{6}{l} & 4+r \end{bmatrix}. \tag{2.52}$$

2.2.3 Neglecting Effects of Shear Deformation and Axial Force

If $N = 0$ and $GA \to \infty (r \to 0)$, the stiffness matrix of beam elements is simplified from Equation (2.52) as

$$[k_{ge}] = [k_{g0}], \tag{2.53}$$

where $[k_{g0}]$ is defined by Equation (2.48). Obviously, one has

$$\psi_1 = \psi_2 = \psi_3 = \psi_4 = 1. \tag{2.54}$$

2.3 EXAMPLES

2.3.1 Bent Frame

In order to illustrate the effects of shear deformation and axial force on the elastic stiffness equation of beam elements, a simple bent frame is considered (see Figure 2.5). The cross-sectional area of the flange plate is equal to that of the web plate, i.e. $A_f = A_w$, for the columns in this bent frame. By employing the stiffness equation of beam elements to the columns of the bent frame and introducing the boundary conditions considering symmetry of the frame, the elastic stiffness equation of the bent frame in terms of the drift and rotation at the top of the columns can be expressed as

$$\frac{EI}{l} \begin{bmatrix} \dfrac{12}{l^2}\psi_1 & \dfrac{6}{l}\psi_2 \\ \dfrac{6}{l}\psi_2 & 4\psi_3 \end{bmatrix} \begin{Bmatrix} \delta \\ \theta \end{Bmatrix} = \begin{Bmatrix} F \\ 0 \end{Bmatrix}, \tag{2.55}$$

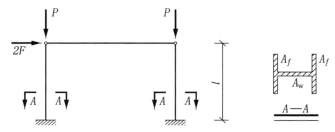

Figure 2.5 A bent frame and its column section

where I is the moment of inertia of the cross section of the column and l is the length of the column.

The solution of Equation (2.55) is

$$\delta = \frac{\psi_3}{12\psi_1\psi_3 - 9\psi_2^2}\frac{Fl^3}{EI}. \tag{2.56}$$

Let δ_0 be the column drift neglecting effects of axial force and shear deformation, i.e.

$$\delta_0 = \frac{Fl^3}{EI}, \tag{2.57}$$

then

$$\frac{\delta}{\delta_0} = \frac{\psi_3}{12\psi_1\psi_3 - 9\psi_2^2}. \tag{2.58}$$

Introduce a parameter C and the slenderness of the cantilever column λ as

$$C = \frac{P}{N_E}, \tag{2.59}$$

$$\lambda = \frac{2l}{\sqrt{I/A}}, \tag{2.60}$$

where $N_E = \pi^2 EI/(2l)^2$ is the Euler load.

Noticing $E/G = 2.6$ for steel, one can obtain

$$\eta = 1 - \frac{\mu P}{GA} = 1 - \frac{76.983C}{\lambda^2}, \tag{2.61}$$

$$(\alpha l) = \sqrt{\frac{Pl^2}{\eta EI}} = \frac{\pi}{2}\sqrt{\frac{C}{\eta}}. \tag{2.62}$$

When $C \neq 0$, $\psi_1 - \psi_3$ in Equation (2.58) are calculated with Equation (2.26), and when $C = 0$ with Equation (2.51), where

$$r = \frac{12\mu EI}{GAl^2} = \frac{374.4}{\lambda^2}. \tag{2.63}$$

Table 2.3 lists the relationship between δ/δ_0 with C and λ. It can be concluded from Table 2.3 that

(1) Both shear deformation and axial force affect the stiffness of beam elements, where the effect of shear deformation relates to the slenderness and the effect of axial force relates to the ratio of axial force to the Euler load, C. As C is inversely proportional to the square of slenderness, the smaller the beam slenderness, the larger the effect of shear deformation and the smaller the effect of axial force on the

Table 2.3 The relationship between λ and C

C	\multicolumn{6}{c}{δ/δ_0}
	$\lambda = 15$	$\lambda = 20$	$\lambda = 30$	$\lambda = 50$	$\lambda = 80$	$\lambda = 120$
0	1.416	1.234	1.104	1.037	1.015	1.007
0.05	1.513	1.308	1.166	1.091	1.067	1.060
0.1	1.625	1.394	1.234	1.155	1.127	1.118
0.3	2.320	1.893	1.622	1.493	1.450	1.436
0.6	6.920	4.200	3.105	2.685	2.555	2.514

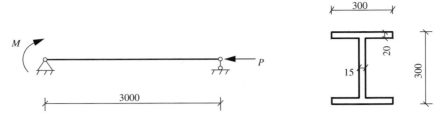

Figure 2.6   A simply supported steel beam

Table 2.4   Rotations at the left end of the beam subjected to different effects

	$C = 0.007$		$C = 0.2$	
Effects considered	Rotation (rad) $\times 10^{-3}$	Relative error (%)	Rotation (rad) $\times 10^{-3}$	Relative error (%)
Idealized beam [a]	1.416	—	1.416	—
Effect of axial force (1)	1.421	0.353	1.645	16.172
Effect of shear deformation (2)	1.496	5.650	1.496	5.650
Effect of initial $y_{0m} = 3$ mm (3) geometric	1.419	0.212	1.419	0.212
imperfection $y_{0m} = 10$ mm (4)	1.425	0.636	1.425	0.636
(1) + (2)	1.502	6.073	1.737	22.670
(1) + (2) + (3)	1.524	7.627	2.561	80.862
(1) + (2) + (4)	1.575	11.229	4.485	216.737

[a] Idealized beam means that effects of axial force, shear deformation and initial geometric imperfection are neglected.

bending stiffness of beam elements. Conversely, the larger the beam slenderness, the smaller the effect of shear deformation and the larger the effect of axial force.

(2) Effects of shear deformation and axial force interact on beam stiffness. The joint effects are larger than the superposition of individual effect of shear deformation and axial force, respectively.

(3) When $\lambda < 30$, the effect of shear deformation on the stiffness of beam elements cannot be neglected, but when $\lambda > 50$ it can be neglected approximately.

(4) When $C > 0.1$, the effect of axial force on the stiffness of beam elements cannot be neglected, but when $C < 0.05$ it can be neglected approximately.

2.3.2 Simply Supported Beam

The rotation at the left end of a simply supported steel beam is examined as shown in Figure 2.6. The elastic and shear modulus of the steel material, E and G, are 206 GPa and 80 GPa, respectively. Only one element is used to represent the entire beam.

The effects of initial geometric imperfection ($y_{0m} = 3$ mm and $y_{0m} = 10$ mm) are involved in the analysis of the beam considering two values of C, $C = 0.007$ and $C = 0.2$. The rotations at the left end of the beam subjected to different effects are listed in Table 2.4. The following observations can be found:

(1) The larger the axial compression force, the smaller the bending stiffness of the beam element.

(2) The initial geometric imperfection may significantly reduce the bending stiffness of the beam subjected to axial compression.

(3) The joint effects of shear deformation, axial force and initial geometric imperfection on stiffness of beam elements are larger than the superposition of their individual effect.

3 Elastic Stiffness Equation of Tapered Beam Element

3.1 TAPERED BEAM ELEMENT

Shear deformation has been verified to significantly influence the structural behaviour of prismatic steel members with I-section under certain conditions, and this conclusion should be true for tapered members as well. It is possible that effects of axial force and shear deformation will act simultaneously on tapered members and influence the stiffness of structures consisting of tapered members. Although there are some works on this topic (Banerjee, 1986; Cleghorn and Tabarrok, 1992; Just, 1977), the aim of this chapter is to derive, for the first time according to the authors' knowledge, the governing equilibrium differential equation of tapered beam elements, including the joint effects, and to propose a method using the Chebyshev polynomial approach to obtain the elemental stiffness matrix.

3.1.1 Differential Equilibrium Equation

The cross section of steel tapered members is usually I-shaped, hot-rolled or welded by three plates. The height of the web is frequently linearly varied, whereas the flanges are symmetric and kept uniform in width along the length direction, as shown in Figure 3.1. For the tapered element described above, the axis of the element remains straight, and the applied forces as well as the corresponding deformations of the element can thus be modelled in the same manner as that shown in Figure 2.1. Following the same procedure as in Section 2.1.1, the equilibrium differential equation of the tapered beam element can be established.

Under the simultaneous action of moment, shear force and axial force (positive for tension and negative for compression), the element deflection consists of two portions. One is induced by the bending deformation and the other by the shear deformation, namely

$$y = y_M + y_Q. \tag{3.1}$$

The curvature of the element caused by bending is

$$y''_M = -\frac{M}{E \cdot I(z)}, \tag{3.2}$$

where $I(z)$ is the moment of inertia of the cross section at distance z from the left end of the element, E is the elastic modulus and M is the cross-sectional moment which can be expressed by

$$M = M_1 - Q_1 \cdot z - N \cdot y. \tag{3.3}$$

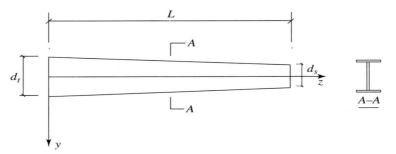

Figure 3.1   A steel tapered member

The slope of the element caused by shearing is

$$y'_Q = \frac{\mu \cdot Q}{G \cdot A(z)} = \frac{\mu}{G \cdot A(z)} \cdot \frac{dM}{dz}, \tag{3.4}$$

where $A(z)$ is the area of the cross section at distance z, Q is the cross-sectional shear force, G is the shear modulus and μ is the shear shape factor varying with the shape of the cross section. For the I-shaped section, μ can be approximately calculated by

$$\mu = \frac{A(z)}{Aw(z)}, \tag{3.5}$$

where $A_w(z)$ is the cross-sectional area of the web at the same cross section for $A(z)$.

Substituting Equation (3.5) into Equation (3.4) gives

$$y'_Q = \frac{1}{G \cdot A_w(z)} \cdot \frac{dM}{dz}, \tag{3.6}$$

and substituting Equation (3.3) into Equation (3.6) leads to

$$y'_Q = \frac{1}{G \cdot A_w(z)} \cdot (-Q_1 - N \cdot y'). \tag{3.7}$$

Differentiating Equation (3.7) gives

$$y''_Q = \frac{1}{G \cdot A_w(z)} \cdot \left[\frac{A'_w(z)}{A_w(z)} \cdot (Q_1 + N \cdot y') - N \cdot y'' \right]. \tag{3.8}$$

Differentiating Equation (3.1) twice and associating Equation (3.2) with Equation (3.8), we obtain

$$y'' = y''_M + y''_Q = -\frac{M_1 - Q_1 \cdot z - N \cdot y}{E \cdot I(z)} + \frac{1}{G \cdot A_w(z)} \cdot \left[\frac{A'_w(z)}{A_w(z)} \cdot (Q_1 + N \cdot y') - N \cdot y'' \right]. \tag{3.9}$$

Equation (3.9) can be simplified as

$$\alpha(z) \cdot y'' - \beta(z) \cdot N \cdot y' - N \cdot y = \beta(z) \cdot Q_1 - (M_1 - Q_1 \cdot z), \tag{3.10}$$

where

$$\alpha(z) = E \cdot I(z) \cdot \gamma(z),$$

$$\beta(z) = E \cdot I(z) \cdot \frac{A'_{\mathrm{w}}(z)}{G \cdot A^2_{\mathrm{w}}(z)},$$

$$\gamma(z) = 1 + \frac{N}{G \cdot A_{\mathrm{w}}(z)}.$$

Equation (3.10) is the governing equation for the equilibrium of tapered beam elements. It should be noted significantly that Equation (3.10) has a general form for any solid or latticed nonprismatic members, other than the forenamed I-shaped sectional tapered members, as long as an appropriate expression is replaced for shear factor μ in Equation (3.5).

3.1.2 Stiffness Equation

Let $\xi = \frac{z}{L}$; Equation (3.10) is converted to nondimensional form by

$$\alpha(\xi) \cdot y'' - \beta(\xi) \cdot L \cdot N \cdot y' - L^2 \cdot N \cdot y = \beta(\xi) \cdot L^2 \cdot Q_1 - L^2 \cdot (M_1 - Q_1 \cdot L \cdot \xi). \tag{3.11}$$

By using the Chebyshev polynomial, the functions $y(\xi), \alpha(\xi)$ and $\beta(\xi)$ can be approached by

$$y(\xi) = \sum_{n=0}^{M} y_n \cdot \xi^n, \tag{3.12a}$$

$$\alpha(\xi) = \sum_{n=0}^{M} \alpha_n \cdot \xi^n, \tag{3.12b}$$

$$\beta(\xi) = \sum_{n=0}^{M} \beta_n \cdot \xi^n. \tag{3.12c}$$

Substituting Equation. (3.12) into Equation (3.11) leads to

$$\sum_{n=0}^{M} \left[\sum_{i=0}^{n} \alpha_i(n+2-i)(n+1-i)y_{n+2-i} \right] \cdot \xi^n$$

$$- L \cdot N \cdot \sum_{n=0}^{M} \left[\sum_{i=0}^{n} \beta_i(n+1-i)y_{n+1-i} \right] \cdot \xi^n - L^2 \cdot N \cdot \sum_{n=0}^{M} y_n \cdot \xi^n \tag{3.13}$$

$$= L^2 \cdot Q_1 \cdot \sum_{n=0}^{M} \beta_n \cdot \xi^n - L^2 \cdot M_1 + L^3 \cdot Q_1 \cdot \xi.$$

According to the principle that the factors at the two sides of Equation (3.13) for the same exponent of ξ should be equal (Eisenberger, 1995), we have

- for $n = 0$:

$$2\alpha_0 \cdot y_2 - L \cdot N \cdot \beta_0 \cdot y_1 - L^2 \cdot N \cdot y_0 = L^2 \cdot Q_1 \cdot \beta_0 - L^2 \cdot M_1; \tag{3.14}$$

- for $n = 1$:

$$6\alpha_0 \cdot y_3 + 2\alpha_1 \cdot y_2 - L \cdot N(2\beta_0 \cdot y_2 + \beta_1 \cdot y_1) - L^2 \cdot N \cdot y_1$$
$$= L^2 \cdot Q_1 \cdot \beta_1 + L^3 \cdot Q_1; \tag{3.15}$$

- for $n \geq 2$:

$$\sum_{i=0}^{n} \alpha_i (n + 2 - i)(n + 1 - i) y_{n+2-i} - L \cdot N$$

$$\cdot \sum_{i=0}^{n} \beta_i (n + 1 - i) y_{n+1-i} - L^2 \cdot N \cdot y_n = L^2 \cdot Q_1 \cdot \beta_n. \tag{3.16a}$$

Rewrite Equation (3.16a) as

$$y_{n+2} = \left[L \cdot N \cdot \sum_{i=0}^{n} \beta_i (n + 1 - i) y_{n+1-i} + L^2 \cdot N \cdot y_n + L^2 \cdot Q_1 \right] \frac{\beta_n - \sum_{i=1}^{n} \alpha_i (n + 2 - i)(n + 1 - i) y_{n+2-i}}{\alpha_0 (n + 2)(n + 1)}. \tag{3.16b}$$

As $\alpha(z)$ and $\beta(z)$ are functions known ahead, the series $\{\alpha_n\}$ and $\{\beta_n\}$ are determinate. Hence, it can be found from Equation (3.16b) that any y_n in series $\{y_n\}(n \geq 4)$ can be expressed by one of the linear combinations of y_0, y_1, y_2, y_3 and Q_1, and series $\{y_n\}$ may be determined when the values of y_0, y_1, y_2, y_3 and Q_1 are obtained.

Consider the following boundary conditions:

- for $\xi = 0$:

$$y(0) = y_0 = 0, \tag{3.17a}$$

$$y'(0) = y_1 = \frac{L}{\gamma(0)} \left[\theta_1 - \frac{Q_1}{G \cdot A_w(0)} \right]; \tag{3.17b}$$

- for $\xi = 1$:

$$y(1) = \sum_{n=0}^{M} y_n = c_1 y_1 + c_2 y_2 + c_3 y_3 + c_4 Q_1 = \delta_2 - \delta_1, \tag{3.18a}$$

$$y'(1) = \sum_{n=0}^{M} n \cdot y_n = c_5 y_1 + c_6 y_2 + c_7 y_3 + c_8 Q_1 = \frac{L}{\gamma(1)} \left[\theta_2 - \frac{Q_1}{G \cdot A_w(1)} \right]. \tag{3.18b}$$

The reason that $y(1)$ and $y'(1)$ in Equation (3.18) are expressed as the linear combinations of y_1, y_2, y_3 and Q_1 is the conclusion obtained from Equation (3.16b) and $y_0 = 0$ from Equation (3.17a).

Letting $y_1 = 1$ and $y_2 = y_3 = Q_1 = 0$, c_1 and c_5 can be determined from Equation (3.18) by

$$c_1 = 1 + \sum_{i=4}^{M} y_n^{(1)}, \tag{3.19a}$$

$$c_5 = 1 + \sum_{i=4}^{M} n \cdot y_n^{(1)}. \tag{3.19b}$$

where $\{y_n^{(1)}\}$ is the series of $\{y_n\}$ determined under the initial condition $y_1 = 1$, $y_2 = y_3 = Q_1 = 0$.

If $y_2 = 1$ and $y_1 = y_3 = Q_1 = 0$, c_2 and c_6 can be determined in the same way. So do c_3 and c_7, c_4 and c_8.

So far y_1, y_2, y_3, Q_1 and M_1 are yet unknown variables. If the boundary deformations of the elements δ_1, δ_2, θ_1 and θ_2 are known, the five equations numbered (3.14), (3.15), (3.17b), (3.18a) and (3.18b) can be combined for solving Q_1 and M_1. The applied forces at the other end of the elements, Q_2 and M_2, may be expressed as a function of Q_1 and M_1 by considering the following equilibrium conditions:

$$Q_2 + Q_1 = 0, \tag{3.20}$$

$$M_2 + M_1 - Q_1 \cdot L - N \cdot (\delta_2 - \delta_1) = 0. \tag{3.21}$$

Then, the stiffness equation of the element is obtained as

$$[k] \cdot \{\delta\} = \{f\}, \tag{3.22}$$

where

$$
\begin{aligned}
\{\delta\} &= [\delta_1, \theta_1, \delta_2, \theta_2]^{\mathrm{T}}, \\
\{f\} &= [Q_1, M_1, Q_2, M_2]^{\mathrm{T}}, \\
[k] &= \begin{bmatrix} -\phi_1 & \phi_2 & \phi_1 & \phi_3 \\ -\phi_4 & \phi_5 & \phi_4 & \phi_6 \\ \phi_1 & -\phi_2 & -\phi_1 & -\phi_3 \\ -\phi_7 & \phi_8 & \phi_7 & \phi_9 \end{bmatrix}.
\end{aligned}
\tag{3.23}
$$

The expressions of $\phi_i(i = 1, 2, \ldots, 9)$ are given in Section 3.3.2.

In theory, the approach described above is accurate. The unique possible error comes computationally from the representation of the realistic deflection y and functions α and β by the Chebyshev polynomial with definite terms, which affects directly not more than the coefficients $c_1 - c_8$. So long as the number of terms for the Chebyshev polynomial, i.e. M, is suitably chosen to make the coefficients $c_1 - c_8$ accurate enough, the satisfactory accuracy of the stiffness matrix of the element can be achieved.

3.2 NUMERICAL VERIFICATION

3.2.1 Symmetry of Stiffness Matrix

The stiffness matrix expressed in Equation (3.23) is of unsymmetrical form. To verify it is actually symmetrical in numerical values, take a steel tapered fixed-hinged beam, shown in Figure 3.2, as an example. The height of the cross section of the beam is varied linearly from 400 to 200 mm. The values of the elements in the stiffness matrix for this beam obtained by the approach proposed hereinbefore considering effects of axial force and shear deformation have the following relations:

$$\phi_2 = 3.390\,034 \times 10^7 = -\phi_4 = 3.390\,029 \times 10^7,$$

$$\phi_3 = 2.018\,891 \times 10^7 = -\phi_7 = 2.018\,896 \times 10^7,$$

$$\phi_6 = 2.308\,029 \times 10^7 = \phi_8 = 2.308\,043 \times 10^7.$$

Obviously, the stiffness matrix is perfectly symmetric.

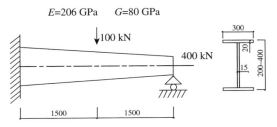

Figure 3.2 A steel tapered fixed-hinged beam (mm)

Table 3.1 Deflection at mid-span δ of a fixed-hinged beam (mm)

Method	Case 1	Case 2	Case 3	Case 4
Proposed (1)	0.4078	0.4091	0.6687	0.6716
FEM (2)	0.4097	0.4110	0.6705	0.6733
$[(2) - (1)]/(2)$ (%)	0.46	0.46	0.27	0.25

Note: Case 1: neither N nor S; case 2: only N; case 3: only S; case 4: both N and S; N = axial force effects; S = shear deformation effects.

3.2.2 Static Deflection

Take the same tapered beam, shown in Figure 3.2, as an example for calculating the deflection at mid-span of the member induced by a lateral point force. The results obtained through representing the whole beam by two tapered Timoshenko–Euler beam elements proposed are compared with those obtained by FEM with stepped representation of the beam in Table 3.1. The beam is divided into 10 segments in FEM. Each segment is modelled by one uniform Timoshenko–Euler beam element (Li and Shen 1995) with the cross section at mid-length of the segment. A compressive axial load with a value of 400 kN is applied at the hinged end of the beam when considering the effect of axial force.

3.2.3 Elastic Critical Load

Figure 3.3 gives a steel tapered cantilever column used by Karabalis (1983) as a numerical example for calculating the elastic critical axial load. The results obtained by Karabalis are compared with those obtained by the approach proposed, using a single tapered Timoshenko–Euler beam element, as in Table 3.2. As approximate geometrical stiffness matrix was employed by Karabalis, the result obtained by the approach proposed is more believable. Moreover, it is reasonable that the elastic critical axial load is reduced when including effects of shear deformation on member stiffness.

3.2.4 Frequency of Free Vibration

The frequencies of the first and second modes of the tapered cantilever beam shown in Figure 3.4 are determined by a single tapered Timoshenko–Euler beam element representation and compared with the results reported by Gupta (1986) and Wekezer (1989) in Table 3.3. The mass matrix used in the computation proposed is cited from Karabalis (1983). As expected, the results obtained by the approach proposed agree very with the values by Gupta (1986) when effects of shear deformation are excluded, but depart slightly on

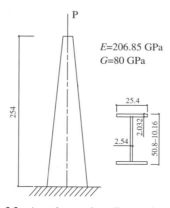

Figure 3.3 A steel tapered cantilever column (mm)

Table 3.2   Elastic critical axial load P_{cr} of the tapered cantilever column (kN)

| | P_{cr} | | |
Case	Proposed (1)	Karabalis (2)	$[(2) - (1)]/(1)(\%)$
1	238.04	241.08	1.28
2	216.62	No result	No comparison

Note: Case 1: without effect of shear deformation; case 2: with effect of shear deformation.

the first frequency and significantly on the second frequency when effects of shear deformation are included. Also, it is important to emphasize that the utilization of approximate shape functions in FEM or stepped representations of the tapered beams yield poor results, as given by Wekezer (1989).

3.2.5   Effect of Term Number Truncated in Polynomial Series

As mentioned above, the number of terms for the Chebyshev polynomial M is the unique factor affecting the accuracy of elemental stiffness computation. Generally, the larger the number M, the more accurate the computation. But a larger M will consume more computation time and thus reduce the advantage of the proposed technique. So making efforts to choose a suitable M, meeting the needs of computational efficiency and accuracy simultaneously, is worthwhile for practical applications. The relative errors of the results of the analyses obtained hereinbefore in Sections 3.2.2–3.2.4 with varying M are shown in Figure 3.5. It is found that a suitable M of around 13 will produce satisfactory results (Li, 2001).

3.2.6   Steel Portal Frame

A gable frame, as shown in Figure 3.6, is analysed as a comprehensive utilization of the proposed method (Li and Li, 2002). This three-bay pitched-roof gable frame comprises tapered external columns (EC), tapered roof beams (RB1, RB2 and RB3) and prismatic internal sway columns (ISC), which have the same sectional dimensions but the height of webs. Table 3.4 lists the section heights of two ends for all the members with linear variation of the section height along the length direction.

The planar behaviour of this gable frame, such as static deflection, natural frequency and elastic critical load, is obtained by both the stepped representation of tapered frame members using general structural analysis software STAAD III and the tapered beam element proposed assuming that failures of out-of-plane and local buckling are prevented. STAAD III can conduct only linear analyses and cannot be used to obtain elastic critical load. Table 3.5 summarizes these results.

A comparison of results in Table 3.5 indicates a general coincidence between the proposed method and STAAD III. It can be found that nonlinearity due to axial force in structural members and shear deformation

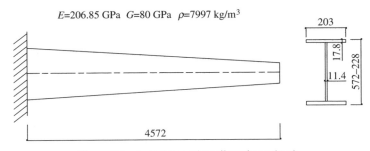

E=206.85 GPa G=80 GPa ρ=7997 kg/m^3

203

17.8

11.4

572–228

4572

Figure 3.4   A steel tapered cantilever beam (mm)

Table 3.3   First and second frequencies, f_1 and f_2, of the tapered cantilever beam

Source	f_1 (Hz)	f_2 (Hz)
Proposed	30.15 (29.27)	149.11 (131.06)
Gupta (1986)	30.533	152.20
Wekezer (1989)	39	233

Note: The values in the parentheses include the effect of shear deformation.

Table 3.4   Linear variation of section height along the length direction for all the members (mm)

		Members				
		EC	ISC	RB1	RB2	RB3
Section height of two ends	d_s	200	300	400	400	300
	d_1	600	300	600	600	600

Table 3.5   Results of the gable frame analysis

Item	Response				Description
Static deflection (mm)	Joint	$\square \to \to$	$\square \downarrow\downarrow$	$\square \to \to$	Vertical loads applied at each
	STAAD III	10.38	11.44	10.62	purlin position with 20 kN,
	Proposed	10.50	11.54	10.71	except at loads eava positions
		(11.21)	(12.08)	(11.45)	with 15 kN; horizontal applied rightly at the top of left EC with 12 kN and right EC with 6 kN
Natural frequency (Hz)	Order	First	Second	Third	No additional mass considered
	STAAD III	2.04	4.71	8.21	but the gable frame's self-mass
	Proposed	1.92	5.03	8.30	
		(1.76)	(5.23)	(8.18)	
Elastic critical load (kN)	Proposed		(2160)		Vertical concentrated loads equally applied at the top of each EC and ISC

Note: The values in the parentheses include effects of axial forces in members (namely geometrical nonlinearity) and shear deformation.

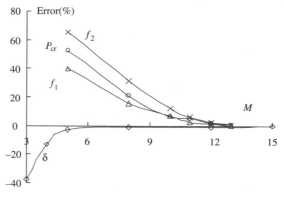

Figure 3.5   Relative errors in computation of δ, f_1, f_2 and P_{cr} versus number of terms for the Chebyshev polynomial M

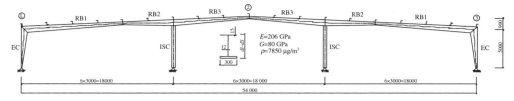

Figure 3.6 A gable frame with tapered components (mm)

leads to significant deviation from linear analysis. Approximately, static deflection deviates about 5 %, and natural frequencies and critical load deviate about 10 %.

3.3 APPENDIX

3.3.1 Chebyshev Polynomial Approach (Rice, 1992)

A function $f(x)$ defined in the range $[-1, 1]$ can be approached approximately by

$$f(x) \approx \frac{1}{2}C_0 + \sum_{j=1}^{M} C_j T_j(x), \tag{3.24}$$

where C_j are the Chebyshev factors being formulated by

$$C_j = \frac{2}{M} \sum_{k=1}^{M} f(x_k) T_j(x_k) = \frac{2}{M} \sum_{k=1}^{M} \left[f\left(\cos\left(\frac{2k-1}{2M}\pi \right) \right) \cdot \cos\left(\frac{j(2k-1)}{2M}\pi \right) \right] \quad (j = 0, 1, \ldots, M). \tag{3.25}$$

For any function defined in the range $[a, b]$, it can be converted to a function in the range $[-1, 1]$ by translating x with

$$y = \frac{[2x - (a+b)]}{(b-a)}. \tag{3.26}$$

If $x \in [a, b]$ and the corresponding converted variable $y \in [-1, 1]$, the Chebyshev polynomial expression for $f(x)$ can be obtained by the Chebyshev iteration formula. It is to find a series of p_j to make

$$f(x) \approx \frac{1}{2}C_0 + \sum_{j=1}^{M} C_j T_j(y) = \sum_{j=0}^{M} q_j \cdot y^j = \sum_{j=0}^{M} p_j \cdot x^j, \tag{3.27}$$

where q_j are temporary variables for determining p_j.

From the Clenshaw formula, we have

$$\sum_{j=0}^{M} q_j \cdot y^j = W_1(y) \cdot y - W_2(y) + \frac{1}{2}C_0, \tag{3.28}$$

where $W_1(y)$ and $W_2(y)$ are polynomials with $M-1$ and $M-2$ exponents, respectively, which can be obtained by the following iterative formula:

$$W_{M+2}(y) = W_{M+1}(y) = 0, \tag{3.29}$$

$$W_j(y) = 2W_{j+1}(y) - W_{j+2}(y) + C_j \quad (j = 0, 1, \ldots, M.). \tag{3.30}$$

When all the factors in $W_1(y)$ and $W_2(y)$ are determined, q_j can be obtained and further used for determining p_j.

3.3.2 Expression of Elements in Equation (3.23)

$$\phi_1 = \frac{\psi_{15}}{\psi_{11}\psi_{15} - \psi_{14}\psi_{12}}, \quad \phi_2 = \frac{\psi_{13}\psi_{15} - \psi_{16}\psi_{12}}{\psi_{11}\psi_{15} - \psi_{14}\psi_{12}}, \quad \phi_3 = -\frac{\psi_{12}\psi_{14}}{\psi_{11}\psi_{15} - \psi_{14}\psi_{12}}, \quad \phi_4 = \frac{1 - \psi_{11}\phi_1}{\psi_{12}},$$

$$\phi_5 = \frac{\psi_{13} - \psi_{11}\phi_2}{\psi_{12}}, \quad \phi_6 = -\frac{\psi_{11}\phi_3}{\psi_{12}}, \quad \phi_7 = \phi_1 L + N - \phi_4, \quad \phi_8 = \phi_2 L - \phi_5, \quad \phi_9 = \phi_3 L - \phi_6,$$

$$\psi_1 = -\frac{g(0) \cdot L}{1 + g(0) \cdot N}, \quad \psi_2 = \frac{L}{1 + g(0) \cdot N}, \quad \psi_3 = -\frac{g(1) \cdot L}{1 + g(1) \cdot N}, \quad \psi_4 = \frac{L}{1 + g(1) \cdot N}$$

$$\psi_5 = \frac{L \cdot \beta_0 (N \cdot \psi_1 + L)}{2\alpha_0}, \quad \psi_6 = -\frac{L^2}{2\alpha_0}, \quad \psi_7 = \frac{L \cdot N \cdot \beta_0 \cdot \psi_2}{2\alpha_0},$$

$$\psi_8 = \frac{2(L \cdot N \cdot \beta_0 - \alpha_1)\psi_5 + L \cdot (\beta_1 + L)(N \cdot \psi_1 + L)}{6\alpha_0}, \quad \psi_9 = \frac{2(L \cdot N \cdot \beta_0 - \alpha_1)\psi_6}{6\alpha_0},$$

$$\psi_{10} = \frac{2(L \cdot N \cdot \beta_0 - \alpha_1)\psi_7 + L \cdot N(\beta_1 + L)\psi_2}{6\alpha_0}, \quad \psi_{11} = c_1\psi_1 + c_2\psi_5 + c_3\psi_8 + c_4,$$

$$\psi_{12} = c_2\psi_6 + c_3\psi_9,$$

$$\psi_{13} = -(c_1\psi_2 + c_2\psi_7 + c_3\psi_{10}), \quad \psi_{14} = c_5\psi_1 + c_6\psi_5 + c_7\psi_8 + c_8 - \psi_3,$$

$$\psi_{15} = c_6\psi_6 + c_7\psi_9,$$

$$\psi_{16} = -(c_5\psi_2 + c_6\psi_7 + c_7\psi_{10}).$$

4 Elastic Stiffness Equation of Composite Beam Element

4.1 CHARACTERISTICS AND CLASSIFICATION OF COMPOSITE BEAM

Concrete slabs are generally laid on the steel beams in multi-storey, high-rise steel buildings. A slab on a beam will bend independently due to vertical floor loads, and a relative shear slip occurs on the interface if there is no connection between them (Figure 4.1). In this case, the concrete slab and the steel beam resist vertical loads jointly but as individual components.

A shear connector can be designed and laid on the slab–beam interface to restrain the relative shear slip (see Figures 4.2 and 4.3), in which case the beam is a concrete–steel composite one and resists vertical floor loads as an integrity (Viest *et al.*, 1997).

Composite beams can be categorized into the following two types according to the performance of shear studs connecting concrete slabs and steel beams:

- *Composite beams with full composite action (Figure 4.2).* Sufficient shear connectors are designed for the fully composite beams so that they can resist the shear force on the interface between concrete slabs and steel beams, and the relative slip is small. The full bending capacity of the composite beams can be ensured in this case.

- *Composite beams with partial composite action (Figure 4.3).* Insufficient shear connectors are designed for the partially composite beams so that they cannot fully resist the shear force on the interface between concrete slabs and steel beams, and the relative slip is relatively large. The full bending capacity of the partially composite beams cannot be achieved. When the number of shear connectors is less than 50 % of that required for fully composite beams, the composite action between concrete slabs and steel beams is actually small, and it is negligible in engineering practice.

A partially composite beam may be a practical option in structural design for the consideration of construction economy, under the condition that the relative slip between the concrete–slab flange and the steel beam is taken into account in the design.

The most efficient and effective way for the analysis of steel frames is the finite element method (FEM) with beam–column members. Inconsistency of degree of freedom (DOF) will occur in finite element analysis of composite frames if two independent axial DOFs are introduced at the two ends of composite beams to consider effects of the relative shear slip (Dissanayeke, Burgess and Davidson, 1995; Faella, Martinelli and Nigro, 2001). To avoid such inconsistency, the elastic stiffness equation of a composite beam element, considering effects of relative slip, is derived based on elastic interaction theory proposed by Newmark, Siess and Viest (1951) through the solution of the governing differential equilibrium equation of the composite element.

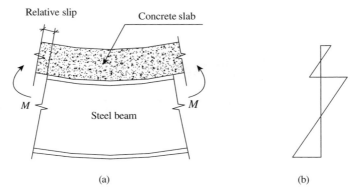

Figure 4.1 Beam without composite action: (a) force and deformation of steel beam and concrete slab; (b) stress distribution along section height

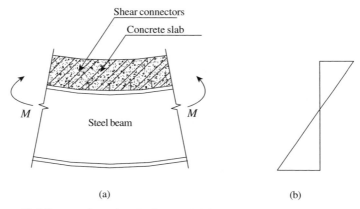

Figure 4.2 Beam with full composite action: (a) force and deformation of steel beam and concrete slab; (b) stress distribution along section height

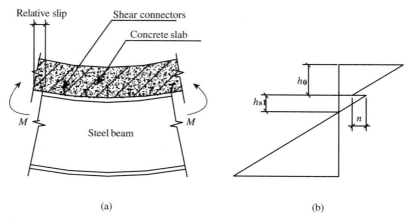

Figure 4.3 Beam with partial composite action: (a) force and deformation of steel beam and concrete slab; (b) stress distribution along section height

4.2 EFFECTS OF COMPOSITE ACTION ON ELASTIC STIFFNESS OF COMPOSITE BEAM

A typical section of a steel–concrete composite beam is illustrated in Figure 4.4. Effects of composite action on the elastic stiffness of composite beams are studied in the following. The plane section is assumed to remain plane in the deformed beam, and an identical elastic modulus for concrete in compression and in tension within elastic scope, namely under the condition of no crushing in compression and no crack in tension, is adopted.

4.2.1 Beam without Composite Action

If there is no composite action between the steel beam and the concrete slab, they will deform individually and have, in the scope of elastic small deformation, the same deflection curves. The strain distribution along the section height of the beam is given in Figure 4.5. The internal moments in steel and concrete sections are

$$M_s = \kappa E_s I_s, \tag{4.1a}$$
$$M_c = \kappa E_c I_c, \tag{4.1b}$$

where κ is the curvature of the common deflection, and $E_s I_s$ and $E_c I_c$ are the bending stiffnesses of steel and concrete sections, respectively.

With the equilibrium of internal and external moments, one has

$$M = M_c + M_s = \kappa(E_c I_c + E_s I_s), \tag{4.2}$$

where M is the external moment applied on the composite section.

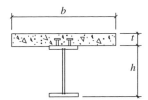

Figure 4.4   A steel–concrete composite beam

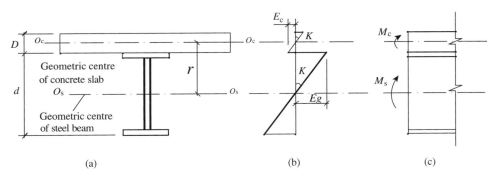

Figure 4.5   Composite beam without composite action: (a) no composite action section; (b) strain distribution along sectional height; (c) internal forces

By Equation (4.2), the bending stiffness of the composite beam without composite action $(EI)^0_{\text{comp}}$ is equal to the algebraic sum of the bending stiffnesses of the steel and concrete components, namely

$$(EI)^0_{\text{comp}} = \sum EI = E_c I_c + E_s I_s. \tag{4.3}$$

A strain difference can be seen along the steel–concrete interface, as shown in Figure 4.1, which is defined as slip strain as

$$\varepsilon_{\text{slip}} = \frac{\mathrm{d}s}{\mathrm{d}x} = \varepsilon_c - \varepsilon_s = \kappa \frac{D}{2} + \kappa \frac{d}{2} = \kappa \left(\frac{D}{2} + \frac{d}{2}\right) = \frac{M}{\sum EI} \cdot r, \tag{4.4}$$

where D and d are the thicknesses of the concrete slab and the height of the steel beam, respectively, and $r = D/2 + d/2$ is the distance from the central axis of the concrete slab to that of the steel beam.

Obviously, a composite beam without composite action behaves, in elastic state, actually as an ordinary beam with a bending stiffness of $\sum EI = E_c I_c + E_s I_s$.

4.2.2 Beam with Full Composite Action

In the case of full composite action, strain is continuous at the steel–concrete interface and linearly distributed along the total section height of the beam, as shown in Figure 4.6, where $C_{a,\infty}$ is the distance from the central axis of the total section to that of the concrete slab. The internal moments in steel and concrete are the same as in Equations (4.1a) and (4.1b). Due to the existence of shear on the steel–concrete interface, the following internal axial compression in concrete, N, and axial tension in steel, T, are produced:

$$N = \varepsilon_c E_c A_c = \kappa \cdot C_{a,\infty} E_c A_c, \tag{4.5}$$

$$T = \varepsilon_s E_s A_s = \kappa \cdot (r - C_{a,\infty}) E_s A_s, \tag{4.6}$$

where $E_s A_s$ and $E_c A_c$ are the axial stiffnesses of the steel and concrete components, respectively.

With view of $N = T$, the height of the neutral axial $C_{a,\infty}$ of the total section can be solved as

$$C_{a,\infty} = \frac{E_s A_s r}{E_c A_c + E_s A_s}. \tag{4.7}$$

The equilibrium of internal and external moments leads to

$$\begin{aligned} M &= M_c + M_s + N \cdot r \\ &= \kappa \cdot (E_c I_c + E_s I_s) + \kappa \cdot C_{a,\infty} E_c A_c r. \end{aligned} \tag{4.8}$$

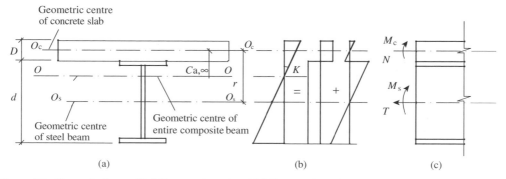

Figure 4.6 Composite beam with full composite action: (a) full composite action section; (b) strain distribution along sectional height; (c) internal forces

Substituting Equation (4.7) into the above equation yields

$$M = \kappa \cdot \left(E_c I_c + E_s I_s + \frac{E_c A_c \cdot E_s A_s}{E_c A_c + E_s A_s} r^2 \right). \tag{4.9}$$

Let

$$\frac{1}{\overline{EA}} = \frac{1}{E_c A_c} + \frac{1}{E_s A_s}, \tag{4.10}$$

then Equation (4.9) becomes

$$M = \kappa \cdot (E_c I_c + E_s I_s + \overline{EA} \cdot r^2) = \kappa \cdot (EI)_{\text{comp}}^{\infty}, \tag{4.11}$$

where $(EI)_{\text{comp}}^{\infty}$ is the bending stiffness of the composite beam with full composite action and is given by

$$(EI)_{\text{comp}}^{\infty} = \sum EI + \overline{EA} \cdot r^2. \tag{4.12}$$

From Equations (4.8) and (4.12), it can be found that the additional axial forces in the steel and concrete components due to composite action lead to an evident increase of bending stiffness $\overline{EA} \cdot r^2$ from $\sum EI$ (bending stiffness without composite action).

4.2.3 Beam with Partial Composite Action

Restrained slip occurs in partial composite action. The strain diagram is given in Figure 4.7. Denote C_c and C_s as the distances from the neutral axes of concrete and steel components to their top surfaces, respectively; the slip strain at the steel–concrete interface can then be expressed as

$$\varepsilon_{\text{slip}} = \varepsilon_c - \varepsilon_s = (D - C_c) \cdot \kappa + C_s \cdot \kappa = (D - C_c + C_s) \cdot \kappa. \tag{4.13}$$

The compression in the concrete slab and the tension in the steel beam are given by

$$N = \kappa \cdot \left(C_c - \frac{D}{2} \right) E_c A_c, \tag{4.14}$$

$$T = \kappa \cdot \left(\frac{d}{2} - C_s \right) E_s A_s. \tag{4.15}$$

The equilibrium of N and T, i.e. $N = T$, results in

$$E_c A_c C_c + E_a A_s C_s = E_c A_c \frac{D}{2} + E_s A_s \frac{d}{2}. \tag{4.16}$$

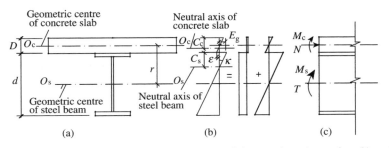

Figure 4.7 Composite beam with partial composite action: (a) partial composite action section; (b) strain distribution along sectional height; (c) internal forces

Combining Equations (4.13) and (4.16), C_c and C_s can be expressed with ε_{slip} and κ:

$$C_c = \frac{1}{E_c A_c} \left[E_s A_s \left(\frac{d}{2} + D \right) + E_c A_c \frac{D}{2} - \frac{\varepsilon_{slip}}{\kappa} E_s A_s \right], \tag{4.17}$$

$$C_s = \frac{1}{E_c A_c} \left[E_s A_s \left(\frac{d}{2} + D \right) - E_c A_c \frac{D}{2} + (E_c A_c - E_s A_s) \frac{\varepsilon_{slip}}{\kappa} \right]. \tag{4.18}$$

Substituting Equation (4.17) back into Equation (4.14) yields

$$N = \overline{EA} r \cdot \kappa - \overline{EA} \varepsilon_{slip}. \tag{4.19}$$

The equilibrium of internal and external moments gives

$$
\begin{aligned}
M &= M_c + M_s + N \cdot r \\
&= \kappa (E_c I_c + E_s I_s) + \overline{EA} r^2 \cdot \kappa - \overline{EA} \cdot r \varepsilon_{slip} \\
&= (EI)_{comp}^{\infty} \cdot \kappa - \overline{EA} \cdot r \varepsilon_{slip} \\
&= (EI)_{comp} \cdot \kappa,
\end{aligned}
\tag{4.20}
$$

where $(EI)_{comp}$ is the bending stiffness of the composite beam with partial composite action and is given by

$$(EI)_{comp} = (EI)_{comp}^{\infty} - \overline{EA} \cdot \frac{r \varepsilon_{slip}}{\kappa}. \tag{4.21}$$

Obviously, the relationship between moment and curvature of the composite beam with partial composite action is no longer linear. In addition to sectional and material parameters, the bending stiffness of the partially composite beam depends also on the slip strain. In Section 4.3, the elastic stiffness equation of the partially composite beam, based on Newmark partial interaction theory, will be derived.

4.3   ELASTIC STIFFNESS EQUATION OF STEEL–CONCRETE COMPOSITE BEAM ELEMENT

4.3.1   Basic Assumptions

The following assumptions are employed in this section:

(1) Both steel and concrete are in elastic state.

(2) The shear stud is also in elastic state, and the shear–slip relationship for single shear stud is

$$Q = K \cdot s, \tag{4.22}$$

   where K is the shear stiffness of a stud with unit N/mm.

(3) The composite action is smeared uniformly on the steel–concrete interface, although the actual shear studs providing composite action are discretely distributed.

(4) The plane section of the concrete slab and the steel beam remains plane independently, which indicates that the strains are linearly distributed along steel and concrete section heights, respectively.

(5) Lift-up of shear studs, namely pull-out of shear studs form the concrete slab, is prevented. The deflection of the steel beam and the concrete slab at the same position along the length is identical, or the steel and concrete components of the composite beam are subjected to the same curvature in deformation.

4.3.2 Differential Equilibrium Equation of Partially Composite Beam

The strains of the concrete and steel components at the interface can be expressed with internal forces as

$$\varepsilon_c = -\frac{N}{E_c A_c} + \frac{M_c}{E_c I_c} \cdot \frac{D}{2}, \tag{4.23a}$$

$$\varepsilon_s = \frac{N}{E_s A_s} - \frac{M_s}{E_s I_s} \cdot \frac{d}{2}, \tag{4.23b}$$

and the slip strain at the steel–concrete interface can then be expressed as

$$\varepsilon_{slip} = \frac{ds}{dx} = \varepsilon_c - \varepsilon_s = -N \cdot \left(\frac{1}{E_c A_c} + \frac{1}{E_s A_s}\right) + \left(\frac{M_c}{E_c I_c} \cdot \frac{D}{2} + \frac{M_s}{E_s I_s} \cdot \frac{d}{2}\right). \tag{4.24}$$

By Equation (4.22), the slip on the interface can be determined with

$$s = \frac{Q}{K}. \tag{4.25}$$

By assumption (3), the shear density transferred by single shear stud on the interface is

$$q = Q/a, \tag{4.26}$$

where a is the spacing of shear studs.

Consider a differential unit of the concrete flange (see Figure 4.8), and the force equilibrium of the unit in horizontal is

$$\frac{dN}{dx} = -q. \tag{4.27}$$

Combining Equations (4.25)–(4.27) gives

$$s = -\frac{a}{K} \cdot \frac{dN}{dx}. \tag{4.28}$$

Then, the slip strain at the steel–concrete interface can also be expressed as

$$\varepsilon_{slip} = \frac{ds}{dx} = -\frac{a}{K} \cdot \frac{d^2 N}{dx^2}. \tag{4.29}$$

Equalling Equation (4.24) to Equation (4.29) results in

$$\frac{a}{K} \frac{d^2 N}{dx^2} - \left(\frac{1}{E_c A_c} + \frac{1}{E_s A_s}\right) N + \frac{M_c}{E_c I_c} \cdot \frac{D}{2} + \frac{M_s}{E_s I_s} \cdot \frac{d}{2} = 0. \tag{4.30}$$

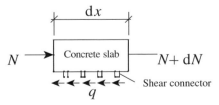

Figure 4.8 Horizontal balance of the concrete unit

By assumptions (4) and (5), the moment–curvature relationship of the steel and concrete components is

$$\frac{M_c}{E_c I_c} = \frac{M_s}{E_s I_s} = \kappa = -y'', \tag{4.31}$$

and it leads to

$$\frac{M_c + M_s}{E_c I_c + E_s I_s} = \kappa = -y''. \tag{4.32}$$

By Equation (4.20), one has

$$N = (M - M_c + M_s)/r. \tag{4.33}$$

Substituting Equations (4.31)–(4.33) into Equation (4.30) leads to the following fourth-order differential equilibrium equation of the partially composite beam:

$$\frac{\mathrm{d}^4 y}{\mathrm{d}x^4} - \frac{K}{a}\left(\frac{1}{E_c A_c} + \frac{1}{E_s A_s} + \frac{r^2}{E_c I_c + E_s I_s}\right)\frac{\mathrm{d}^2 y}{\mathrm{d}x^2} + \frac{1}{E_c I_c + E_s I_s}\cdot\frac{\mathrm{d}^2 M}{\mathrm{d}x^2}$$
$$- \frac{K}{a}\left(\frac{1}{E_c A_c} + \frac{1}{E_s A_s}\right)\cdot\frac{1}{E_c I_c + E_s I_s}M = 0. \tag{4.34}$$

Employing the definition of $(EI)^0_{comp}$ and $(EI)^\infty_{comp}$, we can simplify the above equation as

$$\frac{\mathrm{d}^4 y}{\mathrm{d}x^4} - \frac{k\cdot(EI)^\infty_{comp}}{EA\cdot(EI)^0_{comp}}\frac{\mathrm{d}^2 y}{\mathrm{d}x^2} + \frac{1}{(EI)^0_{comp}}\frac{\mathrm{d}^2 M}{\mathrm{d}x^2} - \frac{k}{EA\cdot(EI)^0_{comp}}M = 0, \tag{4.35}$$

where $k = K/a$ is the shear modulus of the steel–concrete interface (unit: N/mm^2). When $k = \infty$, namely there is no slip on the interface, Equation (4.35) returns to the equation for the composite beam with full composite action, i.e.

$$(EI)^\infty_{comp}\frac{\mathrm{d}^2 y}{\mathrm{d}x^2} + M = 0. \tag{4.36}$$

And when $k = 0$, Equation (4.35) can also return to the equation for the composite beam with none of composite action, i.e.

$$(EI)^0_{comp}\frac{\mathrm{d}^2 y}{\mathrm{d}x^2} + M = 0. \tag{4.37}$$

4.3.3 Stiffness Equation of Composite Beam Element

The typical forces and deformations of the beam element are as in Figure 4.9. The moment at an arbitrary location distance x away from end 1 can be expressed with the end moment M_1 and the end shear Q_1 as

$$M = M_1 - Q_1 x. \tag{4.38}$$

The force balance also determines

$$Q_1 = \frac{1}{l}(M_1 + M_2), \tag{4.39}$$

$$Q_1 = -Q_2. \tag{4.40}$$

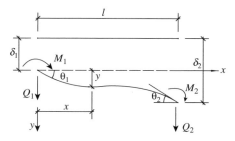

Figure 4.9 The typical forces and deformations of the beam element

Substituting Equation (4.38) into Equation (4.35) yields

$$\frac{d^4 y}{dx^4} - \alpha^2 \frac{d^2 y}{dx^2} - \beta(M_1 - Q_1 x) = 0, \tag{4.41}$$

where α and β are parameters relevant to material properties and section dimensions and are defined as

$$\alpha^2 = \frac{k(EI)^{\infty}_{comp}}{EA \cdot (EI)^0_{comp}}, \tag{4.42}$$

$$\beta = \frac{k}{EA \cdot (EI)^0_{comp}} = \frac{\alpha^2}{(EI)^{\infty}_{comp}}. \tag{4.43}$$

The solution of the above fourth-order differential equation (4.41) is

$$z = y'' = C_1 \cosh \alpha x + C_2 \sinh \alpha x - \frac{1}{(EI)^{\infty}_{comp}}(M_1 - Q_1 x), \tag{4.44}$$

where C_1 and C_2 are integration constants.

Integrating Equation (4.44) twice results in the deflection of the composite beam element with slip as

$$y = \frac{C_1}{\alpha^2} \cosh \alpha x + \frac{C_2}{\alpha^2} \sinh \alpha x - \frac{1}{(EI)^{\infty}_{comp}}\left(\frac{1}{2}M_1 x^2 - \frac{1}{6}Q_1 x^3\right) + C_3 x + C_4, \tag{4.45}$$

where C_3 and C_4 are also integration constants.

Consider the following boundary conditions:

- for $x = 0$:

$$y = 0, \tag{4.46a}$$
$$y' = \theta_1; \tag{4.46b}$$

- for $x = l$:

$$y = \delta_2 - \delta_1, \tag{4.47a}$$
$$y' = \theta_2. \tag{4.47b}$$

Introducing the above boundary conditions into Equation (4.45) yields four simultaneous algebra equations:

$$\frac{C_1}{\alpha^2} + C_4 = 0, \tag{4.48a}$$

$$\frac{C_2}{\alpha} + C_3 = \theta_1, \tag{4.48b}$$

$$\frac{C_1}{\alpha^2} \cosh \alpha l + \frac{C_2}{\alpha^2} \sinh \alpha l - \frac{1}{(EI)_{\text{comp}}^{\infty}} \left(\frac{1}{2} M_1 l^2 - \frac{1}{6} Q_1 l^3 \right) + C_3 l + C_4 = \delta_2 - \delta_1, \tag{4.48c}$$

$$\frac{C_1}{\alpha} \sinh \alpha l + \frac{C_2}{\alpha} \cosh \alpha l - \frac{1}{(EI)_{\text{comp}}^{\infty}} \left(M_1 l - \frac{1}{2} Q_1 l^2 \right) + C_3 = \theta_2. \tag{4.48d}$$

Solving Equations (4.48a)–(4.48d), one has

$$C_1 = \frac{\alpha}{12 + 6\alpha l \sinh \alpha l - 12 \cosh \alpha l} \left[6\alpha(\cosh \alpha l - 1)(\delta_2 - \delta_1) - 6 \sinh \alpha l(\theta_2 - \theta_1) \right.$$
$$\left. - \frac{Q_1 \alpha l^3}{(EI)_{\text{comp}}^{\infty}} (2 + \cosh \alpha l) + \frac{3M_1 \alpha l^2}{(EI)_{\text{comp}}^{\infty}} (1 + \cosh \alpha l) + 6\alpha l(\theta_2 - \theta_1 \cosh \alpha l) - \frac{3l \sinh \alpha l}{(EI)_{\text{comp}}^{\infty}} (2M_1 - Q_1 l) \right], \tag{4.49a}$$

$$C_2 = \frac{\alpha}{12 + 6\alpha l \sinh \alpha l - 12 \cosh \alpha l} \left[-6\alpha \sinh \alpha l(\delta_2 - \delta_1) + 6(\cosh \alpha l - 1)(\theta_2 - \theta_1) \right.$$
$$\left. - \frac{3Q_1 l^2}{(EI)_{\text{comp}}^{\infty}} (\cosh \alpha l - 1) - \frac{6M_1 l}{(EI)_{\text{comp}}^{\infty}} (1 - \cosh \alpha l) - \frac{\alpha l^2 \sinh \alpha l}{(EI)_{\text{comp}}^{\infty}} (3M_1 - Q_1 l) + 6\alpha l \theta_1 \sinh \alpha l \right], \tag{4.49b}$$

$$C_3 = \frac{1}{12 + 6\alpha l \sinh \alpha l - 12 \cosh \alpha l} \left[6(1 - \cosh \alpha l)(\theta_2 + \theta_1) + 6\alpha \sinh \alpha l(\delta_2 - \delta_1) \right.$$
$$\left. + \frac{6}{(EI)_{\text{comp}}^{\infty}} M_1 l(1 - \cosh \alpha l) - \frac{3}{(EI)_{\text{comp}}^{\infty}} Q_1 l^2 (1 - \cosh \alpha l) + \frac{1}{(EI)_{\text{comp}}^{\infty}} \alpha l^2 \sinh \alpha l(3M_1 - Q_1 l) \right], \tag{4.49c}$$

$$C_4 = \frac{1}{\alpha(12 + 6\alpha l \sinh \alpha l - 12 \cosh \alpha l)} \left[6\alpha(1 - \cosh \alpha l)(\delta_2 - \delta_1) + 6\theta_2(\sinh \alpha l - \alpha l) \right.$$
$$+ 6\theta_1(\alpha l \cosh \alpha l - \sinh \alpha l) - \frac{3\alpha}{(EI)_{\text{comp}}^{\infty}} M_1 l^2 (1 + \cosh \alpha l)$$
$$\left. + \frac{\alpha}{(EI)_{\text{comp}}^{\infty}} Q_1 l^3 (2 + \cosh \alpha l) + \frac{3l \sinh \alpha l}{(EI)_{\text{comp}}^{\infty}} (2M_1 - Q_1 l) \right]. \tag{4.49d}$$

In most cases, steel beams are connected to columns fixedly, and when the anchor-hold of negative reinforcement bars in concrete slabs has good performance, it is reasonable to assume that the slip between the steel beams and concrete slabs at the ends of composite beams is negligible, namely

$$s|_{x=0} = 0 \tag{4.50a}$$

and

$$s|_{x=l} = 0. \tag{4.50b}$$

Substituting Equation (4.50) into Equation (4.28) leads to

$$\frac{dN}{dx}\Big|_{x=0} = 0, \tag{4.51a}$$

$$\frac{dN}{dx}\Big|_{x=l} = 0. \tag{4.51b}$$

Combing Equations (4.20), (4.33), (4.38) and (4.44), one has

$$\frac{dN}{dx} = \frac{1}{r}\left[(EI)^0_{\text{comp}}(C_1\alpha \sinh \alpha l + C_2\alpha \cosh \alpha l) + \left(\frac{(EI)^0_{\text{comp}}}{(EI)^\infty_{\text{comp}}} - 1 \right)Q_1 \right]. \tag{4.52}$$

Substituting Equation (4.52) into Equation (4.51) and considering Equations (4.49), (4.39) and (4.40) yields

$$M_1 = \frac{(EI)^0_{\text{comp}}\alpha^3 \sinh \alpha l}{K_s} \cdot 6(\delta_1 - \delta_2) + \left(\frac{3l(EI)^0_{\text{comp}}\alpha^3 \sinh \alpha l}{K_s} + \frac{(EI)^\infty_{\text{comp}}}{l} \right)\theta_1$$

$$+ \left(\frac{3l(EI)^0_{\text{comp}}\alpha^3 \sinh \alpha l}{K_s} - \frac{(EI)^\infty_{\text{comp}}}{l} \right)\theta_2, \tag{4.53a}$$

$$M_1 = \frac{(EI)^0_{\text{comp}}\alpha^3 \sinh \alpha l}{K_s} \cdot 6(\delta_1 - \delta_2) + \left(\frac{3l(EI)^0_{\text{comp}}\alpha^3 \sinh \alpha l}{K_s} + \frac{(EI)^\infty_{\text{comp}}}{l} \right)\theta_1$$

$$+ \left(\frac{3l(EI)^0_{\text{comp}}\alpha^3 \sinh \alpha l}{K_s} + \frac{(EI)^\infty_{\text{comp}}}{l} \right)\theta_2, \tag{4.53b}$$

$$Q_1 = -Q_2 = \frac{M_1 + M_2}{l} = \frac{(EI)^0_{\text{comp}}\alpha^3 \sinh \alpha l}{K_s \cdot l}[12(\delta_2 - \delta_1) + 6l(\theta_1 + \theta_2)], \tag{4.53c}$$

where

$$K_s = \frac{l}{(EI)^\infty_{\text{comp}}}\left[(EI)^0_{\text{comp}}\alpha^3 l \sinh \alpha l + \frac{2}{l^2}((EI)^\infty_{\text{comp}} - (EI)^0_{\text{comp}})(12 + 6\alpha l \sinh \alpha l - 12 \cosh \alpha l) \right]. \tag{4.54}$$

Equation (4.53) can also be expressed with the standard form as

$$M_1 = \frac{(EI)^\infty_{\text{comp}}}{l}\left(4\varphi_2\theta_1 + 2\varphi_3\theta_2 + 6\varphi_1 \frac{\delta_1 - \delta_2}{l} \right), \tag{4.55a}$$

$$M_2 = \frac{(EI)^\infty_{\text{comp}}}{l}\left(2\varphi_3\theta_1 + 4\varphi_2\theta_2 + 6\varphi_1 \frac{\delta_1 - \delta_2}{l} \right), \tag{4.55b}$$

$$Q_1 = -Q_2 = \frac{(EI)^\infty_{\text{comp}}}{l}\left(\frac{4\varphi_2 + 2\varphi_3}{l}\theta_1 + \frac{4\varphi_2 + 2\varphi_3}{l}\theta_2 + 12\varphi_1 \frac{\delta_1 - \delta_2}{l^2} \right), \tag{4.55c}$$

where

$$\varphi_1 = \frac{(EI)^0_{\text{comp}}(\alpha l)^3 \sinh \alpha l}{(EI)^0_{\text{comp}}(\alpha l)^3 \sinh \alpha l + 2((EI)^\infty_{\text{comp}} - (EI)^0_{\text{comp}})(12 + 6\alpha l \sinh \alpha l - 12 \cosh \alpha l)}, \tag{4.56a}$$

$$\varphi_2 = \frac{3}{4} \cdot \frac{(EI)^0_{\text{comp}}(\alpha l)^3 \sinh \alpha l}{(EI)^0_{\text{comp}}(\alpha l)^3 \sinh \alpha l + 2((EI)^\infty_{\text{comp}} - (EI)^0_{\text{comp}})(12 + 6\alpha l \sinh \alpha l - 12 \cosh \alpha l)} + \frac{1}{4}, \tag{4.56b}$$

$$\varphi_3 = \frac{3}{2} \cdot \frac{(EI)^0_{\text{comp}}(\alpha l)^3 \sinh \alpha l}{(EI)^0_{\text{comp}}(\alpha l)^3 \sinh \alpha l + 2((EI)^\infty_{\text{comp}} - (EI)^0_{\text{comp}})(12 + 6\alpha l \sinh \alpha l - 12 \cosh \alpha l)} - \frac{1}{2}. \tag{4.56c}$$

The matrix expression of Equation (4.55) is

$$\frac{(EI)^\infty_{\text{comp}}}{l}
\begin{bmatrix}
\dfrac{12}{l^2}\varphi_1 & \dfrac{4\varphi_2 + 2\varphi_3}{l} & -\dfrac{12}{l^2}\varphi_1 & \dfrac{4\varphi_2 + 2\varphi_3}{l} \\
\dfrac{6}{l}\varphi_1 & 4\varphi_2 & -\dfrac{6}{l}\varphi_1 & 2\varphi_3 \\
-\dfrac{12}{l^2}\varphi_1 & -\dfrac{4\varphi_2 + 2\varphi_3}{l} & \dfrac{12}{l^2}\varphi_1 & -\dfrac{4\varphi_2 + 2\varphi_3}{l} \\
\dfrac{6}{l}\varphi_1 & 2\varphi_3 & -\dfrac{6}{l}\varphi_1 & 4\varphi_2
\end{bmatrix}
\begin{Bmatrix} \delta_1 \\ \theta_1 \\ \delta_2 \\ \theta_2 \end{Bmatrix} =
\begin{Bmatrix} Q_1 \\ M_1 \\ Q_2 \\ M_2 \end{Bmatrix} \tag{4.57}$$

or

$$[k_{ce}]\{\delta_c\} = \{f_c\}, \tag{4.58}$$

where

$$\{\delta_c\} = [\delta_1, \theta_1, \delta_2, \theta_2]^T, \tag{4.59a}$$

$$\{f_c\} = [Q_1, M_1, Q_2, M_2]^T, \tag{4.59b}$$

$$[k_{ce}] = \frac{(EI)^\infty_{\text{comp}}}{l}
\begin{bmatrix}
\dfrac{12}{l^2}\varphi_1 & \dfrac{4\varphi_2 + 2\varphi_3}{l} & -\dfrac{12}{l^2}\varphi_1 & \dfrac{4\varphi_2 + 2\varphi_3}{l} \\
\dfrac{6}{l}\varphi_1 & 4\varphi_2 & -\dfrac{6}{l}\varphi_1 & 2\varphi_3 \\
-\dfrac{12}{l^2}\varphi_1 & -\dfrac{4\varphi_2 + 2\varphi_3}{l} & \dfrac{12}{l^2}\varphi_1 & -\dfrac{4\varphi_2 + 2\varphi_3}{l} \\
\dfrac{6}{l}\varphi_1 & 2\varphi_3 & -\dfrac{6}{l}\varphi_1 & 4\varphi_2
\end{bmatrix}. \tag{4.60}$$

Equation (4.57) or (4.58) is the elastic stiffness equation for the steel–concrete composite beam element with partial composite action and $[k_{ce}]$ is the corresponding elastic stiffness matrix of the element.

The parameters in $[k_{ce}]$ satisfy

$$6\varphi_1 = 4\varphi_2 + 2\varphi_3, \tag{4.61}$$

which indicates that $[k_{ce}]$ is exactly symmetric.

4.3.4 Equivalent Nodal Load Vector

When a composite beam element is subjected to non-nodal loads, the equivalent nodal load vector of the element is necessary for structural analysis using FEM. The equivalent nodal load vectors for three types of non-nodal loads on the composite beam element are to be discussed in this section.

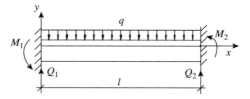

Figure 4.10   A two-end fixed composite beam subject to a uniformly distributed load

The first non-nodal load is the uniformly distributed load applied on a two-end fixed composite beam, as shown in Figure 4.10. By force balance and symmetry, one has

$$Q_1 = Q_2 = \frac{1}{2}ql^2, \tag{4.62}$$

$$M_1 = -M_2. \tag{4.63}$$

And the moment along elemental length is

$$M = Q_1 x - M_1 - \frac{1}{2}qx^2. \tag{4.64}$$

Substituting Equation (4.64) into Equation (4.35) results in the differential equilibrium equation of the two-end fixed composite beam element subjected to the uniform load as

$$\frac{\mathrm{d}^4 y}{\mathrm{d}x^4} - \alpha^2 \frac{\mathrm{d}^2 y}{\mathrm{d}x^2} - \frac{1}{(EI)^0_{\text{comp}}}q - \frac{\alpha^2}{(EI)^\infty_{\text{comp}}}\left(Q_1 x - M_1 - \frac{1}{2}qx^2\right) = 0. \tag{4.65}$$

The solution of Equation (4.65) is

$$y'' = C_1 \cosh \alpha x + C_2 \sinh \alpha x - \frac{q}{\alpha^2}\left(\frac{1}{(EI)^0_{\text{comp}}} - \frac{1}{(EI)^\infty_{\text{comp}}}\right) - \frac{1}{(EI)^\infty_{\text{comp}}}\left(Q_1 x - M_1 - \frac{1}{2}qx^2\right). \tag{4.66}$$

The deflection and rotation can be obtained by integrating Equation (4.66) as

$$\theta = y' = \frac{C_1}{\alpha}\sinh \alpha x + \frac{C_2}{\alpha}\cosh \alpha x - \frac{qx}{\alpha^2}\left(\frac{1}{(EI)^0_{\text{comp}}} - \frac{1}{(EI)^\infty_{\text{comp}}}\right)$$
$$- \frac{1}{(EI)^\infty_{\text{comp}}}\left(\frac{1}{2}Q_1 x^2 - M_1 x - \frac{1}{6}qx^3\right) + C_3, \tag{4.67}$$

$$y = \frac{C_1}{\alpha^2}\cosh \alpha x + \frac{C_2}{\alpha^2}\sinh \alpha x - \frac{1}{2}\frac{qx^2}{\alpha^2}\left(\frac{1}{(EI)^0_{\text{comp}}} - \frac{1}{(EI)^\infty_{\text{comp}}}\right)$$
$$- \frac{1}{(EI)^\infty_{\text{comp}}}\left(\frac{1}{6}Q_1 x^3 - \frac{1}{2}M_1 x^2 - \frac{1}{24}qx^4\right) + C_3 x + C_4, \tag{4.68}$$

where C_1–C_4 are integration parameters.

C_3 and C_4 can be determined with the deflection boundary condition, $y|_{x=0} = y|_{x=l} = 0$, as

$$C_3 = -\frac{C_1}{\alpha^2 l}\cosh \alpha l - \frac{C_2}{\alpha^2 l}\sinh \alpha l + \frac{1}{2}\frac{ql}{\alpha^2}\left(\frac{1}{(EI)^\infty_{\text{comp}}} - \frac{1}{(EI)^\infty_{\text{comp}}}\right)$$
$$+ \frac{1}{(EI)^\infty_{\text{comp}}}\left(\frac{1}{6}Q_1 l^2 - \frac{1}{2}M_1 l - \frac{1}{24}ql^3\right) + \frac{C_1}{\alpha^2 l}, \tag{4.69}$$

$$C_4 = -\frac{C_1}{\alpha^2}. \tag{4.70}$$

C_1 and C_2 can be determined with the slip boundary condition, Equation (4.50), as

$$C_2 = \frac{Q_1}{\alpha} \left(\frac{1}{(EI)^{\infty}_{\text{comp}}} - \frac{1}{(EI)^{0}_{\text{comp}}} \right) = \frac{ql}{2\alpha} \left(\frac{1}{(EI)^{\infty}_{\text{comp}}} - \frac{1}{(EI)^{0}_{\text{comp}}} \right), \tag{4.71}$$

$$C_1 = \frac{ql(1 + \cosh \alpha l)}{2\alpha \sinh \alpha l} \left(\frac{1}{(EI)^{0}_{\text{comp}}} - \frac{1}{(EI)^{\infty}_{\text{comp}}} \right). \tag{4.72}$$

Finally, with the rotation boundary condition $\theta|_{x=0} = 0$, one has

$$\frac{C_2}{\alpha} + C_3 = 0. \tag{4.73}$$

Substituting Equations (4.69), (4.71) and (4.72) into Equation (4.73) leads to

$$M_1 = -M_2 = \frac{1}{12} ql^2. \tag{4.74}$$

Equations (4.62) and (4.74) are the equivalent nodal forces for the uniformly distributed load applied on the composite beam element, which is obviously independent of the composite action of the beam.

As for the other two cases, a concentrated load at mid-span of the beam (see Figure 4.11) and the triangularly distributed load (see Figure 4.12), the equivalent nodal forces can also be determined by a similar procedure as presented hereinabove.

For a concentrated load at mid-span, it can be obtained that

$$Q_1 = Q_2 = \frac{1}{2} Pl, \tag{4.75}$$

$$M_1 = -M_2 = \frac{1}{8} Pl. \tag{4.76}$$

And for distributed triangle loads, it can be obtained that

$$Q_1 = \frac{1}{2} q_0 l - Q_2 = \frac{3}{20} q_0 l$$

$$\cdot \left\{ \begin{array}{c} \dfrac{(EI)^{\infty}_{\text{comp}} - (EI)^{0}_{\text{comp}}}{(EI)^{\infty}_{\text{comp}} \cdot (EI)^{0}_{\text{comp}}} \left[\dfrac{1 - \cosh \alpha l}{\alpha^3 l \sinh \alpha l} \left(\dfrac{40}{3\alpha^2 l} + \dfrac{10l}{3} \right) + \dfrac{1}{\alpha^2} \left(\dfrac{10l}{9} + \dfrac{20}{3\alpha^2 l} \right) \right] - \dfrac{l^3}{12(EI)^{\infty}_{\text{comp}}} \\[4mm] \dfrac{(EI)^{\infty}_{\text{comp}} - (EI)^{0}_{\text{comp}}}{(EI)^{\infty}_{\text{comp}} \cdot (EI)^{0}_{\text{comp}}} \left[\dfrac{2(1 - \cosh \alpha l)}{\alpha^3 \sinh \alpha l} - \dfrac{l}{\alpha^2} \right] + \dfrac{l^3}{12(EI)^{\infty}_{\text{comp}}} \end{array} \right\}, \tag{4.77}$$

$$M_1 = \frac{1}{2} Q_1 l - \frac{1}{24} q_0 l^2, \tag{4.78a}$$

$$M_2 = Q_1 l - M_1 - \frac{1}{6} q_0 l^2. \tag{4.78b}$$

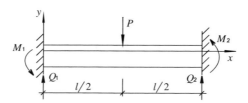

Figure 4.11 A composite beam subjected to a concentrated load at mid-span

EXAMPLE   **49**

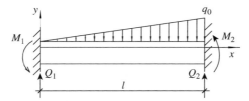

Figure 4.12   A composite beam subjected to a triangularly distributed load

From the above results, it can be found that the equivalent nodal forces of the composite beam element are independent of the shear stiffness on the steel–concrete interface for symmetric load cases (such as uniformly distributed load shown in Figure 4.10 and concentrated load at mid-span shown in Figure 4.11) and have the same form as that of an ordinary beam element. However, for the asymmetric load case (such as distributed triangular loads shown in Figure 4.12), the equivalent nodal forces rely on the shear stiffness on the steel–concrete interface of the composite beam element.

4.4 EXAMPLE

A five-storey steel frame (Baotou Steel & Iron Design and Research Institute, 2000) is selected to illustrate the effectiveness of the stiffness equation of composite beams derived above. The plan and elevation views of the frame are shown in Figure 4.13. The cross-sectional sizes of the frame beams and columns are also given in Figure 4.13, and the material for all the beams and columns is Q235 according to the Chinese standard. This frame was analysed as a pure steel frame, and the composite action from the concrete slab was neglected, under dead and live floor loads, wind and earthquake actions, by Baotou Steel & Iron Design and Research Institute (2000).

The composite beam element developed hereinabove, with various values of steel–concrete interface shear stiffness k, is used to investigate the effects of composite action on resultants and deformations of the steel frame. The effective width of the concrete–slab flange for the composite beams of the frame is determined based on Chinese code GB50017-2003. Alternatively, it is recommended in Chinese code JGJ99-98 that the moment of inertia of the composite beam can be approximately adopted as 1.5 times that of the pure steel beam in high-rise steel buildings when the full composite action between the cast-in-site

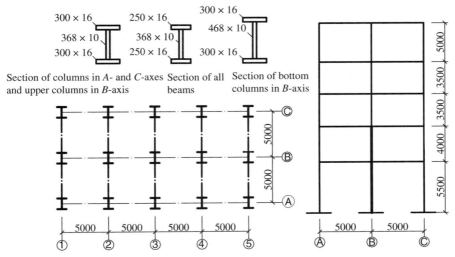

Figure 4.13   The steel frame for example

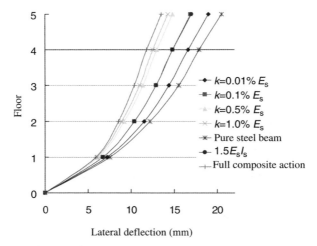

Figure 4.14   Effects of composite action on frame stiffness

concrete slab and the steel beam is considered. The results of various considerations on the composite actions of the frame will be compared as follows.

The values of the steel–concrete interface shear stiffness k are taken as 0.01, 0.1, 0.5 and 1.0 % of the elastic modulus of steel, E_s, and $k = \infty$ represents the fully composite case. It is revealed by numerical analyses that the effect of composite action from the concrete slab is small and can be negligible on internal forces of the frame under vertical and horizontal loads, but this effect is evident on the global lateral stiffness of the frame under horizontal loads and relates to the steel–concrete interface shear stiffness. The effect of composite action on the lateral deflection of the frame subjected to a horizontal earthquake load is given in Figure 4.14, and the relationship between the storey drift and the steel–concrete interface shear stiffness is shown in Figure 4.15.

It can be found from Figure 4.14 that the global frame sway is the greatest when the contribution of concrete slabs is waived and it is the least when the full composite action is considered between concrete slabs and steel beams (the deflection gap between these two extreme cases in this example is 34 %). The deflection curves for partial composite actions are fallen in between these two extreme cases. The results of the approximate consideration of the composite action according to Chinese code JGJ99-98 agree with those of the frame with partially composite beams with a $k$value of 0.1 % E_s.

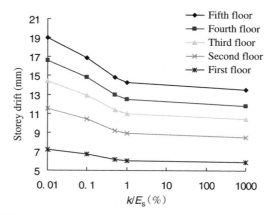

Figure 4.15   Effect of composite action on frame storey drifts

It can be found from Figure 4.15 that the effect of the steel–concrete interface shear stiffness on frame lateral stiffness is significant when k/E_s is less than 1.0 %, otherwise it has little significance. It can also be found that the effect of k on the drifts of the storeys at higher levels is more evident.

4.5 PROBLEMS IN PRESENT WORK

In the derivation of the stiffness equation of the composite beam element with relative slip, an important assumption is made that there is no slip between the concrete slab and the steel beam at the two ends of the element. This assumption is to enhance the stiffness of the composite beam element because there will always be, more or less, a relative slip at the beam ends.

Equation (4.60) can be regressed into

- for $k = \infty$:

$$[k_\infty] = \frac{(EI)^\infty_{\text{comp}}}{l}
\begin{bmatrix}
\dfrac{12}{l^2} & \dfrac{6}{l} & -\dfrac{12}{l^2} & \dfrac{6}{l} \\[6pt]
\dfrac{6}{l} & 4 & -\dfrac{6}{l} & 2 \\[6pt]
-\dfrac{12}{l^2} & -\dfrac{6}{l} & \dfrac{12}{l^2} & -\dfrac{6}{l} \\[6pt]
\dfrac{6}{l} & 2 & -\dfrac{6}{l} & 4
\end{bmatrix};
\tag{4.79}$$

- for $k = 0$:

$$[k_0] =
\begin{bmatrix}
\dfrac{12}{l^3}(EI)^0_{\text{comp}} & \dfrac{6}{l^2}(EI)^0_{\text{comp}} & -\dfrac{12}{l^3}(EI)^0_{\text{comp}} & \dfrac{6}{l^2}(EI)^0_{\text{comp}} \\[10pt]
\dfrac{6}{l^2}(EI)^0_{\text{comp}} & \left(\dfrac{3}{l}(EI)^0_{\text{comp}}+\dfrac{1}{l}(EI)^\infty_{\text{comp}}\right) & -\dfrac{6}{l^2}(EI)^0_{\text{comp}} & \left(\dfrac{3}{l}(EI)^0_{\text{comp}}-\dfrac{1}{l}(EI)^\infty_{\text{comp}}\right) \\[10pt]
-\dfrac{12}{l^3}(EI)^0_{\text{comp}} & -\dfrac{6}{l^2}(EI)^0_{\text{comp}} & \dfrac{12}{l^3}(EI)^0_{\text{comp}} & -\dfrac{6}{l^2}(EI)^0_{\text{comp}} \\[10pt]
\dfrac{6}{l^2}(EI)^0_{\text{comp}} & \left(\dfrac{3}{l}(EI)^0_{\text{comp}}-\dfrac{1}{l}(EI)^\infty_{\text{comp}}\right) & -\dfrac{6}{l^2}(EI)^0_{\text{comp}} & \left(\dfrac{3}{l}(EI)^0_{\text{comp}}+\dfrac{1}{l}(EI)^\infty_{\text{comp}}\right)
\end{bmatrix}.
\tag{4.80}$$

Obviously, Equation (4.60) can be regressed into that for the fully composite beam element, whereas it cannot be regressed into that for the element without any composite action. This error results from the slip boundary condition on the steel–concrete interface assumed in Equation (4.50). By further analysis, when beam bends in asymmetric form (Figure 4.16(a)), it can be derived from Equation (4.80) that

$$\left(\frac{6}{l}\sum EI\right)\theta = M,
\tag{4.81}$$

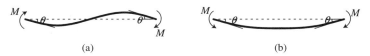

(a)   (b)

Figure 4.16   Two typical deformations of the beam element: (a) asymmetric deformation; (b) symmetric deformation

and when beam bends in symmetric form (Figure 4.16(b)), it can be

$$\left(\frac{2}{l}\overline{EI}\right)\theta = M. \tag{4.82}$$

Equation (4.80) is consistent but Equation (4.81) is not with the bending stiffness relationship of the beam without any composite action. Therefore, good accuracy can be obtained when Equation (4.60) is used for the analysis of frames subjected to horizontal loads because frame beams bend in asymmetric form under horizontal loads. However, error will occur when Equation (4.60) is used for the analysis of frames subjected to vertical loads when beams bend in symmetric form. This is a problem in present research and needs further study.

5 Sectional Yielding and Hysteretic Model of Steel Beam Columns

5.1 YIELDING OF BEAM SECTION SUBJECTED TO UNIAXIAL BENDING

The different states in the yielding process of the beam section subjected to uniaxial bending are shown in Figure 5.1, where σ_s is the steel yielding stress, M_s is the moment at yielding of the sectional edge (termed as initial yielding moment) and M_p is the moment at yielding of the full section (termed as plastic or ultimate yielding moment). M_s and M_p can be calculated as

$$M_s = W_e\sigma_s, \tag{5.1}$$
$$M_p = W_p\sigma_s, \tag{5.2}$$

where W_e is the elastic section modulus and W_p is the plastic section modulus.
   Let

$$\chi_p = \frac{M_p}{M_e} = \frac{W_p}{W_e}, \tag{5.3}$$

which is relevant only to sectional shape and thus can be called as the section plastic shape factor. The values of χ_p for normal symmetric sections are given in Figure 5.2.

5.2 YIELDING OF COLUMN SECTION SUBJECTED TO UNIAXIAL BENDING

Axial forces generally exist in frame columns, and Figure 5.3 illustrates the yielding process of a column section, where M_{sN} and M_{pN} are, respectively, the initial and ultimate yielding moments of the section including the contribution of axial force.
   The condition of initial yielding for a section of any shape can be expressed as

$$\frac{M_{sN}}{W_e} + \frac{N}{A} = \sigma_s. \tag{5.4}$$

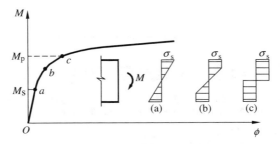

Figure 5.1   Yielding process and moment–curvature relationship of a beam section subjected to uniaxial bending

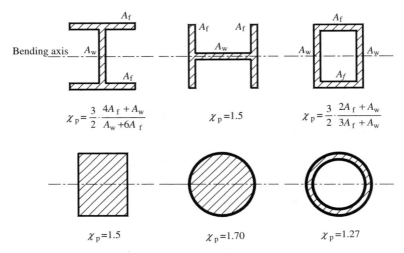

Figure 5.2   Values of plastic shape factors for various symmetric sections

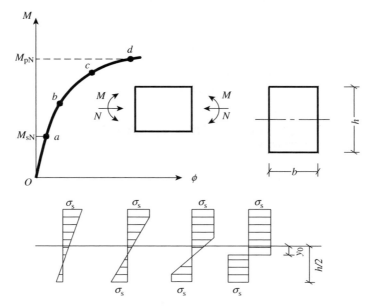

Figure 5.3   Yielding process and moment–curvature relationship of a column section subjected to uniaxial bending

Rewriting it results in

$$M_{\mathrm{sN}} = \left(1 - \frac{N}{N_{\mathrm{p}}}\right) M_{\mathrm{s}}, \tag{5.5}$$

where A is the sectional area, N is the axial force and $N_{\mathrm{p}} = A\sigma_{\mathrm{s}}$ is the yielding load of the section when single axial force is applied.

The ultimate yielding moment can be derived from the equilibrium between sectional internal and external forces on a section in the ultimate state. Considering the ultimate yielding state of a rectangular section (Figure 4), one obtains the equilibrium equations as

$$N = 2\sigma_{\mathrm{s}} b y_0,$$

$$M_{\mathrm{pN}} = \frac{\sigma_{\mathrm{s}} b}{4}(h^2 - 4y_0^2).$$

Eliminating y_0 from the above two equations and noting that $M_{\mathrm{p}} = \sigma_{\mathrm{s}} b h^2/4$ for the rectangular section lead to the expression of M_{pN} as

$$M_{\mathrm{pN}} = \left[1 - \left(\frac{N}{N_{\mathrm{p}}}\right)^2\right] M_{\mathrm{p}}. \tag{5.6}$$

Following the similar procedures as described above, one can calculate M_{pN} of frame columns with biaxial symmetric H sections as follows:

(1) when bending in the major axis:

– for the neutral axis within the web plate, i.e. $0 \leq N/N_{\mathrm{p}} \leq \alpha/(2+\alpha)$,

$$M_{\mathrm{pN}} = \left[1 - \frac{(2+\alpha)^2}{(4+\alpha)\alpha}\left(\frac{N}{N_{\mathrm{p}}}\right)^2\right] M_{\mathrm{p}}; \tag{5.7a}$$

– for the neutral axis within the flange plate, i.e. $\alpha/(2+\alpha) \leq N/N_{\mathrm{p}} \leq 1$,

$$M_{\mathrm{pN}} = \left[1 - \left(\frac{N}{N_{\mathrm{p}}}\right)\right] \frac{2(2+\alpha)}{4+\alpha} M_{\mathrm{p}}; \tag{5.7b}$$

(2) when bending in the minor axis:

– for the neutral axis within the web plate, i.e. $0 \leq N/N_{\mathrm{p}} \leq \alpha/2 + \alpha$,

$$M_{\mathrm{pN}} = M_{\mathrm{p}}; \tag{5.8a}$$

– for the neutral axis within the flange plate, i.e. $\alpha/(2+\alpha) \leq N/N_{\mathrm{p}} \leq 1$,

$$M_{\mathrm{pN}} = \frac{4-\alpha^2}{4}\left[1 - \frac{2\alpha}{2-\alpha}\left(\frac{N}{N_{\mathrm{p}}}\right) - \frac{2+\alpha}{2-\alpha}\left(\frac{N}{N_{\mathrm{p}}}\right)^2\right] M_{\mathrm{p}}, \tag{5.8b}$$

where α is the cross-sectional area ratio of the web to one flange.

The above expressions for H sections bending in the major axis can be used to calculate M_{pN} of box sections except that α is replaced with the cross-sectional area ratio of two web plates to one flange.

For circular or annular sections, M_{pN} is determined by

$$M_{pN} = \frac{3}{4}\left(\sin\varphi - \frac{1}{3}\sin 3\varphi\right)M_p, \tag{5.9a}$$

where φ is solved from

$$1 - \frac{2}{\pi}\left(\varphi - \frac{1}{2}\sin 2\varphi\right) = \frac{N}{N_p}. \tag{5.9b}$$

5.3 YIELDING OF COLUMN SECTION SUBJECTED TO BIAXIAL BENDING

5.3.1 Equation of Initial Yielding Surface

The cross section of frame columns is usually biaxial symmetric because frame columns are often subjected to biaxial bending. The initial yielding surface of a column section can be determined by linear super-imposition of normal stresses within elastic scope and can be written as

$$\left|\frac{M_x}{M_{sx}}\right| + \left|\frac{M_y}{M_{sy}}\right| + \left|\frac{N}{N_p}\right| = 1, \tag{5.10}$$

where M_x and M_y are the bending moments applied about the orthotropic x-axis and y-axis, respectively, N is the axial force applied, and M_{sx} and M_{sy} are the initial yielding moments of the section when M_x and M_y are applied alone, respectively.

Equation (5.10) is the equation of the initial yielding surface for the column section subjected to biaxial bending, which can also be rewritten as

$$\chi_{px}m_x + \chi_{py}m_y + n = 1, \tag{5.11}$$

in which $m_x = |M_x/M_{px}|$, $m_y = |M_y/M_{py}|$, $n = |N/N_p|$, M_{px} and M_{py} are the ultimate yielding moments of the section when M_x and M_y are applied alone, respectivley, and χ_{px} and χ_{py} are the plastic shape factors about the x-axis and y-axis of the section, respectively.

5.3.2 Equation of Ultimate Yielding Surface

The ultimate yielding surface of a column section subjected to biaxial bending moments and axial force can also be defined with m_x, m_y and n. The following describes the derivation of the ultimate yielding surface equation for the rectangular section.

As shown in Figure 5.4, let $y = f(x)$ be the neutral curve in the ultimate state, which indicates that the part of the section above $y = f(x)$ is yielding in tension and that below is yielding in compression. The equilibrium conditions of the section in this ultimate state are governed by

$$N = -\int_{-b/2}^{b/2} 2\sigma_s f(x)\,\mathrm{d}x, \tag{5.12a}$$

$$M_x = -\int_{-b/2}^{b/2} \sigma_s\left[\frac{h^2}{4} - f^2(x)\right]\mathrm{d}x, \tag{5.12b}$$

$$M_y = -\int_{-b/2}^{b/2} 2\sigma_s f(x)\,\mathrm{d}x, \tag{5.12c}$$

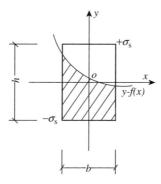

Figure 5.4   Ultimate yielding state of the rectangular section

where $f(x)$ is an undetermined function and can be determined with functional extremum theory. Assume that N and M_y are constant and then $f(x)$ should make M_x maximum. This problem is actually a definite integral inflexion problem with restrain of the definite integral. According to the Lagrange multiplicator method, the functional of the problem can be expressed as

$$H = \sigma_s[h^2/4 - f^2(x)] - \lambda_1[2\sigma_s x f(x)] - \lambda_2[2\sigma_s f(x)]. \qquad (5.13)$$

From the Euler equation,

$$\frac{\partial H}{\partial f(x)} - \frac{d}{dx}\left[\frac{\partial H}{\partial f'(x)}\right] = 0, \qquad (5.14)$$

one obtains

$$f(x) = -\lambda_1 x - \lambda_2. \qquad (5.15)$$

Equation (5.15) indicates that the neutral curve of the rectangular section is a straight line (see Figure 5.5). Substituting Equation (5.15) into Equation (5.12) yields

$$N = 2\sigma_s b \lambda_2, \qquad (5.16a)$$

$$M_x = 2\sigma_s\left[\frac{bh^2}{8} - \frac{\lambda_1^2}{3}\left(\frac{b}{2}\right)^3 - \lambda_2^2\left(\frac{b}{2}\right)\right], \qquad (5.16b)$$

$$M_y = \frac{1}{6}\sigma_s b^3 \lambda_1. \qquad (5.16c)$$

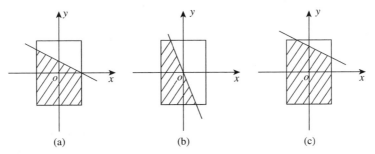

(a)   (b)   (c)

Figure 5.5   Neutral axial positions of the rectangular section

For the rectangular section, the ultimate yielding axial force and ultimate yielding moments are, respectively, obtained by

$$N = \sigma_s bh, \tag{5.17a}$$

$$M_x = \frac{1}{4}\sigma_s bh^2, \tag{5.17b}$$

$$M_y = \frac{1}{4}\sigma_s b^2 h. \tag{5.17c}$$

Combining Equations (5.16) and (5.17) leads to

$$n = \frac{N}{N_p} = \frac{2\lambda_2}{h}, \tag{5.18a}$$

$$m_x = \frac{M_x}{M_{px}} = \frac{8}{bh^2}\left[\frac{bh^2}{8} - \frac{\lambda_1^2}{3}\left(\frac{b}{2}\right)^3 - \lambda_2^2\left(\frac{b}{2}\right)\right], \tag{5.18b}$$

$$m_y = \frac{M_y}{M_{py}} = \frac{2b}{3h}\lambda_1. \tag{5.18c}$$

Eliminating λ_1 and λ_2 from Equation (5.18) results in the explicit equation of the ultimate yielding surface for rectangular sections as

$$n^2 + m_x + \frac{3}{4}m_y^2 = 1. \tag{5.19a}$$

It should be noted that Equation (5.19a) is correct only for the layout of the neutral axis as shown in Figure 5.5(a), and the condition is

$$m_y \leq \frac{2}{3}(1-n) \leq m_x. \tag{5.20a}$$

For the layout of the neutral axis as shown in Figure 5.5(b), the equation of the ultimate yielding surface is

$$n^2 + \frac{3}{4}m_x^2 + m_y = 1, \tag{5.19b}$$

and the corresponding condition is

$$m_x \leq \frac{2}{3}(1-n) \leq m_y. \tag{5.20b}$$

For the layout of the neutral axis as shown in Figure 5.5(c), the equation of the ultimate yielding surface is

$$n + \frac{9}{4}\left[1 - \frac{m_x}{2(1-n)}\right]\left[1 - \frac{m_y}{2(1-n)}\right] = 1, \tag{5.19c}$$

and the condition is

$$m_x \geq \frac{2}{3}(1-n), \qquad m_y \geq \frac{2}{3}(1-n). \tag{5.20c}$$

It can be seen from the above derivation that the equation of the ultimate yielding surface is not unique and depends on the position of the neutral axis.

Following the similar derivation, one can obtain the equations of the ultimate yielding surfaces of biaxial symmetric H and box sections, which are listed in Tables 5.1 and 5.2, respectively. The parameters in such equations are defined in Figures 5.6 and 5.7 for H and box sections, respectively.

Table 5.1 Yielding surface equation for the H section

Order	Position of the neutral axis	Yielding surface equation	Condition
1		$n = \dfrac{2}{(2+\alpha)\beta}\lambda_2(\alpha\beta + 2\lambda_1)$ $m_x = \dfrac{1}{(4+\alpha)\beta}[4\lambda_1 + \alpha\beta(1 - 4\lambda_2^2)]$ $m_y = \dfrac{1}{\beta^2}(\beta^2 - \lambda_1^2 - 4\lambda_1^2\lambda_2^2)$	$0 \le \lambda_2 \le 1/2$ $0 \le \lambda_1 \le \frac{\beta}{1+2\lambda_2}$
2		$n = \dfrac{1}{2+\alpha}\left[2\alpha\lambda_2 + \dfrac{\beta - \lambda_1 + \lambda_1\lambda_2}{\beta}\right]$ $m_x = \dfrac{2}{4+\alpha}\left[1 + \dfrac{\lambda_1 - 2\lambda_1\lambda_2}{\beta} + \alpha\left(\dfrac{1}{2} - 2\lambda_2\right)\right]$ $m_y = \dfrac{1}{\beta^2}\left[\dfrac{\beta^2}{2} - \left(\dfrac{\lambda_1}{2} - \lambda_1\lambda_2\right)^2\right]$	$\lambda_1 \ge \beta/2$ $0 \le \lambda_2 \le 1/2$
3		$n = \dfrac{1}{2+\alpha}\left(\alpha + \dfrac{4\lambda_1\lambda_2}{\beta}\right)$ $m_x = \dfrac{4}{4+\alpha}\dfrac{\lambda_1}{\beta}$ $m_y = \dfrac{1}{\beta^2}(\beta^2 - \lambda_1^2 - 4\lambda_1^2\lambda_2^2)$	$\lambda_2 \ge 1/2$ $0 \le \lambda_1 \le \frac{\beta}{1+2\lambda_2}$
4		$n = \dfrac{1}{2+\alpha}\left[\alpha + \dfrac{\beta - \lambda_1 + \lambda_1\lambda_2}{\beta}\right]$ $m_x = \dfrac{2}{4+\alpha}\left[1 + \dfrac{\lambda_1 - 2\lambda_1\lambda_2}{\beta}\right]$ $m_y = \dfrac{1}{\beta^2}\left[\dfrac{\beta^2}{2} - \left(\dfrac{\lambda_1}{2} - \lambda_1\lambda_2\right)^2\right]$	$\lambda_2 \ge 1/2$ $\dfrac{\beta}{1+2\lambda_2} \le \lambda_1 \le \dfrac{\beta}{2\lambda_2 - 1}$

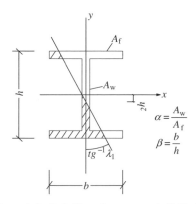

Figure 5.6 Definition of parameters in Table 5.1

Table 5.2 Yielding surface function for the box section

Order	Position of the neutral axis	Yielding surface equation	Condition
1	$\lambda_1 b$ $\lambda_2 h$	$n = 1 - \dfrac{\alpha\lambda_1 + \lambda_2}{1 + \alpha}$ $m_x = \dfrac{\alpha\lambda_1 + \lambda_2(1 - \lambda_1)}{\alpha + 1/2}$ $m_y = \dfrac{\lambda_2 + \alpha\lambda_1(1 - \lambda_1)}{1 + \alpha/2}$ $\frac{1}{1+\alpha}\left[\sqrt{(\alpha + \alpha^2)(1 - n) - (\alpha + \alpha^2/2)m_y}\right.$ $\left. + \sqrt{(1 + \alpha)(1 - n) - (\alpha + 1/2)m_x}\right] - n = 1$	$0 \leq n \leq 1$ $0 \leq m_x \leq \frac{4\alpha+1}{2(2\alpha+1)}$ $0 \leq m_y \leq \frac{4+\alpha}{2(2+\alpha)}$
2	$\lambda_1 h$ λ	$n = \dfrac{1 - \lambda_1 - \lambda_2}{1 + \alpha}$ $m_x = \dfrac{\alpha + \lambda_1(1 - \lambda_1) + \lambda_2(1 - \lambda_2)}{\alpha + 1/2}$ $m_y = \dfrac{\lambda_2 - \lambda_1}{1 + \alpha/2}$ $\dfrac{(1 + \alpha)^2}{2\alpha + 1}n^2 + \dfrac{(1 + \alpha/2)^2}{2\alpha + 1}m_y^2 + m_x = 1$	$0 \leq n \leq \frac{1}{1+\alpha}$ $\frac{2\alpha}{2\alpha+1} \leq m_x \leq 1$ $0 \leq m_y \leq \frac{2}{2+\alpha}$
3	$\lambda_1 b$ $\lambda_2 b$	$n = \dfrac{\alpha(1 - \lambda_1 - \lambda_2)}{1 + \alpha}$ $m_x = \dfrac{\alpha(\lambda_1 - \lambda_2)}{\alpha + 1/2}$ $m_y = \dfrac{1 + \alpha\lambda_1(1 - \lambda_1) + \alpha\lambda_2(1 - \lambda_2)}{1 + \alpha/2}$ $\dfrac{(1 + \alpha)^2}{(2 + \alpha)\alpha}n^2 + \dfrac{(\alpha + 1/2)^2}{(2 + \alpha)\alpha}m_x^2 + m_y = 1$	$0 \leq n \leq \frac{\alpha}{1+\alpha}$ $0 \leq m_x \leq \frac{2\alpha}{2\alpha+1}$ $\frac{2}{2\alpha+1} \leq m_y \leq 1$

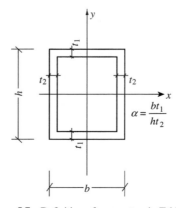

$$\alpha = \frac{bt_1}{ht_2}$$

Figure 5.7 Definition of parameters in Table 5.2

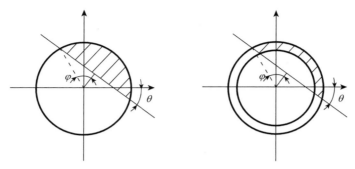

Figure 5.8 Definition of parameters in Equation (5.21)

The similar equations for circular and annular sections in parametric form are

$$n = 1 - \frac{2}{\pi}\left(\varphi - \frac{1}{2}\sin 2\varphi\right), \tag{5.21a}$$

$$m_x = \frac{3}{4}\cos\theta\left(\sin\varphi - \frac{1}{3}\sin 3\varphi\right), \tag{5.21b}$$

$$m_y = \frac{3}{4}\sin\theta\left(\sin\varphi - \frac{1}{3}\sin 3\varphi\right), \tag{5.21c}$$

where the parameters φ and θ are defined in Figure 5.8.

5.3.3 Approximate Expression of Ultimate Yielding Surface

It can be seen from Section 5.3.2 that the exact expression of the ultimate yielding surface for either rectangular, circular sections or H and box sections needs a group of equations, and in some cases the explicit expression cannot be derived. For the sake of convenient application, an approximate expression of the ultimate yielding surface uniformly to all kinds of sections is proposed as

$$\frac{m_x^s}{1 - n^u} + \frac{m_y^t}{1 - n^v} + n^w = 1, \tag{5.22}$$

where s, t, u, v and w are indeterminate parameters and can be determined through the exact equations of the ultimate yielding surface.

Considering the five groups of the control points on the ultimate yielding surface from the exact equations, one can solve the five indeterminate parameters. For example, the five control points may be

- for H section:

$$n = \frac{\alpha}{2(2+\alpha)}, \quad m_x = \frac{16+3\alpha}{4(4+\alpha)}, \quad m_y = 0,$$

$$n = \frac{1+\alpha}{2+\alpha}, \quad m_x = \frac{2}{4+\alpha}, \quad m_y = 0,$$

$$n = \frac{1+\alpha}{2+\alpha}, \quad m_x = 0, \quad m_y = \frac{3}{4},$$

$$n = 0, \quad m_x = \frac{2+\alpha}{4+\alpha}, \quad m_y = \frac{3}{4},$$

$$n = \frac{1+2\alpha}{4(2+\alpha)}, \quad m_x = \frac{3}{4+\alpha}, \quad m_y = \frac{7}{16};$$

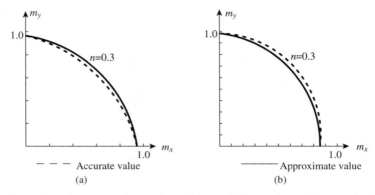

Figure 5.9   Comparison of accurate and approximate ultimate yielding surfaces: (a) box section; (b) H section

- for box section:

$$n = \frac{1}{2(1+\alpha)}, \quad m_x = \frac{8\alpha+3}{8\alpha+4}, \quad m_y = 0,$$

$$n = \frac{\alpha}{2(1+\alpha)}, \quad m_x = 0, \quad m_y = \frac{8+3\alpha}{8+4\alpha},$$

$$n = 0, \quad m_x = \frac{8+3\alpha}{8+4\alpha}, \quad m_y = \frac{1}{2+\alpha},$$

$$n = 0, \quad m_x = \frac{\alpha}{2\alpha+1}, \quad m_y = \frac{8+3\alpha}{8+4\alpha},$$

$$n = \frac{1}{2}, \quad m_x = \frac{1}{2}, \quad m_y = \frac{1}{2};$$

- for circular or annular sections:

$$s = t = 2, \quad u = v,$$
$$n = 0.215, \quad m_x = 0.957, \quad m_y = 0,$$
$$n = 0.609, \quad m_x = 0.650, \quad m_y = 0,$$

where α is defined in Figures 5.6 and 5.7.

A comparison between the approximate and exact equations of the ultimate yielding surfaces for H and box sections is provided in Figure 5.9, where a good coincidence can be found.

5.3.4 Effects of Torsion Moment

Frame columns are possibly subjected to torsion moment, which produces shear stresses to reduce the yielding strength of sections. Therefore, torsion moment affects the equations of the initial and ultimate yielding surfaces for column sections.

Normally, shear stresses resulting from torsion are unevenly distributed over a section, and it is difficult to evaluate the effects of shear stresses on the yielding surfaces of the section. For the sake of simplification, it is assumed that the ratio of the average shear stress over the section, τ, to yielding shear stress, τ_s, is equal to the magnitude of the ratio of the torque applied on the section, M_z, to the ultimate yielding torque, M_{pz}, i.e.

$$\frac{\tau}{\tau_s} = \left| \frac{M_z}{M_{pz}} \right| \tag{5.23}$$

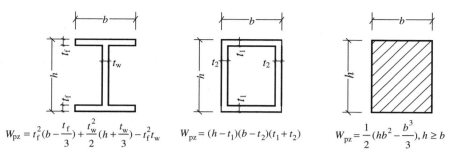

$$W_{pz} = t_f^2(b - \frac{t_f}{3}) + \frac{t_w^2}{2}(h + \frac{t_w}{3}) - t_f^2 t_w \qquad W_{pz} = (h - t_1)(b - t_2)(t_1 + t_2) \qquad W_{pz} = \frac{1}{2}(hb^2 - \frac{b^3}{3}), h \geq b$$

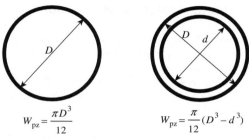

$$W_{pz} = \frac{\pi D^3}{12} \qquad\qquad W_{pz} = \frac{\pi}{12}(D^3 - d^3)$$

Figure 5.10   Values of plastic torsional section modulus

and

$$\tau = m_z \tau_s, \tag{5.24}$$

in which

$$m_z = \left| \frac{M_z}{M_{pz}} \right|, \tag{5.25}$$

$$M_{pz} = W_{pz} \tau_s, \tag{5.26}$$

where W_{pz} is the plastic torsional section modulus and its expression for various sections is given in Figure 5.10.

According to Mises yielding criteria,

$$\sigma^2 + 3\tau^2 = \sigma_s^2 = 3\tau_s^2, \tag{5.27}$$

which leads to

$$\frac{\sigma}{\sigma_s} = \sqrt{1 - \frac{3\tau^2}{\sigma_s^2}} = \sqrt{1 - \frac{\tau^2}{\tau_s^2}} = \sqrt{1 - m_z^2}. \tag{5.28}$$

It indicates that the section yielding stresses decrease due to the existence of torsion moment, and n, m_x and m_y should be replaced with n_T, m_{xT} and m_{yT}, respectively, as

$$n_T = \frac{n}{\sqrt{1 - m_z^2}}, \tag{5.29a}$$

$$m_{xT} = \frac{m_x}{\sqrt{1 - m_z^2}}, \tag{5.29b}$$

$$m_{yT} = \frac{m_y}{\sqrt{1 - m_z^2}}. \tag{5.29c}$$

The equation of the initial yielding surface including effects of torque becomes

$$\chi_{px} m_{xT} + \chi_{py} m_{yT} + n_T = 1,$$ (5.30)

and the approximate equation of the ultimate yielding surface including effects of torque becomes

$$\frac{m_{xT}^s}{1 - n_T^u} + \frac{m_{yT}^t}{1 - n_T^v} + n_T^w = 1.$$ (5.31)

5.4 HYSTERETIC MODEL

5.4.1 Cyclic Loading and Hysteretic Behaviour

Frames are possibly subjected to cyclic loads under earthquake and other dynamic loads. Experiments have shown that after yielding at one loading step, assuming that it is the nth loading, the initial yielding stress of steel in the next unloading and reversal loading, the $(n + 1)$th loading, will be lower than before (see Figure 5.11), i.e. $\sigma_{sn+1} < \sigma_{sn}$, which is known as the Bauschinger effect. If the reversal loading continues, the stress in steel continues to increase till it meets the ultimate yielding stress being higher than before, i.e. $\sigma_{pn+1} > \sigma_{pn}$, which is known as the strain-hardening effect. Let σ_{un} be the stress at the commence of unloading at the nth loading and $\sigma_{un} > \sigma_{pn}$; the ultimate yielding stress at the $(n + 1)$th loading, σ_{pn+1}, can then be approximately equal to σ_{un}, i.e. $\sigma_{pn+1} = \sigma_{un}$. For the first loading, $\sigma_{s1} = \sigma_{p1} = \sigma_s$.

When a steel beam is subjected to cyclic moments, the Bauschinger effect will reflect similarly in the $M - \phi$ relationship (see Figure 5.12), where $M_{sn+1} < M_{sn}$. In the same way and due to the strain-hardening effect, $M_{pn+1} > M_{pn}$ and $M_{pn+1} = M_{un}$. For the first loading, $M_{s1} = M_s$ and $M_{p1} = M_p$.

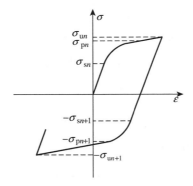

Figure 5.11   Stress–strain curve of steel under cyclic loading

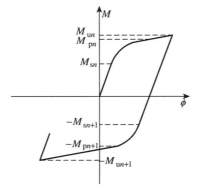

Figure 5.12   Moment–curvature curve under cyclic loading

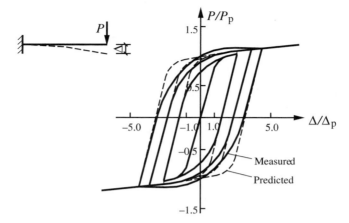

Figure 5.13 Measured versus predicted hysteretic load–deflection curves of a cantilever beam

The solid line in Figure 5.13 is the measured hysteretic load–deflection curve of a cantilever beam subjected to a cyclic load at the free end, where the Bauschinger effect and the strain-hardening effect are evident. The unloading of the beam behaving in elastic stiffness can also be found.

5.4.2 Hysteretic Model of Beam Section

A hysteretic model is the one that describes the relationship between force and displacement of structural members under cyclic loading conditions. A simple model for the hysteretic $M - \phi$ relationship of steel beam sections is a bilinear model, as shown in Figure 5.14, where q is the hardening factor and $q = 0.015$ usually.

The dashed line in Figure 5.13 is the prediction to the hysteretic load–deflection curves of the cantilever beam by the bilinear $M - \phi$ model. Clearly, the bilinear model can represent the basic hysteretic behaviour of a steel beam under cyclic loading, but cannot reflect the nonlinear phase from initial yielding to ultimate yielding very well.

To overcome the above deficit, a nonlinear hysteretic model can be adopted. For this purpose, define the yielding function of a beam section as

$$\Gamma = \left| \frac{M}{M_{\mathrm{p}}} \right| \tag{5.32}$$

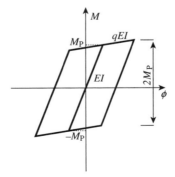

Figure 5.14 Bilinear hysteretic model

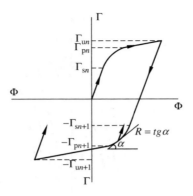

Figure 5.15 Hysteretic $\Gamma - \phi$ curve

and the curvature parameter as

$$\Phi = \left| \frac{\phi}{\phi_{\mathrm{p}}} \right|, \tag{5.33}$$

where ϕ_{p} is the elastic curvature corresponding to M_{p}.

Figure 5.15 gives the nonlinear model to describe the hysteretic $M - \phi$ relationship of beam sections, where Γ_{sn} and Γ_{sn} are the initial and ultimate yielding functions at the nth loading, respectively, Γ_{un} is the yielding function at the nth unloading, and Γ_{sn+1} and Γ_{pn+1} are the initial and ultimate yielding functions at the $(n + 1)$th loading, respectively. Considering the Bauschinger effect and the strain-hardening effect, one can define

$$\Gamma_{sn+1} = \Gamma_{s} - (\Gamma_{bn+1} - 1), \tag{5.34}$$
$$\Gamma_{pn+1} = \Gamma_{bn+1}, \tag{5.35}$$

in which

$$\Gamma_{bn+1} = \begin{cases} \Gamma_{un}, & \Gamma_{un} > \Gamma_{bn}, \\ \Gamma_{bn}, & \Gamma_{un} \leq \Gamma_{bn}, \end{cases} \tag{5.36}$$

where Γ_{bn} and Γ_{bn+1} are the characteristic values of the yielding function at the nth and the $(n+1)$th loading, respectively. For the first loading, $\Gamma_{b1} = 1$.

Γ_{s} is the initial yielding function at the first loading and is defined as

$$\Gamma_{s} = \frac{M_{s}}{M_{p}} = \frac{1}{\chi_{p}}. \tag{5.37}$$

Further, define the recovery force parameter of beam sections as

$$R = \frac{\mathrm{d}\Gamma}{\mathrm{d}\Phi}. \tag{5.38}$$

Then, for the nth loading, one has

$$R = 1, \qquad \Gamma \leq \Gamma_{sn}, \tag{5.39a}$$
$$R = q, \qquad \Gamma \geq \Gamma_{pn}, \tag{5.39b}$$
$$R = 1 - \frac{\Gamma - \Gamma_{sn}}{\Gamma_{pn} - \Gamma_{sn}}(1 - q), \qquad \Gamma_{sn} < \Gamma < \Gamma_{pn}, \tag{5.39c}$$

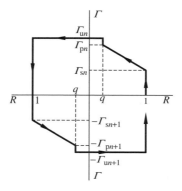

Figure 5.16 Hysteretic $\Gamma - \phi$ curve

and for any unloading

$$R = 1. \tag{5.40}$$

Equations (5.34)–(5.40) construct the nonlinear model for the hysteretic $M - \phi$ relationship of beam sections. The hysteretic behaviour between the yielding function and the recovery force parameter is governed by Equations (5.39) and (5.40), as shown in Figure 5.16.

5.4.3 Hysteretic Model of Column Section Subjected to Uniaxial Bending

If the variation of axial forces in a column is small under cyclic loading, the bilinear hysteretic model illustrated in Figure 5.14 can also be used to represent the hysteretic $M - \phi$ relationship of column sections subjected to uniaxial bending except that M_p should be replaced with M_{pN}. The value of M_{pN} can be determined with Equations (5.6)– (5.9) using the average of axial forces in the process of cyclic loading.

Generally, models in Figure 5.15 or 5.16 can be used to predict the hysteretic $\Gamma - \Phi$ or $\Gamma - R$ relationship of column sections subjected to uniaxial bending, but modify

$$\Gamma = \left| \frac{M}{M_{pN}} \right|, \tag{5.41}$$

$$\Phi = \left| \frac{\phi}{\phi_{pN}} \right| \tag{5.42}$$

and

$$\Gamma_s = \frac{M_{sN}}{M_{pN}}, \tag{5.43}$$

where M_{sN} and M_{pN} are the initial and ultimate yielding moments, respectively, varying with the axial force in the column in the process of loading and ϕ_{pN} is the elastic curvature corresponding to M_{pN}.

5.4.4 Hysteretic Model of Column Section Subjected to Biaxial Bending

The moment–curvature relationship about one axis will be influenced by the bending moment about the other axis for column sections subjected to biaxial bending, which makes it much complex to simply use the

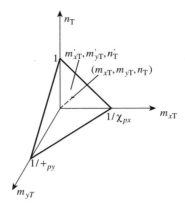

Figure 5.17 Initial yielding surface and the force state point

hysteretic moment–curvature relationship based on uniaxial bending, and the models given in Figure 5.14 or 5.15 become unapplicable.

However, the value of the yielding function Γ can be regarded as an index indicating to what extent the section yields under loads, and the value of the recovery force factor R as an index indicating the sectional stiffness after deformation. The hysteretic model shown in Figure 5.16 can therefore be used to depict the hysteretic $\Gamma - R$ relationship of column sections subjected to biaxial bending. The left-hand side of the ultimate yielding equation in the form of $\Gamma(m_{xT}, \quad m_{yT}, \quad n_T) = 1$ can be defined as the yielding function as

$$\Gamma = \Gamma(m_{xT}, \quad m_{yT}, \quad n_T). \tag{5.44}$$

If the approximate equation for the ultimate yielding, Equation (5.31), is adopted, the yielding function becomes

$$\Gamma = \frac{m_{xT}^s}{1 - n_T^u} + \frac{m_{yT}^t}{1 - n_T^v} + n_T^w. \tag{5.45}$$

The recovery force factor R is calculated with Equations (5.39) and (5.40) as well, where Γ_{sn} and Γ_{pn} are determined according to the recurrence formulae, Equation (5.34)–(5.36), in which Γ_s becomes

$$\Gamma_s = \Gamma(m_{xT}', \quad m_{yT}', \quad n_T'), \tag{5.46}$$

where $(m_{xT}', \quad m_{yT}', \quad n_T')$ is the intersection of the line from the origin to the force state point $(m_{xT}, \quad m_{yT}, \quad n_T)$ with the initial yielding surface in the orthotropic coordinate $m_{xT} - m_{yT} - n_T$, as shown in Figure 5.17.

The validity of the hysteretic model for columns subjected to biaxial bending is obtained with experiments, which will be presented in Chapter 8.

5.5 DETERMINATION OF LOADING AND DEFORMATION STATES OF BEAM–COLUMN SECTIONS

The yielding function can be used to identify the state of beam and column sections loaded as follows:

- if $\Gamma_{t+\Delta t} > \Gamma_t$, it is a loading state,

- if $\Gamma_{t+\Delta t} < \Gamma_t$, it is an unloading state,

- if $\Gamma_{t+\Delta t} = \Gamma_t$, it is a constant load state retaining certain loads,

where Γ_t and $\Gamma_{t+\Delta t}$ are the values of the yielding function at times t and $t + \Delta t$, respectively.
Three deformation states corresponding to the above three loading states are

(1) *loading state*: if $\Gamma_{t+\Delta t} < \Gamma_{sn}$, the section is in elastic state and the recovery force factor $R = 1$, and if $\Gamma_{t+\Delta t} \geq \Gamma_{sn}$, the section is in elasto-plastic state and the recovery force factor R can be determined with Equations (5.39b) and (5.39c);

(2) *unloading state*: the section is in elastic state and the recovery force factor $R = 1$;

(3) *constant load state*: the deformation state of the section at time $t + \Delta t$ is the same as that at time t.

6 Hysteretic Behaviour of Composite Beams

6.1 HYSTERETIC MODEL OF STEEL AND CONCRETE MATERIAL UNDER CYCLIC LOADING

6.1.1 Hysteretic Model of Steel Stress–Strain Relationship

A large amount of effort has been made worldwide to the stress–strain relationship of the steel material under cyclic loading. Several hysteretic models have been developed, where the simplest one is the perfectly elastic–plastic model (see Figure 6.1(a)) ignoring the strain-hardening effect and the Bauschinger effect. The models given in Figure 6.1(b)–(d) can consider both the strain-hardening effect and the Bauschinger effect. The model in Figure 6.1(b) is a bilinear model and the other two are trilinear models with, respectively, softening phase and yielding plateau.

The bilinear model is employed in this chapter because it can capture the principal characteristics of the steel material under cyclic loadings and is convenient to programming. In the bilinear model, q is the hardening factor and generally $q = 0.01$–0.02. If unloading before hardening, the initial elastic modulus E_s is used as loading or unloading stiffness and the Bauschinger effect is ignored, but if unloading after hardening, E_s is used as unloading stiffness and the Bauschinger effect is considered.

6.1.2 Hysteretic Model of Concrete Stress–Strain Relationship

As concrete is a mixture of different materials and there is microcrack in nature, the damage mechanism of concrete is inherently initiated from microcrack developing to macrocrack, and eventually from macrocrack developing to material failure. In the past experiments, it has been found that the skeletal of the hysteretic stress–strain curve of concrete is very close to its uniaxial stress–strain curve. In this chapter, the uniaxial stress–strain curves, well accepted both in compression and in tension, are adopted as the skeletal of the hysteretic curve of the concrete material under cyclic loading. Additionally, unloading and reloading parts are involved in the hysteretic curve constructed to consider the effects of cracking and softening of concrete under cyclic loading.

6.1.2.1 Skeletal curve

Unconfined concrete is considered herein due to little restrains of reinforcement bars to the concrete block in composite beams.

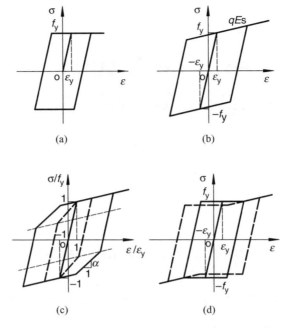

Figure 6.1   Hysteretic stress–strain relationships of the steel material: (a) perfectly elastic–plastic model; (b) bilinear model; (c) trilinear model with softening; (d) trilinear model with yielding

Uniaxial compressive constitution

The uniaxial compressive stress–strain curve of concrete includes stiffening (upward) and softening (downward) phases. The well-known Hognestad formula (Jiang, 1998) is used to construct the stiffening curve and the equations recommended in the Chinese code (GB50010, 2002) for the softening one, i.e.

- for $\varepsilon \leq \varepsilon_{c0}$:

$$\sigma = f_c \left[2 \frac{\varepsilon}{\varepsilon_{c0}} - \left(\frac{\varepsilon}{\varepsilon_{c0}} \right)^2 \right];$$ (6.1)

- for $\varepsilon > \varepsilon_{c0}$:

$$\sigma = f_c \frac{\varepsilon/\varepsilon_{c0}}{\varepsilon/\varepsilon_{c0} + \alpha(\varepsilon/\varepsilon_{c0} - 1)^2},$$ (6.2)

where f_c is the ultimate compressive strength of concrete in uniaxial loading, ε_{c0} is the corresponding strain to f_c and α is a softening parameter relating to material strength, generally ranging from 0.4 to 2.0.

Uniaxial tensile constitution

The following stress–strain relationship is used for uniaxial tension of concrete:

- for $\varepsilon \leq \varepsilon_{t0}$:

$$\sigma = E_0 \varepsilon;$$ (6.3)

- for $\varepsilon > \varepsilon_{t0}$:

$$\sigma = 0.9f_t \exp\left(-0.1\left(\frac{\varepsilon}{\varepsilon_{t0}} - 1\right)\right) + 0.1f_t, \qquad (6.4)$$

where f_t is the ultimate tensile strength of concrete in uniaxial loading, ε_{t0} is the corresponding strain to f_t and E_0 is the tensile modulus of concrete, $E_0 = f_t/\varepsilon_{t0}$.

6.1.2.2 Loading and unloading rule

Unloading in compression and reloading

In the compression state, the unloading and reloading rule (see Figure 6.2) is

(a) when $\varepsilon \le 0.55\varepsilon_{c0}$, the stress–strain unloading and reloading path is determined according to elastic stiffness;

(b) when $\varepsilon > 0.55\varepsilon_{c0}$, the stress–strain unloading and reloading path is determined according the focal point method described below.

The focal points F_1, F_2, F_3 and F_4 are located at the tangent of the skeletal stress–strain curve at origin and their stress coordinates are $-3f_c$, $-f_c$, $-0.75f_c$ and $-0.2f_c$, respectively. Assume that unloading is from an arbitrary point A $(\varepsilon_A, \sigma_A)$, as shown in Figure 6.2. The unloading path is along A–D–B, where point B $(\varepsilon_B, 0)$ is the intersection of line AF_2 with the ε-axis, point D $(\varepsilon_D, \sigma_D)$ is the intersection of line CF_1 and

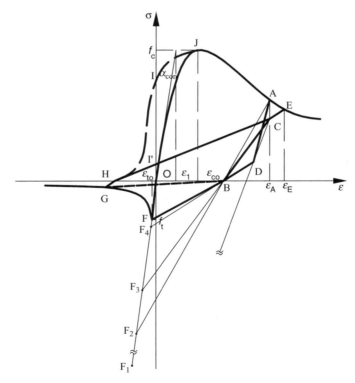

Figure 6.2 Hysteretic stress–strain relationships of the concrete material

line BF_4, and point C $(\varepsilon_C, \sigma_C)$ is that in BF_3 and with a strain value of ε_A. By the rule described above, the coordinates of points B, C and D can be calculated as

$$\varepsilon_B = \frac{f_c \varepsilon_A - \sigma_A \varepsilon_1}{f_c + \sigma_A}, \tag{6.5}$$

$$\sigma_C = \frac{0.75 f_c}{0.75 \varepsilon_1 + \varepsilon_B}(\varepsilon_A - \varepsilon_B), \tag{6.6}$$

$$\varepsilon_D = \frac{3\varepsilon_1 D_2 + \varepsilon_B D_1 - 3 f_c}{D_1 - D_2}, \tag{6.7}$$

$$\sigma_D = (\varepsilon_D - \varepsilon_B) D_1, \tag{6.8}$$

$$D_1 = \frac{0.2 f_c}{\varepsilon_B + 0.2 \varepsilon_1}, \tag{6.9a}$$

$$D_2 = \frac{\sigma_C + 3 f_c}{\varepsilon_A + 3 \varepsilon_1}, \tag{6.9b}$$

where ε_B is the residual strain after unloading and $\varepsilon_1 = \varepsilon_{c0}/2$.

When reloading to compression from point B, the reloading path is along B–C–E, where point E is in the skeletal curve and with a strain value of $1.15\varepsilon_A$, and when reloading in reverse to tension from B, the reloading path depends on the maximum of tensile strain in history, ε_t^*. If $\varepsilon_t^* \leq \varepsilon_{t0}$, namely concrete in tension never cracking, the reloading is along BF, where F (f_t, ε_{t0}) is the point in the skeletal curve corresponding to the maximum tensile stress. However, if $\varepsilon_t^* > \varepsilon_{t0}$, the reloading is along B–G, where G $(\varepsilon_G, \sigma_G)$ is the point corresponding to the maximum tensile stain.

Unloading in tension and reloading

In the tension state, the unloading and reloading rule (see Figure 6.2) is

(a) when $\varepsilon \leq \varepsilon_{t0}$, the stress–strain unloading and reloading path is determined according to elastic stiffness;

(b) when $\varepsilon > \varepsilon_{t0}$, the stress–strain unloading and reloading path is determined by the equations given below.

Assume that unloading is from point G in the softening part of the skeletal curve, as shown in Figure 6.2. The unloading path is along BH and H is the initial point from which the cracking effect is produced, and the strain at point H, ε_H, is

$$\varepsilon_H = \varepsilon_G \left(0.1 + \frac{0.9 \varepsilon_{c0}}{\varepsilon_{c0} + |\varepsilon_G|}\right). \tag{6.10}$$

The path of reloading in reverse depends on the maximum of compressive strain in history, ε_c^*. If $\varepsilon_c^* \leq \varepsilon_{c0}$, namely concrete in compression never crushing, the reloading is along H–I–J, but if $\varepsilon_c^* > \varepsilon_{c0}$, the reloading is along H–I′–C–E.

Point I or I′ is the intersection of reloading path with the σ-axis, and the contact stress corresponding to $\varepsilon = 0$ is

$$\sigma_{con} = 0.3 \sigma_R \left[2 + \frac{|\varepsilon_H|/\varepsilon_{c0} - 4}{|\varepsilon_H|/\varepsilon_{c0} + 2}\right], \tag{6.11}$$

where $\sigma_R = f_c$ for $\varepsilon_c^* \leq \varepsilon_{c0}$ or $\sigma_R = \sigma_A$ for $\varepsilon_c^* > \varepsilon_{c0}$.

Equation for HI or HI′ is

$$\sigma = \sigma_{con}\left(1 - \frac{2 \cdot |\varepsilon|}{|\varepsilon_H| + |\varepsilon|}\right) \qquad (\varepsilon_H \leq \varepsilon < 0). \tag{6.12}$$

Equation for IJ is

$$\sigma = \sigma_{con}(1 - \varepsilon/\varepsilon_{c0}) + \frac{2\varepsilon}{\varepsilon_0 + \varepsilon} f_c \qquad (0 \leq \varepsilon < \varepsilon_{c0}). \qquad (6.13)$$

Equation for I'C is

$$\sigma = \sigma_{con}(1 - \varepsilon/\varepsilon_A) + \frac{2\varepsilon}{\varepsilon_A + \varepsilon} \sigma_C \qquad (0 \leq \varepsilon < \varepsilon_C). \qquad (6.14)$$

If unloading occurs at any point of GI, the unloading path is the direct line from that point to G.

6.2 NUMERICAL METHOD FOR MOMENT–CURVATURE HYSTERETIC CURVES

6.2.1 Assumptions

(1) A plane section of composite beams remains plane when subjected to bending.

(2) The stress–strain relationship for steel and concrete material can be determined according to Figures 6.1(b) and 6.2.

(3) The effect of shear lag in the concrete slab flange of composite beams can be neglected.

(4) No slip occurs at the steel–concrete interface of composite beams, namely full composite action is considered.

6.2.2 Sectional Division

Fibre model is used to analyse the sectional hysteretic behaviour. Strip fibre is divided along the sectional width, as shown in Figure 6.3. Each strip is distributed with uniform normal stress, and the longitudinal reinforcement bar is dealt with as an independent strip.

For the sake of calculation convenience, take the central axis of the steel section as the reference axis. Suppose the strain at the reference axis to be ε_{s0} and the sectional curvature ϕ; the strain of each strip can then be expressed as

$$\varepsilon_i = \varepsilon_{s0} + \phi \cdot y_i, \qquad (6.15)$$

where y_i is the distance from the central point of each strip to the reference axis.

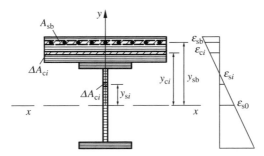

Figure 6.3 Fibre division and strain distribution of the composite section

The stress of each strip can be calculated from the strain given in Equation (6.15) and according to the material constitution of steel and concrete presented in Section 6.1, following which the axial force on each strip can be obtained with

$$N_{ci} = \Delta A_{ci}\sigma_{ci}, \tag{6.16a}$$

$$N_{sb} = A_{sb}\sigma_{sb}, \tag{6.16b}$$

$$N_{si} = \Delta A_{si}\sigma_{si}, \tag{6.16c}$$

where ΔA_{ci}, ΔA_{sb} and ΔA_{si} are the cross-sectional areas of the concrete strip, longitudinal reinforcement bar and steel strip, respectively, and σ_{ci}, σ_{sb} and σ_{si} are strip stresses, respectively.

Then, the axial force N and bending moment M carried by the total section are

$$N = \sum_{i=1}^{n_c} \Delta A_{ci}\sigma_{ci} + \sum_{i=1}^{n_s} \Delta A_{si}\sigma_{si} + A_{sb}\sigma_{sb}, \tag{6.17}$$

$$M = \sum_{i=1}^{n_c} \Delta A_{ci}\sigma_{ci}y_{ci} + \sum_{i=1}^{n_s} \Delta A_{si}\sigma_{si}y_{si} + A_{sb}\sigma_{sb}y_{sb}. \tag{6.18}$$

6.2.3 Calculation Procedure of Moment–Curvature Relationship

In general, a set of values about the bending moment, curvature and strain at the reference axis (reference strain) may be determined for each point in the moment–curvature curve, which satisfy strain compatibility and internal force equilibrium in the section considered. The numerical procedure for determining the moment–curvature curve of composite sections is given in Figure 6.4. To capture the softening (downward) phase in the moment–curvature curve, a curvature-incremental strain-iterative strategy is employed, and the iteration convergence criterion for each incremental analysis is that the axial force on the section considered vanishes, namely

$$N(\varepsilon_{s0}) = 0. \tag{6.19}$$

In strain iteration, a linear interpolation technique is used to speed up the calculation. For this purpose, a convergence function is defined as

$$\psi = N(\varepsilon_{s0}). \tag{6.20}$$

If two trial values of reference strain, ε_{s0}^1 and ε_{s0}^2, are obtained, the next trial value of reference strain is determined with linear interpolation as

$$\varepsilon_{s0}' = \varepsilon_{s0}^1 - \frac{\varepsilon_{s0}^2 - \varepsilon_{s0}^1}{\psi_2 - \psi_1}\psi_1, \tag{6.21}$$

where $\psi_1 = N(\varepsilon_{s0}^1)$ and $\psi_2 = N(\varepsilon_{s0}^2)$.

It should be noted that the double precision format is necessary to set up the variables to preclude the numerical difficulty in iterative calculation. The convergent rate is very good in the iterative procedure proposed, and generally, convergence will be achieved within 10 runs.

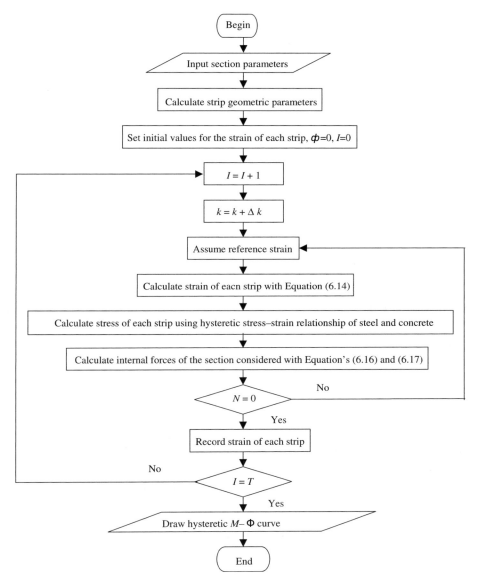

Figure 6.4   Flow chart of moment–curvature calculation

6.3   HYSTERETIC CHARACTERISTICS OF MOMENT–CURVATURE RELATIONSHIPS

A typical hysteretic moment–curvature curve of the composite section is illustrated in Figure 6.5. The characteristics and typical phases of this curve are discussed in this section.

6.3.1   Characteristics of Hysteretic Curves

(1) The hysteretic hoop of the composite beam is plump and stable, without strength or stiffness regression.

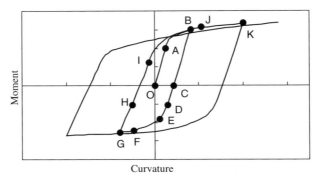

Figure 6.5   Hysteretic curve of the composite section

(2) Asymmetry exists between the positive (sagging) moment zone and the negative (hogging) moment zone in the hysteretic hoop (see Figure 6.5), which results from the asymmetric distribution of the material over the section (see Figure 6.3).

(3) A concrete flange is in compression by positive moment and the hysteretic curve has the characteristics of reinforced concrete beams, whereas the concrete flange is in tension and can be negligible by negative moment and the hysteretic curve has the characteristics of steel beams.

(4) The Bauschinger effect and the strain-hardening effect can be observed in the hysteretic $M–\phi$ curves of composite beam sections.

6.3.2   Typical Phases

(1) *Phase OA*. The moment–curvature relationship in this phase is linear. The steel beam and concrete slab flange are in elastic states. Point A is the initial yielding of the lower steel flange.

(2) *Phase AB*. The moment–curvature relationship in this phase is nonlinear, and the section is in elasto-plastic state. Yielding spreads upwards with the increase of moment, and the neutral axis lifts to keep force balance. The compressive ultimate strength of concrete has not been reached in this phase. The global stiffness of the composite section reduces gradually.

(3) *Phase BC*. Unloading occurs at point B, and the moment–curvature relationship in this phase is linear again. Unloading stiffness is the same as that of initial loading (phase OA). The moment reduces to zero at point C, but the residual positive curvature remains unchanged due to plastic strain.

(4) *Phase CD*. Negative moment is applied in reverse, and the moment–curvature relationship in this phase is linear. The loading stiffness in this phase is equal to the unloading stiffness in phase BC. The concrete in compression is in unloading and at point D, compression in concrete reduces to zero.

(5) *Phase DE*. The moment–curvature relationship in this phase is nonlinear. Tension zone spreads in the concrete flange of the section with the increase of moment, and the contribution of concrete to the stiffness of the composite beam becomes small. The sectional moment resistance is mainly provided by the steel beam and rebar in concrete. Yielding occurs in the compressive lower steel flange at point E.

(6) *Phase EF*. The moment–curvature relationship in this phase is nonlinear. As the plastic zone spreads in the steel beam, the stiffness reduces significantly compared to that in phase DE. At point F, the rebar in tension yields.

(7) *Phase FG*. The moment–curvature relationship in this phase is approximately linear and the slope is small. The section enters the strain-hardening phase. Plastic zones of steel in tension and compression increase.

(8) *Phase GH*. Unloading is from point G, and the moment–curvature relationship in this phase is linear. The unloading stiffness is equal to the loading stiffness in phase DE. Rebars in tension are in elastic unloading. Aggregate locking occurs in the concrete flange of the section at point E.

(9) *Phase HI*. The moment–curvature relationship in this phase is linear. The concrete flange is assumed to come again into compression due to aggregate locking, which gives sectional stiffness an increase so that an antiflexural point appears at H. The unloading stiffness is equal to the loading stiffness in phase OA. The lower part of the steel beam comes in tension after the moment unloads to zero.

(10) *Phase IJ*. Yielding occurs on the outer fibre of the lower steel flange. Even though concrete in compression again can increase sectional stiffness, the entire stiffness reduces with the increase of moment because the plastic zone spreads in the steel beam.

(11) *Phase JK*. The moment–curvature relationship in this phase is linear, but the slope is small. The section enters strain hardening, and the concrete flange crushes when the ultimate compressive strength exceeds.

6.4 PARAMETRIC STUDIES

The skeletal curve of the hysteretic moment–curvature relationship of composite beams is an envelope of many hysteretic hoops. A large amount of numerical studies indicate that the skeletal curve coincides well with the moment–curvature curve of the beam subjected to monotonic loading. Comparisons are given in Figure 6.6 between the hysteretic moment–curvature curve and the monotonic moment–curvature curve for three composite beams.

In this section, parametric studies are given for both skeletal curve and hysteretic curve of composite sections. The typical parameters investigated include height of concrete flange h_c, width of concrete flange B_c, height of steel beam h_s, strength ratio γ, yielding strength of steel f_y and compressive strength of concrete f_{ck}.

6.4.1 Height of Concrete Flange h_c

Five skeletal curves are obtained in Figure 6.7, where results from four different values of h_c as well as from a pure steel beam ($h_c = 0$) are compared. The comparisons reveal that in the positive bending (sagging) moment zone, the initial and ultimate yielding moments increase evidently with the increase of concrete flange height, but in the negative bending (hogging) moment zone, such an increase is small. The bending stiffness and ultimate moment of a composite section are greater than those of the corresponding pure steel section, especially in the positive bending moment zone.

Studies are also conducted on the hysteretic curves of the composite beams with concrete flange heights h_c of 60 and 120 mm. The hysteretic moment–curvature curves, shown in Figure 6.8, demonstrate that the stiffness degradation of composite sections is evident in the positive moment zone, especially in the case of large concrete flange height.

6.4.2 Width of Concrete Flange B_c

The skeletal curves and hysteresis hoops for various widths of concrete flange are shown in Figures 6.9 and 6.10. The effects of concrete flange width on skeletal curves and hysteresis hoops are very similar to those of

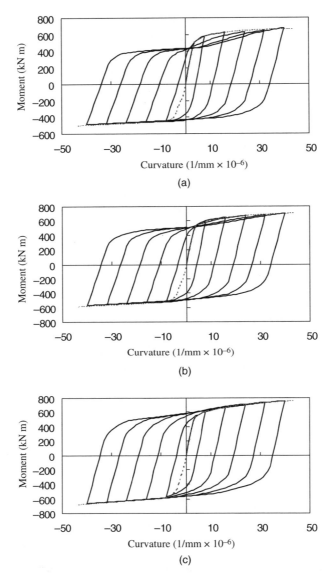

Figure 6.6 Comparison between hysteretic curve and uniaxial moment–curvature curves: (a) $B_c \times h_c = 1400 \, \text{mm} \times 100 \, \text{mm}$, $h_s = 450 \, \text{mm}$, $\gamma = 0.15$, $f_y = 235 \, \text{MPa}$, $f_{ck} = 20 \, \text{MPa}$; (b) $B_c \times h_c = 1400 \, \text{mm} \times 100 \, \text{mm}$, $h_s = 450 \, \text{mm}$, $\gamma = 0.25$, $f_y = 235 \, \text{MPa}$, $f_{ck} = 20 \, \text{MPa}$; (c) $B_c \times h_c = 1400 \, \text{mm} \times 100 \, \text{mm}$, $h_s = 450 \, \text{mm}$, $\gamma = 0.5$, $f_y = 235 \, \text{MPa}$, $f_{ck} = 20 \, \text{MPa}$

concrete flange height. It can be seen from Figure 6.9 that the elastic stiffness and initial yielding moment of composite sections with concrete flange width greater than 1000 mm do not vary much, in both positive and negative moment zones.

6.4.3 Height of Steel Beam h_s

The height of steel beam has great influence on the stiffness and bending strength of composite sections, as shown in Figures 6.11 and 6.12, no matter they are in the positive or negative zone. The results reveal that the

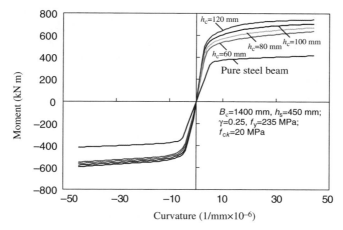

Figure 6.7 Effect of concrete flange height on skeletal curve

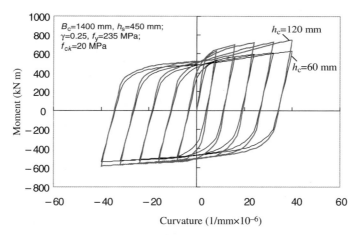

Figure 6.8 Effect of concrete flange height on hysteresis hoops

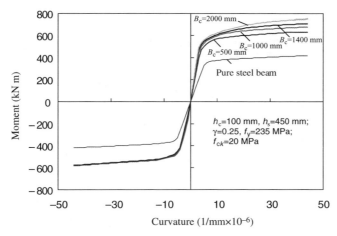

Figure 6.9 Effects of concrete flange width on skeletal curves

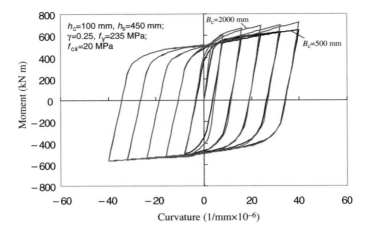

Figure 6.10   Effects of concrete flange width on hysteresis hoops

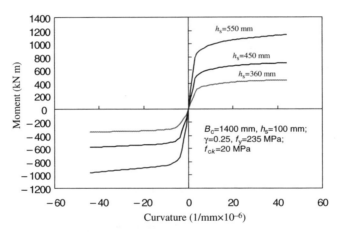

Figure 6.11   Effects of steel beam height on skeletal curves

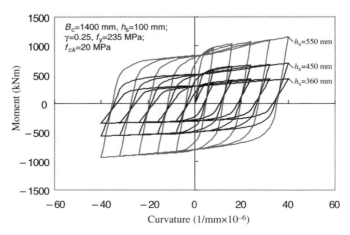

Figure 6.12   Effect of steel beam height on hysteresis hoops

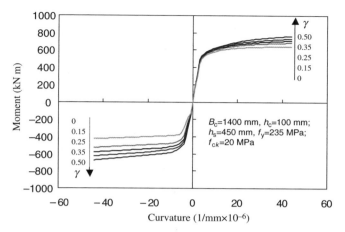

Figure 6.13 Effects of strength ratio on skeletal curves

height of steel beams is a crucial dimension factor to influence the stiffness and strength of composite sections.

6.4.4 Strength Ratio γ

The strength ratio γ is defined as the ratio of longitudinal reinforcement bar strength to steel beam strength, i.e.

$$\gamma = \frac{A_{sb}f_{sb}}{A_s f}, \tag{6.22}$$

where A_{sb} and f_{sb} are the sectional area and design tensile strength of the longitudinal reinforcement bar within the effective breadth of the concrete flange, and A_s and f are the sectional area and design tensile strength of the steel beam, respectively.

The skeletal curves and hysteresis hoops with different strength ratios are given in Figures 6.13 and 6.14. In contrast to the observation above, the effects of strength ratio on skeletal curves in the negative moment zone are greater than those in the positive moment zone. By the results in Figure 6.13, the skeletal curves

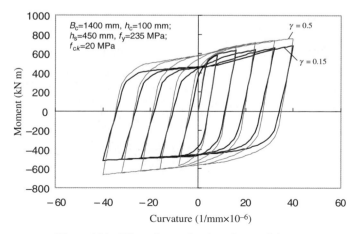

Figure 6.14 Effect of strength ratio on hysteresis hoops

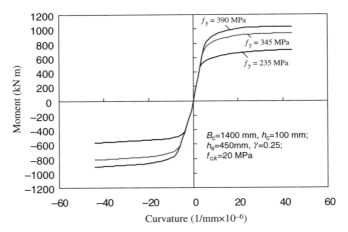

Figure 6.15   Effects of steel yielding strength on skeletal curves

tend to be symmetric with an increase of the strength ratio, namely the stiffness and strength behaviour in the negative moment zone become close to those in the positive zone.

In Figure 6.14, the effects of strength ratio on the hysteretic behaviour in the positive moment zone are evident when the strength ratio is small. But for large strength ratios, the hysteresis hoops of composite sections are close to those of pure steel beams because the concrete contribution is small.

6.4.5   Yielding Strength of Steel f_y

The yielding strength of steel beams has great influence on the bending strength of the corresponding composite beams, as shown in Figures 6.15 and 6.16, no matter they are in the positive or negative zone. However, it does not affect the elastic stiffness because elastic stiffness relates to the elastic modulus of the material, not to the yielding strength.

6.4.6   Compressive Strength of Concrete f_{ck}

By the results illustrated in Figures 6.17 and 6.18, it can be concluded that the compressive strength of concrete has a small effect on the skeletal curves and hysteresis hoops of composite sections.

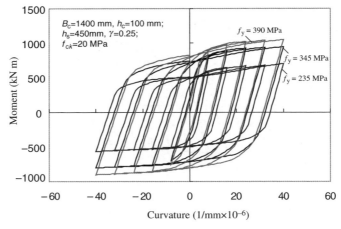

Figure 6.16   Effect of steel yielding strength on hysteresis hoops

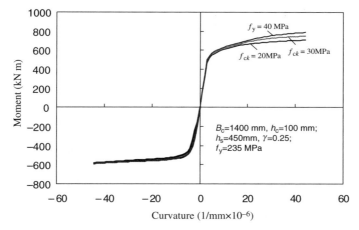

Figure 6.17   Effects of concrete compressive strength on skeletal curves

6.4.7 Summary of Parametric Studies

By parametric studies, the following conclusions can be drawn:

(1) Width and thickness of the concrete flange have an evident influence on the elastic bending stiffness, yielding moment and limit moment of composite beams in positive moment, but have a negligible influence on them when composite beams are in negative moment.

(2) Height of steel beams is the dominant factor to determine the values of bending stiffness and capacity of composite beams, no matter they are in positive or negative bending. But it does not influence the shape of the hysteretic moment–curvature curves much.

(3) Strength ratio is a significant factor relevant to hysteretic behaviour of composite beams. When composite beams are in negative rather than positive moment, the effect of strength ratio is evident. Elastic stiffness, yielding moment and limit moment of composite beams in negative moment increase with strength ratio. The elastic stiffness and yielding moment of composite beams in positive moment are nearly the same, but limit moment increases with the increase of strength ratio.

(4) Strength ratio also affects the shape of hysteretic curves of composite beams. When the strength ratio is small, the effect of concrete on hysteretic curves of composite beams becomes relatively large.

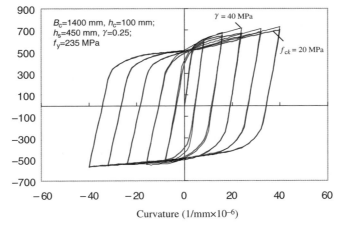

Figure 6.18   Effect of concrete compressive strength on hysteresis hoops

However, the effect of concrete becomes small when the strength ratio increases, and the hysteretic curves of composite beams will become similar to those of steel beams.

(5) Yielding strength of steel and compression strength of concrete have an influence on the bending capacity of composite beams, not on elastic stiffness. The values of such material properties also have a minor influence on the hysteretic shape of composite beams.

6.5 SIMPLIFIED HYSTERETIC MODEL

The numerical method can produce a large amount of skeletal curves and hysteresis hoops for composite sections, but it has low feasibility to be incorporated into a dynamic analysis of steel frames with composite beams, for example, in the appraisal of earthquake resistance of steel frames. A simplified hysteretic model is therefore necessary and can be derived rationally based on the numerical results, which is the task in this section.

6.5.1 Skeletal Curve

A simplified bilinear model is used for the skeletal curve of composite beams, in both positive and negative moment zones, as shown in Figure 6.19. Six parameters are needed to describe the complete bilinear model, which include elastic stiffness K_e^+, yielding moment M_y^+ and hardening stiffness K_p^+ in positive moment and elastic stiffness K_e^-, yielding moment M_y^- and hardening stiffness K_p^- in negative moment.

6.5.1.1 Positive elastic stiffness K_e^+

It has been known from the parametric studies that the effect of longitudinal reinforcement bars on the elastic stiffness of composite beams can be negligible. So Equation (5.13), the expression for the elastic bending stiffness of composite sections with full composite action, can be used to define the positive elastic stiffness, i.e.

$$K_e^+ = (EI)_{comp}^0 + \overline{EA} \cdot r^2. \tag{6.23}$$

6.5.1.2 Positive yielding moment M_y^+

The positive ultimate moment corresponds to the load making the top compressive concrete fibre achieve the maximum compressive strain. Such ultimate moment defined can be calculated by the block-stress assumption over the composite section as shown in Figure 6.20 (for the plastic neutral axis within the concrete flange) and Figure 6.21 (for the plastic neutral axis within the steel beam).

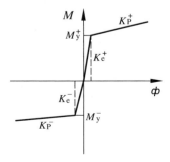

Figure 6.19 Moment–curvature skeletal curve model of the composite section

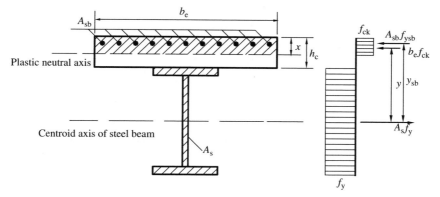

Figure 6.20   Stress distribution in complete yielding state of the section (plastic neutral axis in the concrete flange)

When the plastic neutral axis is located within the concrete flange (see Figure 6.20), namely $A_s f_y \leq b_e h_c f_{ck} + A_{sb} f_{ysb}$, one has

$$M_u^+ = b_e x f_{ck} y + A_{sb} f_{ysb} y_{sb}, \tag{6.24}$$
$$x = (A_s f_y - A_{sb} f_{ysb})/(b_e f_{ck}),$$

where A_s and A_{sb} are the cross-sectional areas of the steel beam and the longitudinal reinforcement bars in the concrete flange, respectively, f_y and f_{ysb} are the yielding strength of the steel beam and longitudinal bars, respectively, f_{ck} is the compressive strength of the concrete material, x is the height of the concrete flange in compression, y is the distance from the centre of the compressive concrete to that of the steel beam and y_{sb} is the distance of the longitudinal bars to the centre of the steel beam.

When the plastic neutral axis is located within the steel beam (see Figure 6.21), namely $A_s f_y \geq b_e h_c f_{ck} + A_{sb} f_{ysb}$, one has

$$M_u^+ = b_e h_c f_{ck} y_1 + A_{sb} y_{sb1} + A_c f_y y_2, \tag{6.25}$$
$$A_c = 0.5(A_s - b_e h_c f_{ck}/f_y - A_{sb} f_{ysb}/f_y), \tag{6.26}$$

where A_c is the cross-sectional area of the steel beam in compression, y_1 is the distance from the centre of the compressive concrete to that of the steel beam in tension, y_{sb1} is the distance of the longitudinal bars to the centre of the steel beam in tension and y_2 is the distance from the centre of the steel beam in compression to that of the steel beam in tension.

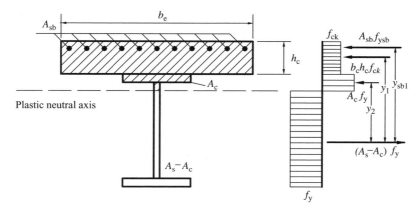

Figure 6.21   Stress distribution in complete yielding state of the section (plastic neutral axis in the steel beam)

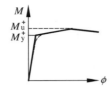

Figure 6.22 The definition of yielding moment

The positive yielding moment is defined as the moment at the intersection of extension lines of the $M-\phi$ elastic curve and hardening curve, as shown in Figure 6.22. Based on numerical results, M_y^+ can be expressed with M_u^+ as

$$M_y^+ = 0.95 M_u^+, \tag{6.27}$$

6.5.1.3 Positive hardening stiffness K_p^+

The positive hardening stiffness K_p^+ can be expressed with the elastic stiffness K_e^+ as

$$K_p^+ = \alpha \cdot K_e^+, \tag{6.28}$$

where α is a hardening factor.

From parametric studies hereinbefore, there is no factor of significance to the hardening stiffness of composite sections. To preserve sufficient strength and ductility, the hardening effect is only considered on the positive moment–resistance of composite sections and the maximum strain of the tensile fibre in the steel beam is limited to $10\,000\mu\varepsilon$. Under such conditions, α can be taken as constant (0.025).

6.5.1.4 Negative elastic stiffness K_e^-

In the negative moment zone, concrete cracks at a very early stage and its contribution to elastic stiffness is small and can be neglected. The elastic stiffness of composite sections can then be determined by ignoring concrete and considering the contribution only from the steel beam and longitudinal bars, as shown in Figure 6.23, i.e.

$$K_e^- = E_s(I_s + A_s \cdot y_4^2 + A_{sb}y_3^2), \tag{6.29}$$

where E_s is the elastic modulus of steel and I_s is the inertial moment of the steel beam about its centre.

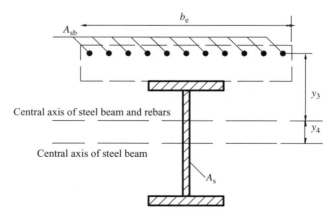

Figure 6.23 The composite section under negative moment

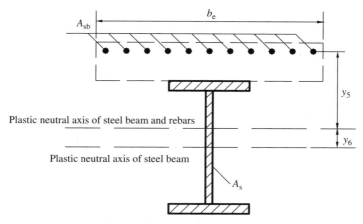

Figure 6.24   Illustration of y_5 and y_6

6.5.1.5 Negative yielding moment M_y^-

In the negative moment zone, the behaviour of composite sections is similar to that of pure steel sections. With reference to the method used for the hysteretic model of steel beams presented in Chapter 5, M_y^- can be expressed as

$$M_y^- = -M_{ys} - A_{sb}f_{ysb}(y_5 + y_6/2), \qquad (6.30)$$

where M_{ys} is the ultimate moment of the steel beam section, $M_{ys} = W_p f_y$, W_p and f_y are the plastic section resistant modulus and yielding strength of steel, respectively, f_{ysb} is the yielding strength of longitudinal bars, and y_5 and y_6 are denoted in Figure 6.24.

6.5.1.6 Negative hardening stiffness K_p^-

The negative hardening stiffness K_p^- can be expressed with the elastic stiffness K_e^- as

$$K_p^- = q \cdot K_e^-, \qquad (6.31)$$

where q is a hardening factor, the value of which can be same as that for steel sections, generally $q = 0.01$–0.02.

6.5.2 Hysteresis Model

The model for the moment–curvature hysteresis hoops of composite sections may be developed on the basis of numerical studies, as illustrated in Figure 6.25.

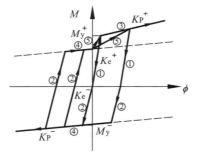

Figure 6.25   Model for moment–curvature hysteresis hoops of the composite section

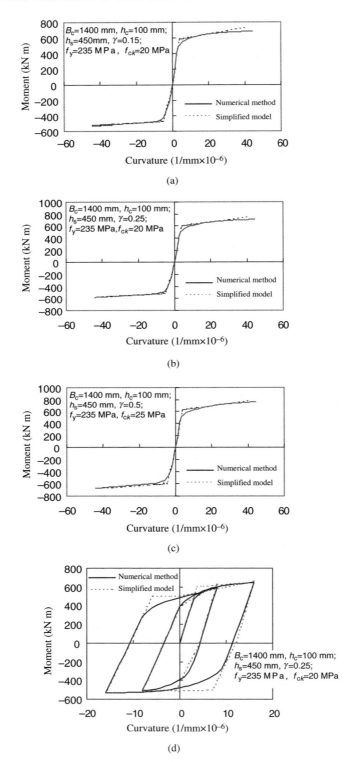

Figure 6.26 Comparisons of skeletal curves and hysteresis hoops between the results of the simplified model and those obtained with the numerical method

6.5.2.1 Loading

The loading path of the hysteresis hoop is determined through the skeletal curves with elastic stiffness in both positive and negative moment zones.

6.5.2.2 Unloading in positive moment zone

In the positive moment zone, the unloading path is along the elastic stiffness K_e^+. After the moment enters the negative zone, loading in reverse is along elastic stiffness K_e^- up to the negative yielding moment. The determination of the negative yielding moment in reverse loading, M_y^{-*}, should account for the Bauschinger effect, where M_y^{-*} depends on whether the positive moment applied on the composite section was greater than the positive yielding moment or not in history. If the positive moment has never reached the positive yielding moment, M_y^{-*} is calculated with Equation (6.30) (actually equal to M_y^-), but if the positive moment has reached the positive yielding moment, M_y^{-*} is that at the intersection of reversal loading path (2) with negative hardening path (4) (see Figure 6.25) passing through the point $(-M_y^-, -\phi_y^-)$, where $-\phi_y^-$ is the curvature corresponding to the negative yielding moment and is given by

$$\phi_y^- = M_y^- / K_e^- . \tag{6.32}$$

6.5.2.3 Unloading in negative moment zone

When the negative moment on the section has never reached the negative yielding moment, the unloading path is along the skeletal curve. If the negative moment has reached the negative yielding moment, the unloading of the moment is according to the elastic stiffness K_e^-, and when the moment enters the positive zone, loading in reverse is also along the elastic stiffness K_e^- up to the positive yielding moment.

The positive yielding moment in reverse loading M_y^{+*} is that at the intersection of reverse loading path (2) with the positive hardening path (4), symmetric and parallel to the negative hardening path (4) (see Figure 6.25). After yielding, the moment increases with the hardening stiffness K_p^-, till the negative curvature vanishes. After the curvature is positive, the moment–curvature path follows the link line (5), which is from the point with zero curvature, the intersection of path (4) in the positive moment zone and

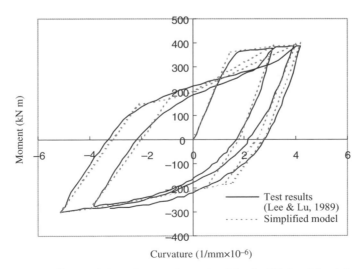

Curvature $(1/\text{mm} \times 10^{-6})$

Figure 6.27 Comparison of hysteresis hoops between the results of the simplified model and those obtained by test

the axis of M, to the unloading point of positive moment last time on the skeletal curve. If the unloading positive moment last time did not exceed the positive yielding moment, the link line (5) is from the zero-curvature point to the point (M_y^+, ϕ_y^+), where ϕ_y^+ is determined by

$$\phi_y^+ = M_y^+ / K_e^+. \tag{6.33}$$

The comparisons of skeletal curves and hysteresis hoops of composite sections between the results of the simplified model presented hereinabove and those obtained with the numerical method are plotted in Figure 6.26(a)–(d). Good agreement can be found in those figures.

The comparison of hysteresis hoops between the results of the simplified model and those obtained by test (Lee and Lu, 1989) is given in Figure 6.27, where good coincidence can be observed as well.

7 Elasto-Plastic Stiffness Equation of Beam Element

When a beam element is in elasto-plastic state, the relationship between the nodal forces and displacements of the element depends on the elemental displacement history. As the elemental displacement history is relevant to the loading history, which is arbitrary, there is no way to establish the stiffness equation with total-quantity form for nodal forces and displacements. However, there is a certain stiffness relationship between incremental nodal forces and displacements with minor step, and it is known as the elasto-plastic incremental stiffness equation.

Assume that equilibrium has been obtained with nodal rotations θ_1 and θ_2 and translations δ_1 and δ_2 corresponding to nodal moments M_1 and M_2 and shears Q_1 and Q_2, as shown in Figure 2.1. If incremental moments dM_1 and dM_2 and shears dQ_1 and dQ_2 are applied, the corresponding incremental rotations $d\theta_1$ and $d\theta_2$ and translations $d\delta_1$ and $d\delta_2$ will occur, as shown Figure 7.1. The incremental forces and displacements can be expressed with vector forms as

$$\{d\delta_g\} = \{d\delta_1, d\theta_1, d\delta_2, d\theta_2\}^T, \tag{7.1}$$

$$\{df_g\} = \{dQ_1, dM_1, dQ_2, dM_2\}^T. \tag{7.2}$$

The elasto-plastic incremental stiffness equation of beam elements can then be expressed as

$$[k_{gp}]\{d\delta_g\} = \{df_g\}, \tag{7.3}$$

where $[k_{gp}]$ is the elasto-plastic tangent stiffness matrix of beam elements.

7.1 PLASTIC HINGE THEORY

The simple or original plastic hinge theory is a straightforward way to establish the elasto-plastic incremental stiffness equation of beam elements, where the following assumptions are raised:

(1) the plastic deformation is concentrated on the two ends of elements;

(2) plastic hinge forms at a section of elements once the moment applied at the section equals to the plastic moment of elements.

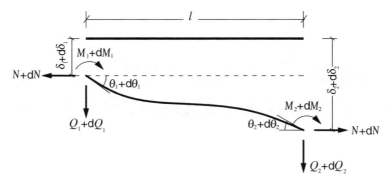

Figure 7.1   Nodal forces and deformations of the beam element

Assumption (1) is the concentrated plastic assumption, which is commonly accepted in the elasto-plastic analysis of frame structures, whereas assumption (2) is based on the ideal elasto-plastic $M - \phi$ relationship, as shown in Figure 7.2.

7.1.1   Hinge Formed at One End of Element

Equation (2.22) can be used for the incremental stiffness equation of a beam element as well, i.e.

$$
\frac{EI}{l}
\begin{bmatrix}
\dfrac{12}{l^2}\psi_1 & \dfrac{6}{l}\psi_2 & -\dfrac{12}{l^2}\psi_1 & \dfrac{6}{l}\psi_2 \\[2mm]
\dfrac{6}{l}\psi_2 & 4\psi_3 & -\dfrac{6}{l}\psi_2 & 2\psi_4 \\[2mm]
-\dfrac{12}{l^2}\psi & -\dfrac{6}{l}\psi_2 & \dfrac{12}{l^2}\psi & -\dfrac{6}{l}\psi_2 \\[2mm]
\dfrac{6}{l}\psi_2 & 2\psi_4 & -\dfrac{6}{l}\psi_2 & 4\psi_3
\end{bmatrix}
\begin{Bmatrix}
d\delta_1 \\ d\theta_1 \\ d\delta_2 \\ d\theta_2
\end{Bmatrix}
=
\begin{Bmatrix}
dQ_1 \\ dM_1 \\ dQ_2 \\ dM_2
\end{Bmatrix},
\tag{7.4}
$$

where $\psi_1 - \psi_4$ are determined with the axial force in the beam element before the incremental axial force is applied.

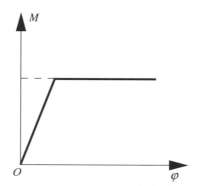

Figure 7.2   Ideal moment–rotation curve of the beam element

Assume that a plastic hinge forms at element end 1, i.e. $M_1 = M_p$, which will lead to $dM_1 = 0$. With static condensation, Equation (7.4) can be rewritten as

$$
\left\{ \begin{array}{c} dQ_1 \\ dQ_2 \\ dM_2 \end{array} \right\} = \frac{EI}{l} \left(\begin{bmatrix} \dfrac{12}{l^2}\psi_1 & -\dfrac{12}{l^2}\psi_1 & \dfrac{6}{l}\psi_2 \\[2mm] -\dfrac{12}{l^2}\psi_1 & \dfrac{12}{l^2}\psi_1 & -\dfrac{6}{l}\psi_2 \\[2mm] \dfrac{6}{l}\psi_2 & -\dfrac{6}{l}\psi_2 & 4\psi_3 \end{bmatrix} \right.
$$

$$
\left. - \begin{bmatrix} \dfrac{6}{l}\psi_2 \\[2mm] -\dfrac{6}{l}\psi_2 \\[2mm] 2\psi_4 \end{bmatrix} (4\psi_3)^{-1} \begin{bmatrix} \dfrac{6}{l}\psi_2 & -\dfrac{6}{l}\psi_2 & 2\psi_4 \end{bmatrix} \right) \left\{ \begin{array}{c} d\delta_1 \\ d\delta_2 \\ d\theta_2 \end{array} \right\}
$$

$$
= \frac{EI}{l} \begin{bmatrix} \dfrac{3}{l^2}\psi_5 & -\dfrac{3}{l^2}\psi_5 & \dfrac{3}{l}\psi_6 \\[2mm] -\dfrac{3}{l^2}\psi_5 & \dfrac{3}{l^2}\psi_5 & -\dfrac{3}{l}\psi_6 \\[2mm] \dfrac{3}{l}\psi_6 & -\dfrac{3}{l}\psi_6 & 3\psi_7 \end{bmatrix} \left\{ \begin{array}{c} d\delta_1 \\ d\delta_2 \\ d\theta_2 \end{array} \right\}, \tag{7.5}
$$

where

$$
\psi_5 = \frac{1}{\psi_3}\left(4\psi_1\psi_3 - 3\psi_2^2\right), \tag{7.6a}
$$

$$
\psi_6 = \frac{\psi_2}{\psi_3}\left(2\psi_3 - \psi_4\right), \tag{7.6b}
$$

$$
\psi_7 = \frac{1}{3\psi_3}\left(4\psi_3^2 - \psi_4^2\right). \tag{7.6c}
$$

As

$$
\psi_2 = \frac{2\psi_3 + \psi_4}{3}, \tag{7.7}
$$

substituting it into Equation (7.6b) and noting Equation (7.6c) yields

$$
\psi_6 = \psi_7. \tag{7.8}
$$

With Equation (7.8), the incremental stiffness equation of the beam element with a plastic hinge formed at end 1 can be written as

$$
[k_{gp1}]\{d\delta_g\} = \{df_g\}, \tag{7.9}
$$

where

$$
[k_{ge1}] = \frac{EI}{l} \begin{bmatrix} \dfrac{3}{l^2}\psi_5 & 0 & -\dfrac{3}{l^2}\psi_5 & \dfrac{3}{l}\psi_6 \\[2mm] 0 & 0 & 0 & 0 \\[2mm] -\dfrac{3}{l^2}\psi_5 & 0 & \dfrac{3}{l^2}\psi_5 & -\dfrac{3}{l}\psi_6 \\[2mm] \dfrac{3}{l}\psi_6 & 0 & -\dfrac{3}{l}\psi_6 & 3\psi_6 \end{bmatrix}. \tag{7.10}
$$

A similar stiffness equation of the element when a plastic hinge forms at end 2 can be

$$[k_{gp2}]\{d\delta_g\} = \{df_g\}, \tag{7.11}$$

where

$$[k_{ge2}] = \frac{EI}{l} \begin{bmatrix} \dfrac{3}{l^2}\psi_5 & \dfrac{3}{l}\psi_6 & -\dfrac{3}{l^2}\psi_5 & 0 \\[2mm] \dfrac{3}{l}\psi_6 & 3\psi_6 & -\dfrac{3}{l}\psi_6 & 0 \\[2mm] -\dfrac{3}{l^2}\psi_5 & -\dfrac{3}{l}\psi_6 & \dfrac{3}{l^2}\psi_5 & 0 \\[2mm] 0 & 0 & 0 & 0 \end{bmatrix}. \tag{7.12}$$

ψ_5 and ψ_6 in Equations (7.10) and (7.12) are calculated according to Equations (7.6a) and (7.6b), respectively. If the effects of axial force are neglected, substituting Equation (2.40) into Equation (7.6) results in

$$\psi_5 = \psi_6 = \frac{4}{4 + r}. \tag{7.13}$$

If the effects of shear deformation are neglected, and only the first term of $(\alpha l)^2$ in series expansions of ψ_5 and ψ_6 is reserved, substituting Equation (2.27) into Equation (7.6) leads to

$$\psi_5 = 1 + \frac{6}{15}(\alpha l)^2 = 1 + \frac{6}{15}\frac{Nl^2}{EI}, \tag{7.14a}$$

$$\psi_6 = 1 + \frac{1}{15}(\alpha l)^2 = 1 + \frac{1}{15}\frac{Nl^2}{EI}. \tag{7.14b}$$

Substituting Equation (7.14) into Equations (7.10) and (7.12) yields

$$[k_{gp1}] = [k_{ge1}] + [k_{gG1}], \tag{7.15a}$$

$$[k_{gp2}] = [k_{ge2}] + [k_{gG2}], \tag{7.15b}$$

where

$$[k_{ge1}] = \frac{EI}{l} \begin{bmatrix} \dfrac{3}{l^2} & 0 & -\dfrac{3}{l^2} & \dfrac{3}{l} \\[2mm] 0 & 0 & 0 & 0 \\[2mm] -\dfrac{3}{l^2} & 0 & \dfrac{3}{l^2} & -\dfrac{3}{l} \\[2mm] \dfrac{3}{l} & 0 & -\dfrac{3}{l} & 3 \end{bmatrix}, \tag{7.16a}$$

$$[k_{ge2}] = \frac{EI}{l} \begin{bmatrix} \dfrac{3}{l^2} & \dfrac{3}{l} & -\dfrac{3}{l^2} & 0 \\[2mm] \dfrac{3}{l} & 3 & -\dfrac{3}{l} & 0 \\[2mm] -\dfrac{3}{l^2} & -\dfrac{3}{l} & \dfrac{3}{l^2} & 0 \\[2mm] 0 & 0 & 0 & 0 \end{bmatrix}, \tag{7.16b}$$

$$[k_{gG1}] = N \begin{bmatrix} \dfrac{6}{5l} & 0 & -\dfrac{6}{5l} & \dfrac{1}{5} \\[2mm] 0 & 0 & 0 & 0 \\[2mm] -\dfrac{6}{5l} & 0 & \dfrac{6}{5l} & -\dfrac{1}{5} \\[2mm] \dfrac{1}{5} & 0 & -\dfrac{1}{5} & \dfrac{l}{5} \end{bmatrix}, \tag{7.17a}$$

$$[k_{gG2}] = N \begin{bmatrix} \dfrac{6}{5l} & \dfrac{1}{5} & -\dfrac{6}{5l} & 0 \\ \dfrac{1}{5} & \dfrac{l}{5} & -\dfrac{1}{5} & 0 \\ -\dfrac{6}{5l} & -\dfrac{1}{5} & \dfrac{6}{5l} & 0 \\ 0 & 0 & 0 & 0 \end{bmatrix}. \tag{7.17b}$$

In the above equations, $[k_{ge1}]$ and $[k_{ge2}]$ are the elastic stiffness matrices for the beam element with hinge at end 1 and end 2, respectively, which exclude effects of both shear deformation and axial forces, and $[k_{gG1}]$ and $[k_{gG2}]$ are the geometrical stiffness matrices for the beam element with hinge at end 1 and end 2, respectively.

If effects of both shear deformation and axial force are neglected, Equation (7.15) is simplified to

$$[k_{gp1}] = [k_{ge1}], \tag{7.18a}$$
$$[k_{gp2}] = [k_{ge2}]. \tag{7.18b}$$

7.1.2 Hinge Formed at Both Ends of Element

If plastic hinges form at both end 1 and end 2 of the element, the equilibrium conditions before and after incremental forces applied are (see Figures 2.1 and 7.1)

$$Q_2 = -Q_1 = N\frac{\delta_2 - \delta_1}{l}, \tag{7.19a}$$

$$Q_2 + dQ_2 = -(Q_1 + dQ_1) = (N + dN)\frac{(\delta_2 + d\delta_2) - (\delta_1 + d\delta_1)}{l}. \tag{7.19b}$$

Substituting Equation (7.19a) into Equation (7.19b) and omitting high-order minor quantities lead to

$$dQ_2 = -dQ_1 = \frac{N}{l}(d\delta_2 - d\delta_1). \tag{7.20}$$

Equation (7.20) can be rewritten in matrix form as

$$[k_{gp3}]\{d\delta_g\} = \{df_g\}, \tag{7.21}$$

where

$$[k_{ge3}] = N \begin{bmatrix} \dfrac{1}{l} & 0 & \dfrac{1}{l} & 0 \\ 0 & 0 & 0 & 0 \\ \dfrac{1}{l} & 0 & \dfrac{1}{l} & 0 \\ 0 & 0 & 0 & 0 \end{bmatrix}. \tag{7.22}$$

7.2 CLOUGH MODEL

Clough, Benuska and Wilson (1965) proposed a model based on plastic hinge theory and bilinear $M - \phi$ relationship of beam sections, which actually introduces the hardening effect of material to the simple plastic hinge model. The beam element of the Clough model is assumed to be superpositioned with two parallel components working together, as shown in Figure 7.3, where one is an idealized elasto-plastic component and the other an infinitely elastic component.

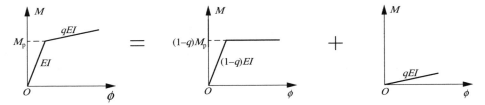

Figure 7.3 Clough model

When the bending moment at one end of the beam element is equal to or larger than the sectional plastic moment M_p and the element is at the same time in loading state, plastic hinge will form at the same end of the idealized elasto-plastic component. As the two components of the element work together, the stiffness of the beam element can be the summation of these two components assumed. That is if end 1 of the element is yielding ($M_1 \geq M_p$ and $M_1 dM_1 \geq 0$):

$$[k_{gp}] = (1 - q)[k_{gp1}] + q[k_{ge}]; \tag{7.23a}$$

• if end 2 of the element is yielding ($M_2 \geq M_p$ and $M_2 dM_2 \geq 0$):

$$[k_{gp}] = (1 - q)[k_{gp2}] + q[k_{ge}]; \tag{7.23b}$$

• if both ends of the element are yielding:

$$[k_{gp}] = (1 - q)[k_{gp3}] + q[k_{ge}], \tag{7.23c}$$

where $[k_{gp1}]$, $[k_{gp2}]$ and $[k_{gp3}]$ are the stiffness matrices of the beam element, determined with plastic hinge theory; $[k_{ge}]$ is the elastic stiffness matrix of the beam element and q is the hardening factor in the bilinear $M - \phi$ model.

When the beam element is subjected to cyclic loading, the stiffness matrices of the beam element can be calculated by combining the Clough model with the bilinear $M - \phi$ hysteretic model.

7.3 GENERALIZED CLOUGH MODEL

The realistic $M - \phi$ relationship of steel beams can be represented with the curve shown in Figure 7.4, which is different from the bilinear model, and nonlinearity occurs in the phase from initial yielding to ultimate yielding. To determine the nonlinear $M - \phi$ relationship of beam elements, the authors proposed a generalized Clough model in 1990.

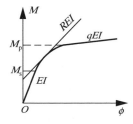

Figure 7.4 Realistic moment–curvature curve of the steel beam

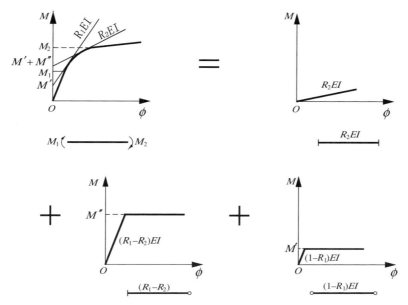

Figure 7.5 Generalized Clough model ($R_1 \geq R_2$)

In the generalized Clough model, a beam element is assumed to be superpositioned with three parallel components working together, as shown in Figure 7.5. The first is a two-end clamped component, the second is a hinged-clamped component and the third is a two-end hinged component. It can be seen from Figure 7.5 that the generalized Clough model is a transient one and corresponds to the loading state of the beam element at the time considered.

The tangent stiffness matrices of the beam element can be expressed with the generalized Clough model as

$$[k_{\mathrm{gp}}] = R_2[k_{\mathrm{ge}}] + (R_1 - R_2)[k_{\mathrm{gp2}}] + (1 - R_1)[k_{\mathrm{gp3}}], \quad \text{for} \quad R_1 \geq R_2, \tag{7.24a}$$

or

$$[k_{\mathrm{gp}}] = R_1[k_{\mathrm{ge}}] + (R_2 - R_1)[k_{\mathrm{gp1}}] + (1 - R_2)[k_{\mathrm{gp3}}], \quad \text{for} \quad R_1 \leq R_2, \tag{7.24b}$$

where R_1 and R_2 are the recovery force parameters for end 1 and end 2 of the element, respectively, which are relevant to the moments at the both ends of the element, M_1 and M_2, and the deformation state of the element, and can be calculated according to Equations (5.39) and (5.40).

7.4 ELASTO-PLASTIC HINGE MODEL

Giberson (1969) proposed an end-spring model (see Figure 7.6) to establish the elasto-plastic stiffness equation of a beam element after its ends yield. Based on the model by Giberson, Li (1988) proposed the elasto-plastic hinge model. The principal assumptions involved in the elasto-plastic hinge model are as follows:

(1) The rotation of the elemental end section always includes elastic and plastic portions, namely the elemental end is constantly an elasto-plastic hinge.

(2) The plastic rotation of the elemental end can be represented by a virtual rotating spring, which is dependent only on the bending moment at the same end.

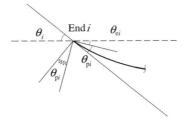

Figure 7.6 Elasto-plastic spring model

By assumption (1), the incremental deformations of the elements can be written as

$$\{d\delta_g\} = \{d\delta_{ge}\} + \{d\delta_{gp}\}, \tag{7.25}$$

where $\{d\delta_{ge}\}$ and $\{d\delta_{gp}\}$ are the elastic and plastic deformation components.

Dividing the incremental deformations by the two ends results in

$$\{d\delta_g\} = \left[\{d\delta_{g1}\}^T, \{d\delta_{g2}\}^T\right]^T, \tag{7.26}$$

where

$$\{d\delta_{g1}\} = [d\delta_1, d\theta_1]^T, \tag{7.27a}$$
$$\{d\delta_{g2}\} = [d\delta_2, d\theta_2]^T. \tag{7.27b}$$

Note that

$$\{d\delta_{gp1}\} = [0, d\theta_{p1}]^T = [g]\{1\}d\theta_{p1}, \tag{7.28a}$$
$$\{d\delta_{gp2}\} = [0, d\theta_{p2}]^T = [g]\{1\}d\theta_{p2}, \tag{7.28b}$$

where

$$[g] = \begin{bmatrix} 0 & 0 \\ 0 & 1 \end{bmatrix}, \tag{7.29}$$
$$[1] = [1 \ 1]^T. \tag{7.30}$$

Dividing the incremental forces with two components $\{df_{gt}\}$ and $\{df_{gn}\}$, one has

$$\{df_g\} = \{df_{gt}\} + \{df_{gn}\}, \tag{7.31}$$

where

$$\{df_{gt}\} = \left[\{df_{gt1}\}^T, \{df_{gt2}\}^T\right], \tag{7.32a}$$

$$\{df_{gt1}\} = [dQ_1, 0]^T, \tag{7.32b}$$

$$\{df_{gt2}\} = [dQ_2, 0]^T, \tag{7.32c}$$

$$\{df_{gn}\} = \left[\{df_{gn1}\}^T, \{df_{gn2}\}^T\right], \tag{7.33a}$$

$$\{df_{gn1}\} = [0, dM_1]^T, \tag{7.33b}$$

$$\{df_{gn2}\} = [0, dM_2]^T. \tag{7.33c}$$

Evidently, $\{df_{gt}\}$ and $\{df_{gn}\}$ are the components of $\{df_g\}$ orthotropic to each other. The relationship between the incremental forces $\{df_{gn}\}$ and incremental elastic deformations $\{d\delta_{ge}\}$ *isconstantly*

$$\{df_g\} = [k_{ge}]\{d\delta_{ge}\}. \tag{7.34}$$

Rewrite Equation (7.34) in partitioned matrix form as

$$\left\{ \begin{array}{c} \{df_{g1}\} \\ \{df_{g2}\} \end{array} \right\} = \left[\begin{array}{cc} [k_{gebb}] & [k_{gebt}] \\ [k_{getb}] & [k_{gebb}] \end{array} \right] \left\{ \begin{array}{c} \{d\delta_{ge1}\} \\ \{d\delta_{ge1}\} \end{array} \right\} \tag{7.35}$$

or

$$\{df_{g1}\} = [k_{gebb}]\{d\delta_{ge1}\} + [k_{gebt}]\{d\delta_{ge2}\}, \tag{7.36a}$$
$$\{df_{g2}\} = [k_{getb}]\{d\delta_{ge1}\} + [k_{gebb}]\{d\delta_{ge2}\}. \tag{7.36b}$$

By assumption (2), one has

$$\{df_{gn}\} = [k_n]\{d\delta_{gp}\}, \tag{7.37}$$

where

$$[k_n] = \left[\begin{array}{cc} [k_{nb}] & 0 \\ 0 & [k_{nt}] \end{array} \right]. \tag{7.38}$$

Then, $\{df_{gn}\}$ can be partitioned as

$$\{df_{gn1}\} = [k_{nb}]\{d\delta_{gp1}\}, \tag{7.39a}$$
$$\{df_{gn2}\} = [k_{nt}]\{d\delta_{gp2}\}, \tag{7.39b}$$

where

$$[k_{nb}] = \left[\begin{array}{cc} 0 & 0 \\ 0 & B_b \end{array} \right], \quad [k_{nt}] = \left[\begin{array}{cc} 0 & 0 \\ 0 & B_t \end{array} \right], \tag{7.40}$$

$$B_b = \alpha_1 k_{ge22}, \quad B_t = \alpha_2 k_{ge44}. \tag{7.41}$$

k_{ge22} and k_{ge44} are the elements of the second row and second column and fourth row and fourth column, respectively, in the stiffness matrix $[k_{ge}]$, and α_1 and α_2 are the elasto-plastic hinge factors for the two ends of the beam element.

The elasto-plastic hinge factor can be understood as the stiffness factor of the virtual rotating spring and should satisfy the following two conditions:

(1) If the section considered is in elastic state, $\alpha = \infty$ and the plastic incremental rotation of the section is zero.

(2) If the section considered is in ideal plastic state (without hardening), $\alpha = 0$ and the plastic incremental rotation of the section is arbitrary.

To satisfy the above conditions, α can be defined as

$$\alpha_i = \frac{R_i}{1 - R_i}, \quad i = 1, 2, \tag{7.42}$$

where R_1 and R_2 are the recovery force parameters defined in Equation (5.38) and can be calculated according to Equations (5.39) and (5.40).

From Equations (7.25), (7.31), (7.36) and (7.39), one has

$$\{df_{gt1}\} = \{df_{g1}\} - \{df_{gn1}\} = [k_{gebb}]\{d\delta_{g1}\} + [k_{gebt}]\{d\delta_{g2}\} - ([k_{gebb}] + [k_{nb}])\{d\delta_{gp1}\} - [k_{gebt}]\{d\delta_{gp2}\},$$

$$= [k_{gebb}]\{d\delta_{g1}\} + [k_{gebt}]\{d\delta_{g2}\} - ([k_{gebb}] + [k_{nb}])\{d\delta_{gp1}\} - [k_{gebt}]\{d\delta_{gp2}\}, \tag{7.43a}$$

$$\{df_{gt2}\} = \{df_{g2}\} - \{df_{gn2}\}$$

$$= [k_{getb}]\{d\delta_{g1}\} + [k_{gett}]\{d\delta_{g2}\} - [k_{getb}]\{d\delta_{gp1}\} - ([k_{gett}] + [k_{nt}])\{d\delta_{gp2}\}. \tag{7.43b}$$

As vectors $\{d\delta_{gp1}\}$ and $\{d\delta_{gp2}\}$ are orthotropic with $\{df_{gt1}\}$ and $\{dt_{gt2}\}$, which can be known from Equations (7.28) and (7.32), one has

$$\{d\delta_{gp1}\}^T\{df_{gt1}\} = d\theta_{p1}\{1\}^T[g]^T\{df_{gt1}\} = 0, \tag{7.44a}$$

$$\{d\delta_{gp2}\}^T\{df_{gt2}\} = d\theta_{p2}\{1\}^T[g]^T\{df_{gt2}\} = 0. \tag{7.44b}$$

The solutions of $d\theta_{p1}$ and $d\theta_{p2}$ and the expressions of the stiffness matrix of the element will be relevant to the state of the elemental ends, which is discussed in the following.

7.4.1 Both Ends Yielding

When both ends of the element yield, $d\theta_{p1} \neq 0$ and $d\theta_{p2} \neq 0$, with which Equation (7.44) becomes

$$\{1\}^T[g]^T\{df_{gt1}\} = 0, \tag{7.45a}$$

$$\{1\}^T[g]^T\{df_{gt2}\} = 0. \tag{7.45b}$$

Substituting Equation (7.43) into Equation (7.45) yields

$$\{1\}^T[g]^T([k_{gebb}] + [k_{nb}])[g]\{1\}d\theta_{p1} + \{1\}^T[g]^T[k_{gebt}][g]\{1\}d\theta_{p2}$$
$$= \{1\}^T[g]^T[k_{gebb}]\{d\delta_{g1}\} + \{1\}^T[g]^T[k_{gebt}]\{d\delta_{g2}\}, \tag{7.46a}$$

$$\{1\}^T[g]^T[k_{getb}][g]\{1\}d\theta_{p1} + \{1\}^T[g]^T([k_{gett}] + [k_{nt}])[g]\{1\}d\theta_{p2}$$
$$= \{1\}^T[g]^T[k_{getb}]\{d\delta_{g1}\} + \{1\}^T[g]^T[k_{gett}]\{d\delta_{g2}\}. \tag{7.46b}$$

The solution of the above equations about $d\theta_{p1}$ and $d\theta_{p2}$ is

$$\begin{Bmatrix} d\theta_{p1} \\ d\theta_{p2} \end{Bmatrix} = \begin{bmatrix} k_{bb} & k_{bt} \\ k_{tb} & k_{tt} \end{bmatrix}^{-1} \begin{bmatrix} [H_{bb}] & [H_{bt}] \\ [H_{tb}] & [H_{tt}] \end{bmatrix} \begin{Bmatrix} d\delta_{g1} \\ d\delta_{g2} \end{Bmatrix}, \tag{7.47}$$

where

$$k_{bb} = \{1\}^T[g]^T([k_{gebb}] + [k_{nb}])[g]\{1\}, \tag{7.48a}$$

$$k_{tt} = \{1\}^T[g]^T([k_{gett}] + [k_{nt}])[g]\{1\}, \tag{7.48b}$$

$$k_{bt} = \{1\}^T[g]^T[k_{gebt}][g]\{1\}, \tag{7.48c}$$

$$k_{tb} = \{1\}^T[g]^T[k_{getb}][g]\{1\}, \tag{7.48d}$$

$$[H_{bb}] = \{1\}^T[g]^T[k_{gebb}], \tag{7.49a}$$

$$[H_{tt}] = \{1\}^T[g]^T[k_{gett}], \tag{7.49b}$$

$$[H_{bt}] = \{1\}^T[g]^T[k_{gebt}], \tag{7.49c}$$

$$[H_{tb}] = \{1\}^T[g]^T[k_{getb}]. \tag{7.49d}$$

Let

$$[G] = \begin{bmatrix} [g] & 0 \\ 0 & [g] \end{bmatrix}, \tag{7.50}$$

$$[E] = \begin{bmatrix} \{1\} & 0 \\ 0 & \{1\} \end{bmatrix}, \tag{7.51}$$

$$[L_{bt}] = \begin{bmatrix} k_{bb} & k_{bt} \\ k_{tb} & k_{tt} \end{bmatrix}^{-1}. \tag{7.52}$$

From Equations (7.28), (7.47) and (7.49), one has

$$\{d\delta_{gp}\} = [G][E][L_{bt}][E]^{T}[G]^{T}[k_{ge}]\{d\delta_g\}, \tag{7.53}$$

and substituting Equations (7.25) and (7.53) into Equation (7.34) yields

$$\{df_g\} = ([k_{ge}] - [k_{ge}][G][E][L_{bt}][E]^{T}[G]^{T}[k_{ge}])\{d\delta_g\}. \tag{7.54}$$

Thus, the stiffness matrix when both ends of the element are yielding is

$$[k_{gp}] = [k_{ge}] - [k_{ge}][G][E][L_{bt}][E]^{T}[G]^{T}[k_{ge}]. \tag{7.55}$$

7.4.2 Only End 1 Yielding

When only end 1 of the element yields, $d\theta_{p1} \neq 0$ and $d\theta_{p2} = 0$, with which Equations (7.43a), (7.44a) and (7.45a) become

$$\{1\}^{T}[g]^{T}([k_{gebb}] + [k_{nb}])[g]\{1\}d\theta_{p1} = \{1\}^{T}[g]^{T}[k_{gebb}]\{d\delta_{g1}\} + \{1\}^{T}[g]^{T}[k_{gebt}]\{d\delta_{g2}\}, \tag{7.56}$$

Solution of Equation (7.56) about $d\theta_{p1}$ gives

$$d\theta_{p1} = \frac{1}{k_{bb}}(\{1\}^{T}[g]^{T}[k_{gebb}]\{d\delta_{g1}\} + \{1\}^{T}[g]^{T}[k_{gebt}]\{d\delta_{g2}\}) \tag{7.57}$$

or

$$\begin{Bmatrix} d\theta_{p1} \\ d\theta_{p2} \end{Bmatrix} = \begin{bmatrix} 1/k_{bb} & 0 \\ 0 & 0 \end{bmatrix}\begin{bmatrix} [H_{bb}] & [H_{bt}] \\ [H_{tb}] & [H_{tt}] \end{bmatrix}\begin{Bmatrix} d\delta_{g1} \\ d\delta_{g2} \end{Bmatrix}. \tag{7.58}$$

Let

$$[L_b] = \begin{bmatrix} 1/k_{bb} & 0 \\ 0 & 0 \end{bmatrix}. \tag{7.59}$$

Then, the stiffness matrix of the element when end 1 yields can be expressed as

$$[k_{gp}] = [k_{ge}] - [k_{ge}][G][E][L_b][E]^{T}[G]^{T}[k_{ge}]. \tag{7.60}$$

7.4.3 Only End 2 Yielding

When only end 2 of the element yields, $d\theta_{p1} = 0$ and $d\theta_{p2} \neq 0$. Similar to the discussion in Section 7.4.2, let

$$[L_t] = \begin{bmatrix} 0 & 0 \\ 0 & 1/k_{tt} \end{bmatrix}. \tag{7.61}$$

Then, the stiffness matrix of the element when end 2 yields can be expressed similarly as

$$[k_{gp}] = [k_{ge}] - [k_{ge}][G][E][L_t][E]^T[G]^T[k_{ge}]. \tag{7.62}$$

7.4.4 Summary

As a summary, the form of the stiffness matrix of the beam element obtained with the elasto-plastic hinge model can be expressed in the following unified form as

$$[k_{gp}] = [k_{ge}] - [k_{ge}][G][E][L][E]^T[G]^T[k_{ge}], \tag{7.63}$$

where matrix $[L]$ can be determined with

(a) when no yielding occurring at both ends of the element:

$$[L] = \begin{bmatrix} 0 & 0 \\ 0 & 0 \end{bmatrix}; \tag{7.64a}$$

(b) when yielding occurring at only end 1 of the element:

$$[L] = \begin{bmatrix} 1/k_{bb} & 0 \\ 0 & 0 \end{bmatrix}; \tag{7.64b}$$

(c) when yielding occurring at only end 2 of the element:

$$[L] = \begin{bmatrix} 0 & 0 \\ 0 & 1/k_{tt} \end{bmatrix}; \tag{7.64c}$$

(d) when yielding occurring at both ends of the element:

$$[L] = \begin{bmatrix} k_{bb} & k_{bt} \\ k_{tb} & k_{tt} \end{bmatrix}^{-1}. \tag{7.64d}$$

7.5 COMPARISON BETWEEN ELASTO-PLASTIC HINGE MODEL AND GENERALIZED CLOUGH MODEL

For the sake of comparison, let $l = 1$ and $EI = 1$ for the beam element, and ignore the effects of axial force and shear deformation in the following discussion.

7.5.1 Only End 1 Yielding

The stiffness matrix of the element based on the elasto-plastic hinge model can be determined with Equations (7.63) and (7.64b) as

$$[k_{gp}]_{B1} = \begin{bmatrix} 3 + 9R_1 & 6R_1 & -3 - 9R_1 & 3 + 3R_1 \\ 6R_1 & 4R_1 & -6R_1 & 2R_1 \\ -3 - 9R & -6R_1 & 3 + 9R_1 & -3 - 9R \\ 3 + 3R_1 & 2R_1 & -3 - 9R & 4R_1 \end{bmatrix}. \tag{7.65}$$

From Equation (7.65), we may find that when end 1 of the element is in elastic state, i.e. $R_1 \to 1$, then $[k_{gp}]_{B1} \to [k_{ge}]$, and when end 1 yields fully and the element is perfectly elasto-plastic, then $[k_{gp}]_{B1}$ is absolutely equal to the result from the simple plastic hinge theory. The above agreements indicate that $[k_{gp}]_{B1}$ satisfies the continuity condition on the yielding at one end of the element.

The stiffness matrix from the generalized Clough model when only end 1 yields is

$$[k_{gp}]_{C1} = R_1 \begin{bmatrix} 12 & 6 & -12 & 6 \\ 6 & 4 & -6 & 2 \\ -12 & -6 & 12 & -6 \\ 6 & 2 & -6 & 4 \end{bmatrix} + (1 - R_1) \begin{bmatrix} 3 & 0 & -3 & 0 \\ 0 & 0 & 0 & 0 \\ -3 & 0 & 3 & -3 \\ 3 & 0 & -3 & 3 \end{bmatrix}. \tag{7.66}$$

Comparison of Equations (7.65) and (7.66) shows coincidence between the elasto-plastic hinge model and the generalized Clough model.

7.5.2   Both Ends Yielding

The stiffness matrix of the element based on the elasto-plastic hinge model can be determined with Equations (7.63) and (7.64d) as

$$[k_{gp}]_{B3} = \frac{1}{D}$$

$$\cdot \begin{bmatrix} 12R_1 + 12R_2 + 24R_1R_2 & 12R_1 + 12R_1R_2 & -12R_1 - 12R_2 - 24R_1R_2 & 12R_2 + 12R_1R_2 \\ 12R_1 + 12R_1R_2 & 12R_1 + 4R_1R_2 & -12R_1 - 12R_1R_2 & 8R_1R_2 \\ -12R_1 - 12R_2 - 24R_1R_2 & -12R_1 - 12R_1R_2 & 12R_1 + 12R_2 + 24R_1R_2 & -12R_2 - 12R_1R_2 \\ 12R_2 + 24R_1R_2 & 8R_1R_2 & -R_2 - 12R_1R_2 & 12R_2 + 4R_1R_2 \end{bmatrix}.$$
$$\tag{7.67}$$

where $D = 3 + R_1 + R_2 - R_1R_2$.

It can be seen from Equation (7.67) that when end 2 of the element is elastic, i.e. $R_2 \to 1$, then $[k_{gp}]_{B3} \to [k_{gp}]_{B1}$; when both ends are elastic, i.e. $R_1 \to 1$, $R_2 \to 1$, then $[k_{gp}]_{B3} \to [k_{ge}]$; and when both ends enter the same plastic state, namely $R_1 = R_2 = R$, then

$$[k_{gp}]_{B3} = \frac{R}{3 + 2R - R^2} \begin{bmatrix} 24 + 24R & 12 + 12R & -24_2 - 24R & 12 + 12R \\ 12 + 12R & 12 + 4R & -12 - 12R & 8R \\ -24 - 24R & -12 - 12R & 24 + 24R & -12 - 12R \\ 12 + 12R & 8R & -12 - 12R & 12 + 4R \end{bmatrix}. \tag{7.68}$$

If both ends are fully yielding and the element is perfectly elasto-plastic, i.e. $R = 0$, then with Equation (7.68), $[k_{gp}]_{B3} = 0$. It can be found from the above discussions that the stiffness of the beam element by the elasto-plastic hinge model can be continuous along with arbitrary deformation states.

The significant limitation of the Clough model is that there is a stiffness break when beam elements enter elasto-plastic state from elastic state. On the contrary, the generalized Clough model overcomes this point and keeps the stiffness of beam elements changing smoothly between elastic and elasto-plastic states. However, comparison of Equations (7.67) and (7.24) indicates disagreement between the stiffness of the beam element by elasto-plastic hinge model and that by the generalized Clough model. For the sake of convenient comparison, assume the plastic states at both ends of the element are the same, namely $R_1 = R_2 = R$, then the stiffness of the beam element by the generalized Clough model is

$$[k_{\text{gp}}]_{C3} = R \begin{bmatrix} 12 & -6 & -12 & 6 \\ 6 & 4 & -6 & 2 \\ -12 & -6 & 12 & -6 \\ 6 & 2 & -6 & 4 \end{bmatrix}. \tag{7.69}$$

The following reasons may cause the difference between Equations (7.68) and (7.69):

(1) Two or three parallel components are assumed to be working together in the Clough and generalized Clough models. However, compatibility between deformations of the idealized elastic component and those of the idealized elasto-plastic component is hardly satisfied. On the contrary, the elasto-plastic hinge model is deformation compatible.

(2) When one end (end 1) is elastic and the other end yields, the moment transfer factor from the elastic end to the yielding end is $2R_1/(3 + R_1)$ in the beam element. This value can be considered as the capability to accept moment at the yielding end from the other end, which is regardless with the other end being elastic or plastic. The moment transfer factor by the Clough model for both ends yielding is 0.5, whereas that by the generalized Clough model depends on the yielding extent at the other end and is $2R_1/(3 + R_1)$ (if $R_2 \geq R_1$) or 0.5 (if $R_1 \geq R_2$). In the elasto-plastic hinge model, however, this moment transfer factor is constantly equal to $2R_1/(3 + R_1)$, which indicates its advantage over the Clough or generalized Clough model.

7.5.3 Numerical Example

Consider the frame as shown in Figure 7.7, where the moment–curvature relationship of the frame column is assumed to be

$$M = 1 - e^{\phi}, \tag{7.70}$$

and , $M_p = 1$, $\phi_p = M_p/EI = 1$. So, the recovery force parameter of the column section is

$$R = \frac{d(M/M_p)}{d(\phi/\phi_p)} = \frac{dM}{d\phi} = e^{-\phi} = 1 - M. \tag{7.71}$$

Noting that the yielding extent at both column ends is constantly equal and the structure is symmetric, the relationship between the drift and the horizontal force at the top of the frame column can be determined by the elasto-plastic hinge model as

$$\frac{R}{3 + 2R - R^2}(24 + 24R)d\delta_B = dF$$

or

$$d\delta_B = \left(\frac{1}{8R} - \frac{1}{24}\right)dF. \tag{7.72}$$

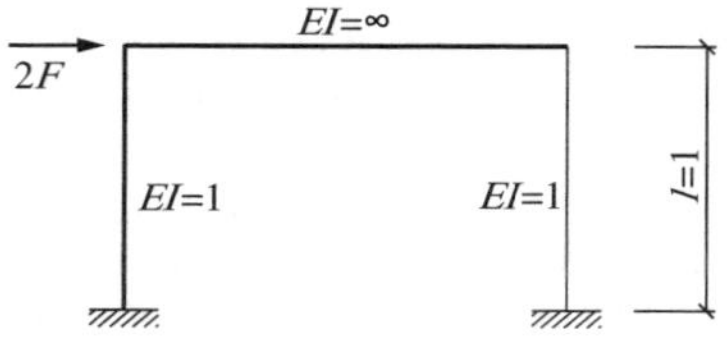

Figure 7.7 Rigid frame with a beam of infinite stiffness

Table 7.1 $(\delta_B/\delta_C) - M$ relationship

M	0	0.1	0.3	0.5	0.7	0.9	1
δ_B/δ_C	1.000	1.025	1.079	1.139	1.209	1.304	1.500

As the moment at the top of the column is $M = \frac{1}{2}F$, substituting Equation (7.71) into Equation (7.72) results in

$$d\delta_B = \frac{1}{4}\left(\frac{1}{1-M} - \frac{1}{3}\right)dM. \tag{7.73}$$

The solution for δ_B of the above equation is

$$\delta_B = \frac{1}{4}\left(\ln\frac{1}{1-M} - \frac{M}{3}\right). \tag{7.74}$$

Similarly, the relationship between the drift and the horizontal force at the top of the column can also be obtained with the generalized Clough model as

$$12Rd\delta_C = dF \tag{7.75}$$

or

$$d\delta_C = \frac{1}{6}\left(\frac{dM}{1-M}\right). \tag{7.76}$$

The solution for δ_C of the above equation is

$$\delta_C = \frac{1}{6}\ln\frac{1}{1-M}. \tag{7.77}$$

Dividing δ_B by δ_C yields

$$\frac{\delta_B}{\delta_C} = \frac{1}{2}\left[3 + \frac{M}{\ln(1-M)}\right]. \tag{7.78}$$

Table 7.1 lists the values of δ_B/δ_C varying with M, where it can be found that when both ends of the beam element yield, the stiffness determined by the elasto-plastic hinge model is smaller than that by the generalized Clough model.

7.6 EFFECTS OF RESIDUAL STRESSES AND TREATMENT OF TAPERED ELEMENT

7.6.1 Effects of Residual Stresses on Plasticity Spread Along Element Section

No matter hot-rolled or welded steel members, residual stresses with significant magnitude exist (see Figure 7.8) in beam and column members with the shape of H or box in section for multi-storey, high-rise steel buildings because the flange and web plates are normally quite thick. Generally, tensile stresses will be produced at the cross of flange and web plates or where cooling speed is relatively slow after welding or hot rolling, and self-balanced compression stresses will be produced at other area of sections. Due to the

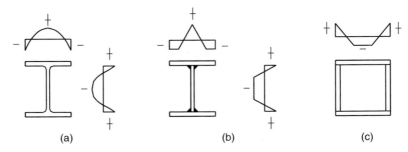

Figure 7.8 Typical distribution of residual stresses for steel members: (a) hot-rolled H-type section; (b) welded H-type section; (c) welded box section

existence of initially compressive stresses, the part of a section in compression may go into plasticity in advance although the ultimate plastic strength of the section is the same. The effects of residual stresses on $M - \phi$ relationship of steel sections in bending are given in Figure 7.9, where OCB and ODB curves correspond to those without and with effects of residual stresses, respectively, and OEB is a simplified one of OCB. The material property of steel in Figure 7.9 is assumed to be idealized elasto-plastic.

The realistic distribution of residual stresses along the section of steel members is complicated, which relates not only to the processes of cutting, welding and rolling but also to the plate thickness and sectional shape. It is deemed to be difficult to exactly consider the effects of residual stresses on the plastic spread over the section. In practice, a more straightforward and efficient way is to modify the equation of the initial yielding surface of sections.

The typical equations for the initial yielding surfaces of wide flange H sections are

- without residual stresses:

$$\frac{N}{N_y} + \frac{\chi_p M}{M_p} = 1.0; \tag{7.79}$$

- with residual stresses:

$$\frac{N}{0.8N_y} + \frac{\chi_p M}{0.9M_p} = 1.0, \tag{7.80}$$

where χ_p is the plastic shape factor of sections. The corresponding curves for Equations (7.79) and (7.80) are shown in Figure 7.10.

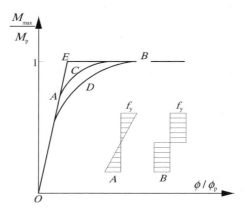

Figure 7.9 Effect of residual stresses on the moment–curvature relationship of steel sections

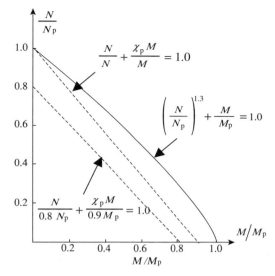

Figure 7.10   Initial and ultimate yielding curves of the H section

7.6.2   Effects of Residual Stresses on Plasticity Spread Along Element Length

When the ratio of axial force to squash load is large for a member in compression, residual stresses can influence the plasticity distribution along element length. A transient elastic modulus concept, namely the concept of tangent modulus, is proposed to take this effect into account.

Two types of tangent modulus based on different column strength equations have been proposed. The CRC column strength equations (Galambos, 1988) can be employed in deriving the tangent modulus. The ratio of the tangent modulus to the elastic modulus E_t/E is proposed to be (Galambos, 1988)

$$\frac{E_t}{E} = 1.0, \quad \text{for} \quad N \le 0.5N_y, \tag{7.81a}$$

$$\frac{E_t}{E} = \frac{4N}{N_y}\left(1 - \frac{N}{N_y}\right), \quad \text{for} \quad N > 0.5N_y. \tag{7.81b}$$

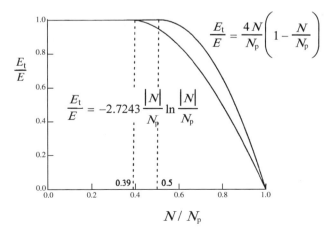

Figure 7.11   Tangent modulus with and without consideration of initial geometric imperfection

The LRFD column strength equation can also be used to derive the tangent modulus, where the effects of residual stresses and the initial geometric deflection of the compressive member are included. The ratio of the tangent modulus to the elastic modulus, E_t/E, is given by (AISC, 1994)

$$\frac{E_t}{E} = 1.0, \quad \text{for} \quad N \leq 0.39N_p, \tag{7.82a}$$

$$\frac{E_t}{E} = -2.7243\frac{N}{N_p}\ln\left[\frac{N}{N_p}\right], \quad \text{for} \quad N > 0.39N_p. \tag{7.82b}$$

As the LRFD equation includes the effect of initial geometric deflection, the tangent modulus by LRFD is less than that by CRC, at the same axial force. A comparison of these two tangent modulus proposals is given in Figure 7.11.

7.6.3 Treatment of Tapered Element

Plastic hinge theory, Clough model, generalized Clough model and elasto-plastic hinge model based on centralized plasticity at the ends of elements can be used to establish the elasto-plastic stiffness equation for tapered elements. Because the elasto-plastic hinge model is the most generalized one for the elasto-plastic beam element, Equation (7.63) can be used in practice to calculate the elasto-plastic stiffness matrix $[k_{gP}]$ of the tapered element where the elastic stiffness matrix $[k_{ge}]$ can be obtained with the methods proposed in Chapter 3.

The ratio of the axial force to the squash load is different at the two ends of a tapered element with the same axial force. If the effects of residual stresses should be involved, the following equations can be used to determine the modified tangent modulus E_t:

$$E_t = \frac{A_1}{A_1 + A_2}E_{t1} + \frac{A_2}{A_1 + A_2}E_{t2}, \tag{7.83}$$

where A_1 and A_2 are sectional areas of the two ends of the tapered element, and E_{t1} and E_{t2} are the tangent moduli determined by Equation (7.81) or (7.82).

7.7 BEAM ELEMENT WITH PLASTIC HINGE BETWEEN TWO ENDS

To eliminate the needs to divide a frame member into two or more elements to model the effects of distributed loads on the member in the second-order inelastic analysis of the frame, Chen and Chan (1995) proposed an elemental stiffness equation for a beam element with mid-span plastic hinge. This method is refined in this section to consider a possible plastic hinge at any position between the elemental ends.

Referring to Figure 7.12, an internal node C between elemental ends is inserted so that the element is divided into two parts, the lengths of which are L_a and L_b, respectively. Assume the maximum bending

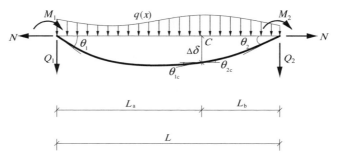

Figure 7.12   A beam element with plastic hinge within two ends

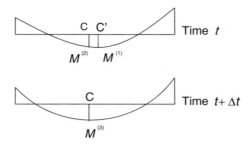

Figure 7.13 Position of the maximum moment within two ends

moment $M^{(1)}$ at time t is at position C′ and the maximum bending moment $M^{(3)}$ at time $t + dt$ is at position C (see Figure 7.13). For derivation of incremental stiffness matrix of the element during $t \rightarrow t + dt$, a virtual state of moment $M^{(2)}$ is conceived, which is the bending moment at the same position of $M^{(3)}$ at the time t. The incremental stiffness relationship of each part of the element can be expressed in the standard form as

- for the part of L_a:

$$
\begin{Bmatrix} dQ_1 \\ dM_1 \\ dQ_{1c} \\ dM_{1c} \end{Bmatrix} = [K_{pa}] \begin{Bmatrix} d\delta_1 \\ d\theta_1 \\ d\delta_{1c} \\ d\theta_{1c} \end{Bmatrix} = \begin{bmatrix} a_{11} & a_{12} & a_{13} & a_{14} \\ & a_{22} & a_{23} & a_{24} \\ & & a_{33} & a_{34} \\ & & & a_{44} \end{bmatrix} \begin{Bmatrix} d\delta_1 \\ d\theta_1 \\ d\delta_{1c} \\ d\theta_{1c} \end{Bmatrix} ; \tag{7.84a}
$$

- for the part of L_b:

$$
\begin{Bmatrix} dQ_{2c} \\ dM_{2c} \\ dQ_2 \\ dM_2 \end{Bmatrix} = [K_{pb}] \begin{Bmatrix} d\delta_{2c} \\ d\theta_{2c} \\ d\delta_2 \\ d\theta_2 \end{Bmatrix} = \begin{bmatrix} b_{11} & b_{12} & b_{13} & b_{14} \\ & b_{22} & b_{23} & b_{24} \\ & & b_{33} & b_{34} \\ & & & b_{44} \end{bmatrix} \begin{Bmatrix} d\delta_{2c} \\ d\theta_{2c} \\ d\delta_2 \\ d\theta_2 \end{Bmatrix} , \tag{7.84b}
$$

where $[K_{pa}]$ and $[K_{pb}]$ are the elasto-plastic stiffness matrices for the parts of L_a and L_b of the element, respectively, and a_{ij} and b_{ij} $(i, j = 1, 2, 3, 4)$ are the corresponding elements in such matrices.

It can be seen from Figure 7.12 that the two parts of the elements share the same deformation components at their junction, namely $d\delta_{1c} = d\delta_{2c} = d\delta_c$ and $d\theta_{1c} = d\theta_{2c} = d\theta_c$. Combining Equations (7.84a) and (7.84b), one has

$$
\begin{Bmatrix} dQ_1 \\ dM_1 \\ dQ_2 \\ dM_2 \\ dQ_{1c} + dQ_{2c} \\ dM_{1c} + dM_{2c} \end{Bmatrix} = \begin{bmatrix} a_{11} & a_{12} & 0 & 0 & a_{13} & a_{14} \\ & a_{22} & 0 & 0 & a_{23} & a_{24} \\ & & b_{33} & b_{34} & b_{13} & b_{23} \\ & & & b_{44} & b_{14} & b_{24} \\ & & & & a_{33} + b_{11} & a_{34} + b_{12} \\ & & & & & a_{44} + b_{22} \end{bmatrix} \begin{Bmatrix} d\delta_1 \\ d\theta_1 \\ d\delta_2 \\ d\theta_2 \\ d\delta_c \\ d\theta_c \end{Bmatrix} . \tag{7.85}
$$

For the purpose of static condensation to eliminate the freedom degree of the displacements of internal node, the above stiffness matrix is partitioned into internal and external degrees of freedom as

$$
\begin{Bmatrix} df_e \\ df_i \end{Bmatrix} = \begin{bmatrix} k_{ee} & k_{ei} \\ k_{ei}^T & k_{ii} \end{bmatrix} \begin{Bmatrix} d\delta_e \\ d\delta_i \end{Bmatrix} , \tag{7.86}
$$

where $\{df_e\}$ and $\{df_i\}$ are the elemental end and internal force vectors, respectively, and $\{d\delta_e\}$ and $\{d\delta_i\}$ are the elemental end and internal deformation vectors, respectively. Their expressions are as follows:

$$\{df_e\} = [dQ_1, dM_1, dQ_2, dM_2]^T,$$

$$\{d\delta_e\} = [d\delta_1, d\theta_1, d\delta_2, d\theta_2]^T,$$

$$\{df_i\} = [dQ_{1c} + dQ_{2c}, dM_{1c} + dM_{2c}]^T,$$

$$\{df_e\} = [dQ_1 \; dM_1 \; dQ_2 \; dM_2]^T,$$

$$\{d\delta_e\} = [d\delta_1 \; d\theta_1 \; d\delta_2 \; d\theta_2]^T,$$

$$\{df_i\} = [dQ_{1c} + dQ_{2c} \; dM_{1c} + dM_{2c}]^T,$$

$$\{d\delta_i\} = [d\delta_c \; d\theta_c]^T,$$

$$k_{ee} = \begin{bmatrix} a_{11} & a_{12} & 0 & 0 \\ a_{12} & a_{22} & 0 & 0 \\ 0 & 0 & b_{33} & b_{34} \\ 0 & 0 & b_{34} & b_{44} \end{bmatrix}, \tag{7.87a}$$

$$k_{ei} = \begin{bmatrix} a_{13} & a_{14} \\ a_{23} & a_{24} \\ b_{13} & b_{23} \\ b_{14} & b_{24} \end{bmatrix}, \tag{7.87b}$$

$$k_{ii} = \begin{bmatrix} a_{33} + b_{11} & a_{34} + b_{12} \\ a_{34} + b_{12} & a_{44} + b_{22} \end{bmatrix}. \tag{7.87c}$$

As no external forces are applied at internal node C, namely $\{df_i\} = \{0\}$, $\{d\theta_i\}$ in Equation (7.86) can be expressed with $\{d\theta_e\}$. The stiffness equation condensed off the internal displacement vector is

$$(k_{ee} - k_{ei} k_{ii}^{-1} k_{ei}^T)\{d\delta_e\} = \{df_e\}. \tag{7.88}$$

In the above derivation, it is assumed that the internal plastic hinge occurs at position of C at time , and the moment increases from $M^{(2)}$ at t to $M^{(3)}$ at $t + dt$. But actually in the duration $t \rightarrow t + dt$, the moment change should have been from $M^{(1)}$ at position of C′ to $M^{(3)}$ at position of C. A stiffness matrix modification $([k_{ee} - k_{ei} k_{ii}^{-1} k_{ei}^T]_{Ct} - [k_{ee} - k_{ei} k_{ii}^{-1} k_{ei}^T]_{C't})$ may be superimposed to approximately take the effect from position change of internal plastic hinge into account. The subscripts in the stiffness matrix modification indicate the position and the time of maximum bending moment.

Assume that the internal plastic hinge occurs at the position of maximum bending moment between the two ends. The position of the maximum bending moment between the two ends of the element, position C, varies in the loading process. Hence, the rational way to trace the internal plastic hinge is to calculate the position of the maximum bending moment at each loading step after elemental yielding. Two common internal loading patterns for beam elements are concentrated load and uniformly distributed load, as shown in Figure 7.14.

If one concentrated load is applied within the beam span, the position of the maximum moment within span is certainly the loading position. But if a uniformly distributed load is applied, the position of the maximum moment within span is changeable. The condition of the maximum moment within the beam span is

$$\frac{dM(x)}{dx} = 0 \quad \text{or} \quad Q(x) = 0. \tag{7.89}$$

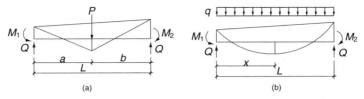

Figure 7.14 Load patterns within beam span: (a) concentrated load case; (b) uniformly distributed load case

The shear at end 1 can be expressed as

$$Q_1 = \frac{M_1 - M_2}{L} + \frac{1}{2}qL. \tag{7.90}$$

And letting the shear be equal to zero yields the position of the maximum moment desired:

$$x = \frac{M_1 - M_2}{qL} + \frac{1}{2}L. \tag{7.91}$$

As for the beam element with both concentrated load and uniformly distributed load within span, one can divide this element into two segments at the position where the concentrated load is applied. The maximum moment position of each segment can be determined according the method for the uniformly distributed load case as mentioned above. With comparison of the maximum moments of two segments of the element induced by the uniformly distributed load and the bending moment where the concentrated load is applied, the real maximum moment of this beam element can be obtained with the maximum of the above three moments.

7.8 SUBDIVIDED MODEL WITH VARIABLE STIFFNESS FOR COMPOSITE BEAM ELEMENT

7.8.1 Subdivided Model

The composite beams in steel frames resist vertical loads not only from frame slabs, but also subjected to several types of horizontal loads. Figure 7.15 gives typical bending moment diagrams of the steel frame storeys under three different loading conditions, i.e. vertical loads, horizontal loads and combination of vertical and horizontal loads. It can be seen from Figure 7.15 that the positive (sagging) and negative (hogging) moments will apply at the ends of composite beams repeatedly when horizontal loads induced by earthquakes occur. This gives new challenges to elasto-plastic analysis of steel frames comprising composite beams.

A subdivided model with variable stiffness is proposed in this section to construct a practical analysis model for elasto-plastic analysis of composite beams. The following assumptions are adopted:

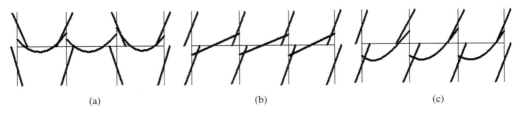

Figure 7.15 Typical moment diagrams of frame storeys: (a) subjected to vertical load; (b) subjected to horizontal load; (c) subjected to both vertical and horizontal loads

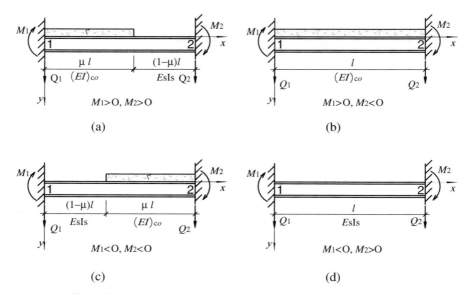

Figure 7.16   Subdivided model with variable stiffness for composite beams

(1) ignore the tensile strength of concrete, namely $f_{ct} = 0$;

(2) involve contribution of rebars within effective width of concrete flange of composite beams;

(3) the portion of a beam in the positive moment zone is considered as a composite beam;

(4) the portion of a beam in the negative moment zone is regarded as a pure steel beam.

Figure 7.16 illustrates four cases of the composite beam divided into subelements with different stiffness. The positive moment is in the clockwise direction.

In Figure 7.16(a), the moments at the left and right ends of the element are positive, which indicates a linear variation of moment within the span of the beam. Concrete flange is in compression in the part of the beam close to the left end and in tension in the part of the beam close to the right end.

In Figure 7.16(b), the moment at the left end is positive and that at the right end is negative (in sagging bending). The concrete flange of the beam is always in compression.

The moments at the left and right ends of the beam, as shown in Figure 7.16(c), are negative, and the curvature reverses within the span. The concrete flange of the beam is in tension in the part close to the left end and in compression in the part close to the right end.

In Figure 7.16(d), the moment at the left end is negative and that at the right end is positive (in hogging bending). The concrete flange of the composite beam is therefore always in tension.

7.8.2   Stiffness Equation of Composite Beam Element

To use the stiffness equations of prismatic beam elements for establishing the stiffness equation of a composite beam element, an internal node A is inserted at the position where curvature reverses. The composite beam with two segments of different curvatures is shown in Figure 7.17, where the force and deformation of the two segments are also illustrated.

The stiffness equation of subelement 1 with the concrete flange in compression is

$$
\begin{bmatrix}
h_{11} & h_{12} & h_{13} & h_{14} \\
h_{12} & h_{22} & h_{23} & h_{24} \\
h_{13} & h_{23} & h_{33} & h_{34} \\
h_{14} & h_{24} & h_{34} & h_{44}
\end{bmatrix}
\begin{Bmatrix}
\delta_1 \\
\theta_1 \\
\delta_A \\
\theta_A
\end{Bmatrix}
=
\begin{Bmatrix}
Q_1 \\
M_1 \\
Q_A \\
M_A
\end{Bmatrix}
\tag{7.92a}
$$

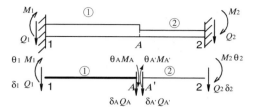

Figure 7.17   Subdivided composite beam element

or

$$[K]^{(1)}\{\delta\}^{(1)} = \{f\}^{(1)}. \tag{7.92b}$$

The stiffness equation of subelement 2 with the concrete flange in tension and out of action is

$$\begin{bmatrix} g_{11} & g_{12} & g_{13} & g_{14} \\ g_{12} & g_{22} & g_{23} & g_{24} \\ g_{13} & g_{23} & g_{33} & g_{34} \\ g_{14} & g_{24} & g_{34} & g_{44} \end{bmatrix} \begin{Bmatrix} \delta_{A'} \\ \theta_{A'} \\ \delta_A \\ \theta_A \end{Bmatrix} = \begin{Bmatrix} Q_{A'} \\ M_{A'} \\ Q_2 \\ M_2 \end{Bmatrix}. \tag{7.93a}$$

or

$$[K]^{(2)}\{\delta\}^{(2)} = \{f\}^{(2)}. \tag{7.93b}$$

It should be noted that the elemental stiffness matrices $[K]^{(1)}$ and $[K]^{(2)}$ can denote the elastic stiffness matrices or incremental elasto-plastic stiffness matrices of subelements 1 and 2.

To condense the degree of freedom of the internal node A, employ the deformation compatibility at node A as

$$\delta_A = \delta_{A'}, \tag{7.94a}$$
$$\theta_A = \theta_{A'} \tag{7.94b}$$

and the equilibrium condition as

$$Q_A + Q_{A'} = 0, \tag{7.95a}$$
$$M_A + M_{A'} = 0, \tag{7.95b}$$

By Equations (7.92) and (7.93), one has

$$Q_A = h_{13}\delta_1 + h_{23}\theta_1 + h_{33}\delta A + h_{34}\theta A, \tag{7.96a}$$
$$Q_{A'} = g_{11}\delta A + g_{12}\theta A + g_{13}\delta_2 + g_{14}\theta_2, \tag{7.96b}$$
$$M_A = h_{14}\delta_1 + h_{24}\theta_1 + h_{34}\delta A + h_{44}\theta A, \tag{7.96c}$$
$$M_{A'} = g_{12}\delta A + g_{22}\theta A + g_{23}\delta_2 + g_{24}\theta_2. \tag{7.96d}$$

Substituting Equation (7.96) into Equation (7.95) yields

$$(h_{33} + g_{11})\delta_A + (h_{34} + g_{12})\theta_A = -h_{13}\delta_1 - h_{23}\theta_1 - g_{13}\delta_2 - g_{14}\theta_2, \tag{7.97a}$$
$$(h_{34} + g_{12})\delta_A + (h_{44} + g_{22})\theta_A = -h_{14}\delta_1 - h_{24}\theta_1 - g_{23}\delta_2 - g_{24}\theta_2. \tag{7.97b}$$

Solving Equation (7.97), one obtains

$$
\left\{ \begin{array}{c} \delta A \\ \theta A \end{array} \right\} = \begin{bmatrix} h_{33} + g_{11} & h_{34} + g_{12} \\ h_{34} + g_{12} & h_{44} + g_{22} \end{bmatrix}^{-1} \begin{bmatrix} -h_{13} & -h_{23} & -g_{13} & -g_{14} \\ -h_{14} & -h_{24} & -g_{23} & -g_{24} \end{bmatrix} \left\{ \begin{array}{c} \delta_1 \\ \theta_1 \\ \delta_2 \\ \theta_2 \end{array} \right\}.
\tag{7.98}
$$

Let

$$
G = \det \left(\begin{bmatrix} h_{33} + g_{11} & h_{34} + g_{12} \\ h_{34} + g_{12} & h_{44} + g_{22} \end{bmatrix} \right).
\tag{7.99}
$$

The nodal forces of node 1 can be expressed with the nodal deformations with Equation (7.92a) as

$$
\begin{bmatrix} h_{11} & h_{12} & 0 & 0 \\ h_{12} & h_{22} & 0 & 0 \end{bmatrix} \left\{ \begin{array}{c} \delta_1 \\ \theta_1 \\ \delta_2 \\ \theta_2 \end{array} \right\} + \begin{bmatrix} h_{13} & h_{14} \\ h_{23} & h_{24} \end{bmatrix} \left\{ \begin{array}{c} \delta A \\ \theta A \end{array} \right\} = \left\{ \begin{array}{c} Q_1 \\ M_1 \end{array} \right\}.
\tag{7.100}
$$

Substituting Equation (7.98) into Equation (7.100) yields

$$
\left(\begin{bmatrix} h_{11} & h_{12} & 0 & 0 \\ h_{12} & h_{22} & 0 & 0 \end{bmatrix} + \begin{bmatrix} h_{13} & h_{14} \\ h_{23} & h_{24} \end{bmatrix} \begin{bmatrix} h_{33} + g_{11} & h_{34} + g_{12} \\ h_{34} + g_{12} & h_{44} + g_{22} \end{bmatrix}^{-1} \begin{bmatrix} -h_{13} & -h_{23} & -g_{13} & -g_{14} \\ -h_{14} & -h_{24} & -g_{23} & -g_{24} \end{bmatrix} \right) \left\{ \begin{array}{c} \delta_1 \\ \theta_1 \\ \delta_2 \\ \theta_2 \end{array} \right\} = \left\{ \begin{array}{c} Q_1 \\ M_1 \end{array} \right\}.
\tag{7.101}
$$

The nodal forces of node 2 can be expressed with the nodal deformations with Equation (7.92b) as

$$
\begin{bmatrix} g_{13} & g_{23} \\ g_{14} & g_{24} \end{bmatrix} \left\{ \begin{array}{c} \delta_{A'} \\ \theta_{A'} \end{array} \right\} + \begin{bmatrix} 0 & 0 & g_{33} & g_{34} \\ 0 & 0 & g_{34} & g_{44} \end{bmatrix} \left\{ \begin{array}{c} \delta_1 \\ \theta_1 \\ \delta_2 \\ \theta_2 \end{array} \right\} = \left\{ \begin{array}{c} Q_2 \\ M_2 \end{array} \right\}.
\tag{7.102}
$$

Substituting Equation (7.98) into Equation (7.102) yields

$$
\left(\begin{bmatrix} 0 & 0 & g_{33} & g_{34} \\ 0 & 0 & g_{34} & g_{44} \end{bmatrix} + \begin{bmatrix} g_{13} & g_{23} \\ g_{14} & g_{24} \end{bmatrix} \begin{bmatrix} h_{33} + g_{11} & h_{34} + g_{12} \\ h_{34} + g_{12} & h_{44} + g_{22} \end{bmatrix}^{-1} \begin{bmatrix} -h_{13} & -h_{23} & -g_{13} & -g_{14} \\ -h_{14} & -h_{24} & -g_{23} & -g_{24} \end{bmatrix} \right) \left\{ \begin{array}{c} \delta_1 \\ \theta_1 \\ \delta_2 \\ \theta_2 \end{array} \right\} = \left\{ \begin{array}{c} Q_2 \\ M_2 \end{array} \right\}.
\tag{7.103}
$$

Combining Equations (7.101) and (7.103) leads to the elemental stiffness equation as

$$
\begin{bmatrix} k_{11} & k_{12} & k_{13} & k_{14} \\ & k_{22} & k_{23} & k_{24} \\ & & k_{33} & k_{34} \\ \text{symm} & & & k_{44} \end{bmatrix} \left\{ \begin{array}{c} \delta_1 \\ \theta_1 \\ \delta_2 \\ \theta_2 \end{array} \right\} = \left\{ \begin{array}{c} Q_1 \\ M_1 \\ Q_2 \\ M_2 \end{array} \right\}.
\tag{7.104a}
$$

or

$$
[K]\{\delta\} = \{f\},
\tag{7.104b}
$$

where $[K]$ is the stiffness matrix of the composite beam element and the elements in the matrix are obtained with

$$k_{11} = h_{11} + \frac{1}{G}[-h_{13}^2(h_{44} + g_{22}) + 2h_{13}h_{14}(h_{34} + g_{12}) - h_{14}^2(h_{33} + g_{11})],$$

$$k_{22} = h_{22} + \frac{1}{G}[-h_{23}^2(h_{44} + g_{22}) + 2h_{23}h_{24}(h_{34} + g_{12}) - h_{24}^2(h_{33} + g_{11})],$$

$$k_{33} = g_{33} + \frac{1}{G}[-g_{13}^2(h_{44} + g_{22}) + 2g_{13}g_{23}(h_{34} + g_{12}) - g_{23}^2(h_{33} + g_{11})],$$

$$k_{44} = g_{44} + \frac{1}{G}[-g_{14}^2(h_{44} + g_{22}) + 2g_{14}g_{24}(h_{34} + g_{12}) - g_{24}^2(h_{33} + g_{11})],$$

$$k_{12} = h_{12} + \frac{1}{G}[-h_{13}h_{23}(h_{44} + g_{22}) + h_{13}h_{24}(h_{34} + g_{12}) + h_{14}h_{23}(h_{34} + g_{12}) - h_{14}h_{24}(h_{33} + g_{11})],$$

$$k_{13} = \frac{1}{G}[-h_{13}g_{13}(h_{44} + g_{22}) + h_{13}g_{23}(h_{34} + g_{12}) + h_{14}g_{13}(h_{34} + g_{12}) - h_{14}g_{23}(h_{33} + g_{11})],$$

$$k_{14} = \frac{1}{G}[-h_{13}g_{14}(h_{44} + g_{22}) + h_{13}g_{24}(h_{34} + g_{12}) + h_{14}g_{14}(h_{34} + g_{12}) - h_{14}g_{24}(h_{33} + g_{11})],$$

$$k_{23} = \frac{1}{G}[-h_{23}g_{13}(h_{44} + g_{22}) + h_{23}g_{23}(h_{34} + g_{12}) + h_{24}g_{13}(h_{34} + g_{12}) - h_{24}g_{23}(h_{33} + g_{11})],$$

$$k_{24} = \frac{1}{G}[-h_{23}g_{14}(h_{44} + g_{22}) + h_{23}g_{24}(h_{34} + g_{12}) + h_{24}g_{14}(h_{34} + g_{12}) - h_{24}g_{24}(h_{33} + g_{11})],$$

$$k_{34} = g_{34} + \frac{1}{G}[-g_{13}g_{14}(h_{44} + g_{22}) + g_{13}g_{24}(h_{34} + g_{12}) + g_{23}g_{14}(h_{34} + g_{12}) - g_{23}g_{24}(h_{33} + g_{11})].$$

7.9 EXAMPLES

7.9.1 A Steel Portal Frame with Prismatic Members

The steel portal frame with prismatic members shown in Figure 7.18 is taken as a benchmark example for the plastic analysis of steel frames, the plastic zone and plastic hinge solution of which can be found in references (Chen, Li and Xia, 1985; Toma and Chen, 1992; Vogel, 1985). For verification purpose, this frame is also analysed using the prismatic beam element proposed in this chapter. The material properties of steel are $E = 205\text{kN}/\text{mm}^2$ and $\sigma = 235\text{N}/\text{mm}^2$.

When the effects of initial geometric imperfection and strain hardening are neglected, the ultimate load factor (the ratio of the limit load-bearing capacity to the reference load shown in Figure 7.18) obtained with the elasto-plastic hinge model proposed in this chapter is $\lambda = 0.96$, whereas the result by the plastic zone method (Chen, 1993) is $\lambda = 0.97$. When the effects of initial geometric imperfection and strain hardening

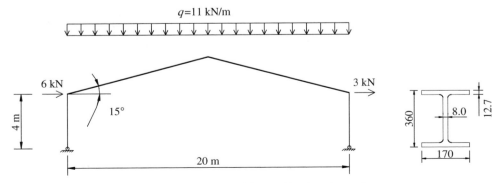

Figure 7.18   A steel portal frame with prismatic members

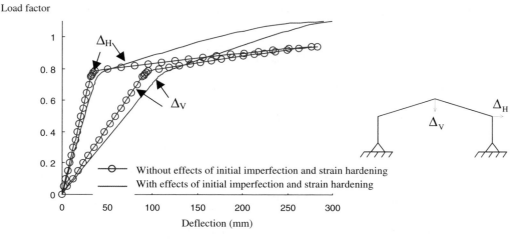

Figure 7.19   Load–deflection curves of the prismatic portal frame

are considered in the analysis, the result from this chapter is $\lambda = 1.10$, whereas $\lambda = 1.07$ by the plastic zone method. The load–deflection curves are plotted in Figure 7.19.

Considering moment resistance is the dominant factor for the load-bearing capacity of steel portal frames, all of the members used in this example frame have compact sections so that their plastic rotation capacity is very good, which makes the effect of strain hardening significant on the ultimate capacity of the frame. As the strain-hardening model used in this chapter neglects the yielding plateau and that used in reference (Chen, 1993) considers it, it is rational that the result obtained by the method proposed in this chapter is slightly greater than that in reference when the strain-hardening effect is considered.

7.9.2   A Steel Portal Frame with Tapered Members

A steel portal frame with tapered members, as shown in Figure 7.20, is analysed in reference with the plastic zone method (Chen, 2000b). Such an analysis is repeated with the tapered beam element proposed in

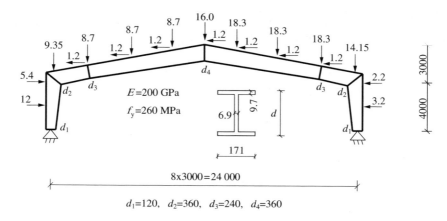

$d_1=120, \quad d_2=360, \quad d_3=240, \quad d_4=360$

Load unit: kN, size unit: mm

Figure 7.20   A steel portal frame with tapered members

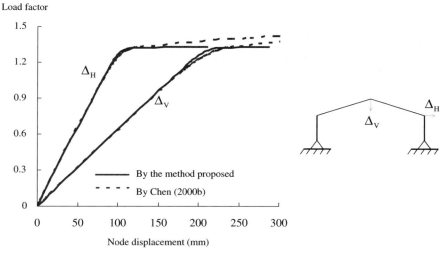

Figure 7.21   Load–displacement curves of the tapered steel portal frame

this chapter, where each tapered column and beam of the frame is represented with a tapered beam element. The load–displacement curves are plotted and compared in Figure 7.21.

As the tapered element model proposed can consider the effects of strain hardening, shear deformation, residual stresses and initial geometric imperfection, the ultimate load factors obtained with consideration of such individual and joint effects are listed in Table 7.2. The hardening factor $q = 0.02$ and shear modulus $G = 80$ GPa are adopted for the analysis.

It can be seen from Table 7.2 that the effects of shear deformation, residual stresses and initial geometric imperfection are negligible on the load-bearing capacity of normal single-storey steel portal frames. But the effect of strain hardening is significant if the plastic deformation capacity of the frame members can be ensured.

7.9.3 Vogel Portal Frame

Vogel portal frame (Vogel, 1985) had received a wide study as a benchmark frame (Avery and Mahendran, 2000a, 2000b, 2000c, 2000d; Chen, 1993; Chen and Kim, 1997; Kim and Lü, 1992; Toma and Chen, 1992). The frame size, material properties and load information are illustrated in Figure 7.22, and the frame member sizes are listed in Table 7.3. The horizontal displacement of rightupper corner (node A) versus load factor curve by the elasto-plastic hinge model presented in this chapter is compared with that by Toma and Chen (1992) with the plastic zone method in Figure 7.23. The ultimate load factor obtained by the method proposed is $\lambda = 1.03$, whereas that by Toma and Chen (1992) is $\lambda = 1.022$.

Table 7.2   Load factors of the steel portal frame obtained by the method proposed

Without effects of initial geometric imperfection, strain hardening and shear deformation	1.306
With the effect of residual stresses	1.300
With the effect of initial geometric imperfection	1.298
With the effect of shear deformation	1.306
With the effect of strain hardening	1.402
With the effect of initial geometric imperfection, strain hardening and shear deformation	1.380

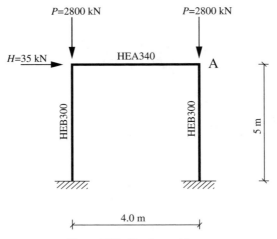

Figure 7.22   Vogel portal frame

Table 7.3   Member sizes and sectional properties of the Vogel portal frame

Section properties	d (mm)	b_f (mm)	t_w (mm)	t_f (mm)	A (mm^2)	I ($\times 10^6$ mm^4)	S ($\times 10^3$ mm^3)
HEA340	330	300	9.5	16.5	13 300	276.9	1850
HEB300	300	300	11.0	19.0	14 900	251.7	1869

7.9.4   Vogel Six-Storey Frame

Vogel six-storey frame (Vogel, 1985) usually appears in benchmark study of planar steel frames (Avery and Mahendran, 2000a, 2000b, 2000c, 2000d; Chan and Chui, 1997; Chen, 1993; Chen and Kim, 1997; Kim and Lü, 1992; Morteza, Torkamani and Sonmez, 2001; Toma and Chen, 1992). The frame size, material properties and load information are illustrated in Figure 7.24, and the frame member sizes are listed in Table 7.4. The horizontal displacement of rightupper corner (node A) versus load factor curve by the elasto-plastic hinge model presented in this chapter is compared with that by Toma and Chen (1992) with the plastic zone method in Figure 7.25. The ultimate load factor obtained by the method proposed is $\lambda = 1.15$,

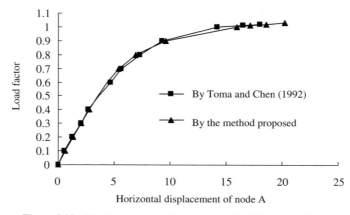

Figure 7.23   Displacement–load factor curve of the Vogel portal frame

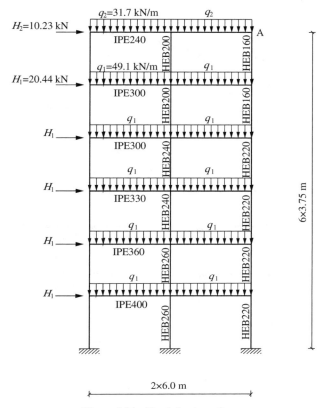

Figure 7.24 Vogel six-storey frame

whereas that by Toma and Chen (1992) is $\lambda = 1.18$. The axial force and moment diagrams in the ultimate state are shown in Figure 7.26, where the final plastic hinge distribution is dotted in the moment diagram.

7.9.5 A Single-Storey Frame with Mid-Span Concentrated Load

A simple portal frame with mid-span concentrated load at the beam is selected as an example. Figure 7.27 gives the frame size and load information. The elastic modulus and yielding strength of the material for the

Table 7.4 Member sizes and sectional properties of the Vogel six-storey frame

Section properties	d (mm)	b_f (mm)	t_w (mm)	t_f (mm)	A (mm^2)	I ($\times 10^6$ mm^4)	S ($\times 10^3$ mm^3)
HEA340	330	300	9.5	16.5	13 300	276.9	1850
HEB160	160	160	8.0	13.0	5430	24.92	354
HEB200	200	200	9.0	15.0	7810	56.96	643
HEB220	220	220	9.5	16.0	9100	80.91	827
HEB240	240	240	10.0	17.0	10 600	112.6	1053
HEB260	260	260	10.0	17.5	11 800	149.2	1283
HEB300	300	300	11.0	19.0	14 900	251.7	1869
IPE240	240	120	6.2	9.8	3910	38.92	367
IPE300	300	150	7.1	10.7	5380	83.56	628
IPE330	330	160	7.5	11.5	6260	117.7	804
IPE360	360	170	8.0	12.7	7270	162.7	1019
IPE400	400	180	8.6	13.5	8450	231.3	1307

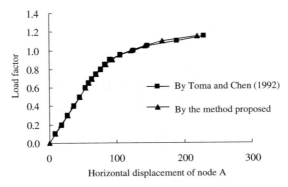

Figure 7.25 Displacement–load factor curve of the Vogel six-storey frame

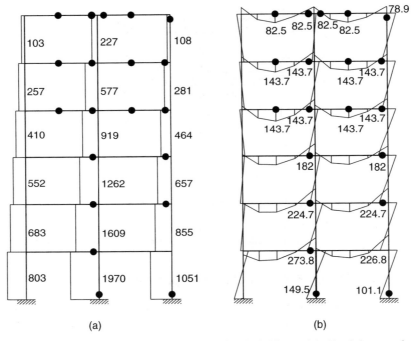

(a)   (b)

Figure 7.26 Resultant axial force and moment distribution with plastic hinges of the Vogel six-storey frame: (a) axial forces (kN); (b) moments (kN m)

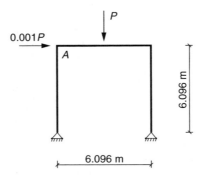

Figure 7.27 Single-storey frame with a concentrated load at the beam

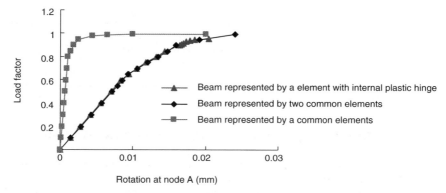

Figure 7.28 Rotation–load factor curves

frame are $E = 206$ GPa and $f_y = 235$ MPa, respectively. USA wide flange sections W18 × 50 and W12 × 65 are selected for the beam and column of the frame, respectively. The rotation versus load factor curves by different element strategies are shown in Figure 7.28. In conventional element treatment where the beam is represented with one common element (without consideration of possible plastic hinge between ends) and the equivalent force and moment of the mid-span concentrated load are enforced at the two end nodes of the element, redistribution of internal forces due to formation of the mid-span plastic hinge within the beam is ignored, and large error occurs for the rotation at node A as shown in Figure 7.28.

To consider the effect of the mid-span plastic hinge, two common beam elements may be used with a common node set at the location of the concentrated load. Alternatively, the beam with mid-span concentrated load can be represented by a single beam element with plastic hinge within span, which is obviously good for computational efficiency in the analysis of the frames.

7.9.6 A Single-Storey Frame with Distributed Load

A simple portal frame with distributed load at the beam is analysed. Figure 7.29 gives the frame size and load information. The elastic modulus and yielding strength of the material for the frame are $E = 206$ GPa and $f_y = 345$ MPa, respectively. USA wide flange section W18 × 31 is selected for all the members of the frame. The horizontal displacement versus load factor curve obtained with a single element with plastic hinge between two ends of the beam of the frame is given in Figure 7.30. The results by representation with two elements for the beam (Chen and Chan, 1995) are also shown in Figure 7.30 for comparison. The change of position of the maximum moment in the beam is checked and tabulated in Table 7.5. It can be found from Table 7.5 that the position of the maximum moment in the beam of the single-storey frame does not change much during loading.

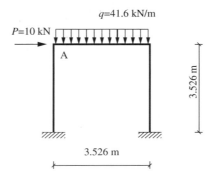

Figure 7.29 Single-storey frame with a distributed load at the beam

Table 7.5   Variation of the location (away from the left end of the beam) of the maximum moment with load factors

Load factor	0.3	0.6	0.9	1.2	1.5	1.8	2.1	2.4	2.7
Location (m)	1.6611	1.6608	1.6606	1.6604	1.6602	1.6600	1.6597	1.6595	1.6592
Load factor	3.0	3.1	3.2	3.3	3.4	3.5	3.6	3.7	3.8
Location (m)	1.6595	1.6666	1.6825	1.6975	1.7115	1.7243	1.7350	1.7440	1.7524

7.9.7   A Four-Storey Frame with Mid-Span Concentrated Load

The structure examined is a four-storey frame with mid-span concentrated loads as shown in Figure 7.31. Table 7.6 gives the frame member size. The horizontal displacement versus load factor curves obtained both by analysis with the elements with internal hinge proposed in this chapter and with the normal elements through dividing the frame beam into two elements (Shu and Shen 1993) are shown in Figure 7.32. The ultimate load factor obtained with the proposed elements is $\lambda = 1.03$, whereas that with normal elements (Shu and Shen, 1993) is $\lambda = 0.99$. The sequence of plastic hinges formed in the frame is illustrated in Figure 7.33.

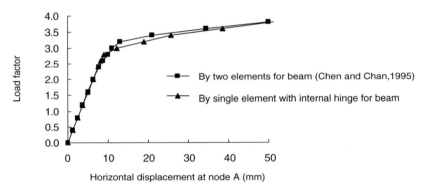

Figure 7.30   Displacement–load factor curve of the single-storey frame

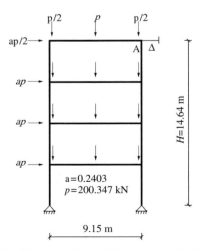

Figure 7.31   Four-storey frame with concentrated loads at beams

Table 7.6 Member sizes of the four-storey frame

Member section	H (mm)	B (mm)	t_w (mm)	t_f (mm)	A (mm^2)	I ($\times 10^6$ mm^4)
W16 × 40	406.7	177.5	7.9	12.7	7610	215
W10 × 60	259.6	256	10.7	17.3	11 400	142
W12 × 79	314.5	306.8	11.9	18.8	15 000	276

Table 7.7 Member sizes of the composite frame

Member section	H (mm)	B (mm)	t_w (mm)	t_f (mm)	A (mm^2)	I ($\times 10^6$ mm^4)
W12 × 27	304	165	6.02	10.16	5062	84.0
W12 × 50	310	205	9	16	8484	157.9

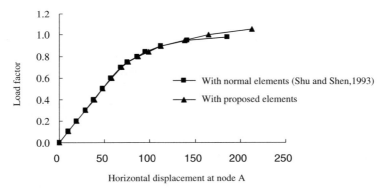

Figure 7.32 Displacement–load factor curve of the four-storey frame

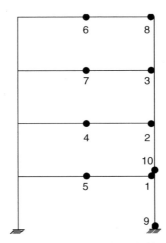

Figure 7.33 Sequence of plastic hinges

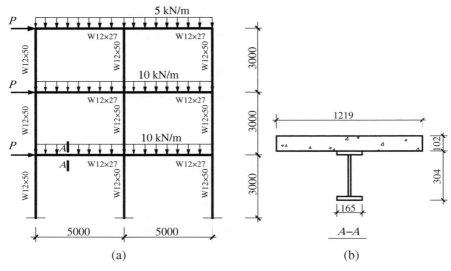

Figure 7.34   A two-span three-storey composite frame: (a) geometry and loading; (b) steel–concrete composite beam

7.9.8   A Two-Span Three-Storey Composite Frame

Consider a two-span three-storey frame (Figure 7.34(a)) with steel columns and steel–concrete composite beams (Figure 7.34(b)). The sizes of steel beams (W12 × 27) and steel columns (W12 × 50) are given in Table 7.7, and the elastic modulus of steel, $E_s = 200$ GPa, steel yielding strength, $f_y = 252.4$ MPa, concrete compressive strength, $f_c = 16$ MPa, and concrete–steel interface shear stiffness $k = 0.1 \% E_s$ are adopted for the analysis of the frame.

Effects of concrete flange cracking of the composite beam in action of negative bending moment on the behaviour of the frame may be considered with the composite beam element presented here-inabove. A comparison of the horizontal load versus lateral deflection curves obtained with different elemental strategies is given in Figure 7.35. From the results in Figure 7.35, it is clear that the composite action coming from concrete slabs can enhance the sway stiffness and ultimate capacity of steel frames.

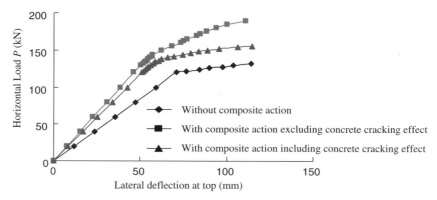

Figure 7.35   Load–deflections curves of the composite frame

8 Elastic and Elasto-Plastic Stiffness Equations of Column Element

8.1 FORCE AND DEFORMATION OF COLUMN ELEMENT

Generally, three forces N_z, Q_x and Q_y and three moments M_x, M_y and M_z are applied at each end of a column element, as shown in Figure 8.1, where all the directions of the forces and moments are positive. Correspondingly, the deformation components of the column element include three translations, δ_x, δ_y and δ_z, and three rotations, θ_x, θ_y and θ_z. Denote the force and deformation vectors of the column element as $\{f_c\}$ and $\{\delta_c\}$, then

$$\{f_c\} = \left[\{f_{c1}\}^T, \quad \{f_{c2}\}^T \right]^T, \tag{8.1}$$

$$\{\delta_c\} = \left[\{\delta_{c1}\}^T, \quad \{\delta_{c2}\}^T \right]^T, \tag{8.2}$$

where

$$\{f_{c1}\} = \left[N_{z1}, \quad Q_{x1}, \quad M_{y1}, \quad Q_{y1}, \quad M_{x1}, \quad M_{z1} \right]^T, \tag{8.3a}$$

$$\{f_{c2}\} = \left[N_{z2}, \quad Q_{x2}, \quad M_{y2}, \quad Q_{y2}, \quad M_{x2}, \quad M_{z2} \right]^T, \tag{8.3b}$$

$$\{\delta_{c1}\} = \left[\delta_{z1}, \quad \delta_{x1}, \quad \theta_{y1}, \quad \delta_{y1}, \quad \theta_{x1}, \quad \theta_{z1} \right]^T, \tag{8.4a}$$

$$\{\delta_{c2}\} = \left[\delta_{z2}, \quad \delta_{x2}, \quad \theta_{y2}, \quad \delta_{y2}, \quad \theta_{x2}, \quad \theta_{z2} \right]^T. \tag{8.4b}$$

8.2 ELASTIC STIFFNESS EQUATION OF COLUMN ELEMENT SUBJECTED TO BIAXIAL BENDING

Under the precondition of small deflection, it is independent between the lateral deflections with axial and torsional deformations for column elements. The elastic stiffness equation of a column element is therefore obtained by combining the stiffness equation of the corresponding beam element with axial and torsional stiffness relationship, i.e.

$$\{f_c\} = [k_{ce}]\{\delta_c\} \tag{8.5a}$$

Advanced Analysis and Design of Steel Frames   Guo-Qiang Li and Jin-Jun Li
© 2007 John Wiley & Sons, Ltd

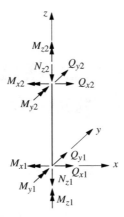

Figure 8.1   Forces on the column element

or

$$\begin{bmatrix} \{f_{c1}\} \\ \{f_{c2}\} \end{bmatrix} = \begin{bmatrix} [k_{cebb}] & [k_{cebt}] \\ [k_{cetb}] & [k_{cett}] \end{bmatrix} \begin{bmatrix} \{\delta_{c1}\} \\ \{\delta_{c2}\} \end{bmatrix}, \tag{8.5b}$$

in which $[k_{ce}]$ is the elastic stiffness matrix of the spatial column element and $[k_{cebb}]$, $[k_{cebt}]$, $[k_{cetb}]$ and $[k_{cett}]$ are the partitioned matrices of $[k_{ce}]$ given by

$$[k_{cebb}] = \begin{bmatrix} \dfrac{EA}{l} & 0 & 0 & 0 & 0 & 0 \\ 0 & \dfrac{12EI_y}{l^3}\psi_{1y} & \dfrac{6EI_y}{l^2}\psi_{2y} & 0 & 0 & 0 \\ 0 & \dfrac{6EI_y}{l^2}\psi_{2y} & \dfrac{4EI_y}{l^3}\psi_{3y} & 0 & 0 & 0 \\ 0 & 0 & 0 & \dfrac{12EI_x}{l^3}\psi_{1x} & \dfrac{6EI_y}{l^2}\psi_{2x} & 0 \\ 0 & 0 & 0 & \dfrac{6EI_y}{l^2}\psi_{2x} & \dfrac{4EI_x}{l^3}\psi_{3x} & 0 \\ 0 & 0 & 0 & 0 & 0 & \dfrac{GI_z}{l} \end{bmatrix}, \tag{8.6a}$$

$$[k_{cett}] = \begin{bmatrix} \dfrac{EA}{l} & 0 & 0 & 0 & 0 & 0 \\ 0 & \dfrac{12EI_y}{l^3}\psi_{1y} & \dfrac{6EI_y}{l^2}\psi_{2y} & 0 & 0 & 0 \\ 0 & -\dfrac{6EI_y}{l^2}\psi_{2y} & \dfrac{4EI_y}{l^3}\psi_{3y} & 0 & 0 & 0 \\ 0 & 0 & 0 & \dfrac{12EI_x}{l^3}\psi_{1x} & -\dfrac{6EI_y}{l^2}\psi_{2x} & 0 \\ 0 & 0 & 0 & -\dfrac{6EI_y}{l^2}\psi_{2x} & \dfrac{4EI_x}{l^3}\psi_{3x} & 0 \\ 0 & 0 & 0 & 0 & 0 & \dfrac{GI_z}{l} \end{bmatrix}, \tag{8.6b}$$

$$
[k_{\text{cetb}}] = [k_{\text{cebt}}]^{\text{T}} =
\begin{bmatrix}
-\dfrac{EA}{l} & 0 & 0 & 0 & 0 & 0 \\[2ex]
0 & -\dfrac{12EI_y}{l^3}\psi_{1y} & -\dfrac{6EI_y}{l^2}\psi_{2y} & 0 & 0 & 0 \\[2ex]
0 & \dfrac{6EI_y}{l^2}\psi_{2y} & \dfrac{4EI_y}{l^3}\psi_{4y} & 0 & 0 & 0 \\[2ex]
0 & 0 & 0 & -\dfrac{12EI_x}{l^3}\psi_{1x} & -\dfrac{6EI_y}{l^2}\psi_{2x} & 0 \\[2ex]
0 & 0 & 0 & \dfrac{6EI_y}{l^2}\psi_{2x} & \dfrac{4EI_x}{l^3}\psi_{4x} & 0 \\[2ex]
0 & 0 & 0 & 0 & 0 & -\dfrac{GI_z}{l}
\end{bmatrix},
\tag{8.6c}
$$

where I_x and I_y are the inertial moments about the x and y axes of the element section, respectively, I_z is the torsional inertial moment of the section, and $\psi_{1x} - \psi_{4x}$ and ψ_{1y}. It should be noted that for opened sections, such as the H section, the torsional stiffness, ψ_{4y} are the factors defined in Equation (2.21) or (2.26), calculated with the inertial GI_z/l, in Equation (8.5) is relatively small due to warping action. In the analysis of steel moment of section and the shear shape factors about the x and y axes, respectively. frames, the structural torsional stiffness is provided in majority not by that of frame columns themselves, but by the 'planar frame action' (see Equation (15.64)). So the error from the torsional stiffness in the column stiffness equation can be neglected in the frame analysis.

8.3 ELASTO-PLASTIC STIFFNESS EQUATIONS OF COLUMN ELEMENT SUBJECTED TO BIAXIAL BENDING

When the column element is in elasto-plastic state, its stiffness will change continuously. The incremental method can be used to develop the elasto-plastic tangent stiffness equation based on the elasto-plastic hinge model.

Similarly, the deformations at the element ends may be divided into elastic and plastic portions as

$$
\{d\delta_c\} = \{d\delta_{ce}\} + \{d\delta_{cp}\},
\tag{8.7}
$$

where

$$
\{d\delta_{ce}\} = \left[\{d\delta_{ce1}\}^{\text{T}}, \quad \{d\delta_{ce2}\}^{\text{T}}\right]^{\text{T}},
\tag{8.8}
$$

$$
\{d\delta_{cp}\} = \left[\{d\delta_{cp1}\}^{\text{T}}, \quad \{d\delta_{cp2}\}^{\text{T}}\right]^{\text{T}},
\tag{8.9}
$$

$$
\{d\delta_{ce1}\} = \left[d\delta_{z1e}, \quad d\delta_{x1e}, \quad d\theta_{y1e}, \quad d\delta_{y1e}, \quad d\theta_{x1e}, \quad d\theta_{z1e}\right]^{\text{T}},
\tag{8.10a}
$$

$$
\{d\delta_{ce2}\} = \left[d\delta_{z2e}, \quad d\delta_{x2e}, \quad d\theta_{y2e}, \quad d\delta_{y2e}, \quad d\theta_{x2e}, \quad d\theta_{z2e}\right]^{\text{T}},
\tag{8.10b}
$$

$$
\{d\delta_{cp1}\} = \left[d\delta_{z1p}, \quad d\delta_{x1p}, \quad d\theta_{y1p}, \quad d\delta_{y1p}, \quad d\theta_{x1p}, \quad d\theta_{z1p}\right]^{\text{T}},
\tag{8.11a}
$$

$$
\{d\delta_{cp2}\} = \left[d\delta_{z2p}, \quad d\delta_{x2p}, \quad d\theta_{y2p}, \quad d\delta_{y2p}, \quad d\theta_{x2p}, \quad d\theta_{z2p}\right]^{\text{T}}.
\tag{8.11b}
$$

Also, the forces at the element ends can be divided into two portions as

$$
\{df_c\} = \{df_{ct}\} + \{df_{cn}\},
\tag{8.12}
$$

where

$$
\{df_{ct}\} = \left[\{df_{ct1}\}^{\text{T}}/, \quad \{df_{ct2}\}^{\text{T}}\right]^{\text{T}},
\tag{8.13}
$$

$$
\{df_{cn}\} = \left[\{df_{cn1}\}^{\text{T}}/, \quad \{df_{cn2}\}^{\text{T}}\right]^{\text{T}},
\tag{8.14}
$$

$$\{df_{ct1}\} = \begin{bmatrix} 0, & dQ_{x1}, & 0, & dQ_{y1}, & 0, & 0 \end{bmatrix}^{T}, \tag{8.15a}$$

$$\{df_{ct2}\} = \begin{bmatrix} 0, & dQ_{x2}, & 0, & dQ_{y2}, & 0,0 \end{bmatrix}^{T}, \tag{8.15b}$$

$$\{df_{cn1}\} = \begin{bmatrix} dN_{z1}, & 0, & dM_{y1}, & 0, & dM_{x1}, & dM_{z1} \end{bmatrix}^{T}, \tag{8.16a}$$

$$\{df_{cn2}\} = \begin{bmatrix} dN_{z2}, & 0, & dM_{y2}, & 0, & dM_{x2}, & dM_{z2} \end{bmatrix}^{T}. \tag{8.16b}$$

According to the rule of plasticity flow, one has

$$\{d\delta_{cp1}\} = [g_1]\{1\}\lambda_1, \tag{8.17a}$$

$$\{d\delta_{cp2}\} = [g_2]\{1\}\lambda_2, \tag{8.17b}$$

in which

$$[g_1] = \text{diag}\left[\frac{\partial \Gamma_1}{\partial N_{z1}}, \quad 0, \quad \frac{\partial \Gamma_1}{\partial M_{y1}}, \quad 0, \quad \frac{\partial \Gamma_1}{\partial M_{x1}}, \quad \frac{\partial \Gamma_1}{\partial M_{z1}}\right], \tag{8.18a}$$

$$[g_2] = \text{diag}\left[\frac{\partial \Gamma_2}{\partial N_{z2}}, \quad 0, \quad \frac{\partial \Gamma_2}{\partial M_{y2}}, \quad 0, \quad \frac{\partial \Gamma_2}{\partial M_{x2}}, \quad \frac{\partial \Gamma_2}{\partial M_{z2}}\right], \tag{8.18b}$$

$$[1] = [1, 1, 1, 1, 1, 1]^{T}. \tag{8.19}$$

where Γ_1 and Γ_2 are the yielding functions at the two ends of the column element, defined in Equation (5.45), and λ_1 and λ_2 are the proportional factors of the plastic deformations at the two ends of the element.

The relationship between incremental forces $\{df_c\}$ and incremental elastic deformations $\{d\delta_{ce}\}$ is constantly

$$\{df_g\} = [k_{ce}]\{d\delta_{ge}\}. \tag{8.20}$$

Rewriting Equation (8.20) in partitioned form leads to

$$\{df_{c1}\} = [k_{cebb}]\{d\delta_{ce1}\} + [k_{cebt}]\{d\delta_{ce2}\}, \tag{8.21a}$$

$$\{df_{c2}\} = [k_{cetb}]\{d\delta_{ce1}\} + [k_{cebb}]\{d\delta_{ce2}\}. \tag{8.21b}$$

According to the elasto-plastic hinge model, the plastic deformation at the end of the element is merely relevant to the forces at the same end of the element. Then,

$$\{df_{cn1}\} = [k_{nb}]\{d\delta_{cp1}\}, \tag{8.22a}$$

$$\{df_{cn2}\} = [k_{nt}]\{d\delta_{cp2}\} \tag{8.22b}$$

or

$$\{df_{cn}\} = [k_n]\{d\delta_{cp}\}, \tag{8.23}$$

where

$$[k_n] = \begin{bmatrix} [k_{nb}] & 0 \\ 0 & [k_{nt}] \end{bmatrix},$$

$$[k_{nb}] = \text{diag}\begin{bmatrix} B_{b1}, & B_{b2}, & B_{b3}, & B_{b4}, & B_{b5}, & B_{b6} \end{bmatrix},$$

$$[k_{nt}] = \text{diag}\begin{bmatrix} B_{t1}, & B_{t2}, & B_{t3}, & B_{t4}, & B_{t5}, & B_{t6} \end{bmatrix}. \tag{8.24}$$

Equation (8.17) indicates that the plastic deformation components of a plastic hinge are proportional to the same factor so that, noting Equation (8.22), the force components corresponding to the plastic deformation components should also be proportional to the same factor. Hence,

$$
\begin{aligned}
B_{bi} &= \alpha_1 k_{ceii}, & i &= 1, 2, 3, 4, 5, 6 \\
B_{ti} &= \alpha_2 k_{cejj}, & j &= 7, 8, 9, 10, 11, 12,
\end{aligned}
\tag{8.25}
$$

where, α_1 and α_2 are the elasto-plastic hinge factors for the two ends of the element, and k_{ceii} is the element of the ith row and the ith column of the stiffness matrix $[k_{ce}]$.

When the elemental section is in elastic state, all the plastic deformation components of the section are zero, and then it is required that $\alpha = \infty$, and when the section is in idealized plastic state, these plastic deformation components can be arbitrary in value and then $\alpha = 0$. To satisfy the above conditions, α can be defined as

$$
\alpha_i = \frac{R_i}{1 - R_i}, \qquad i = 1, 2,
\tag{8.26}
$$

where R_1 and R_2 are the recovery force factors defined in Equation (5.38) for the two ends of the element and can be calculated according to Equations (5.39) and (5.40).

From Equations (8.12), (8.21) and (8.22), one has

$$
\begin{aligned}
\{df_{ct1}\} = \{df_{c1}\} - \{df_{cn1}\} &= [k_{cebb}]\{d\delta_{c1}\} + [k_{cebt}]\{d\delta_{c2}\} \\
&- ([k_{cebb}] + [k_{nb}])[g_1]\{1\}\lambda_1 - [k_{cebt}][g_2]\{1\}\lambda_2,
\end{aligned}
\tag{8.27a}
$$

$$
\begin{aligned}
\{df_{ct2}\} = \{df_{c2}\} - \{df_{cn2}\} &= [k_{cetb}]\{d\delta_{c1}\} + [k_{cett}]\{d\delta_{c2}\} \\
&- [k_{cetb}][g_1]\{1\}\lambda_1 - ([k_{cett}] + [k_{nt}])[g_2]\{1\}\lambda_2.
\end{aligned}
\tag{8.27b}
$$

It is known from Equations (8.11) and (8.15) that vectors $\{d\delta_{cp1}\}$ and $\{d\delta_{cp2}\}$ are orthotropic with $\{df_{ct1}\}$ and $\{df_{ct2}\}$, respectively. Hence,

$$
\{d\delta_{cp1}\}^T\{df_{ct1}\} = \lambda_1(k_{bb}\lambda_1 + k_{bt}\lambda_2 - [H_{11}]\{d\delta_{c1}\} - [H_{12}]\{d\delta_{c2}\}) = 0,
\tag{8.28a}
$$

$$
\{d\delta_{cp2}\}^T\{df_{ct2}\} = \lambda_2(k_{tb}\lambda_1 + k_{tt}\lambda_2 - [H_{21}]\{d\delta_{c1}\} - [H_{22}]\{d\delta_{c2}\}) = 0,
\tag{8.28b}
$$

where

$$
k_{bb} = \{1\}^T[g_1]^T([k_{cebb}] + [k_{nb}])[g_1]\{1\},
\tag{8.29a}
$$

$$
k_{bt} = \{1\}^T[g_1]^T[k_{cebt}][g_2]\{1\},
\tag{8.29b}
$$

$$
k_{tb} = \{1\}^T[g_2]^T[k_{cetb}][g_1]\{1\},
\tag{8.29c}
$$

$$
k_{tt} = \{1\}^T[g_2]^T([k_{cett}] + [k_{nt}])[g_2]\{1\},
\tag{8.29d}
$$

$$
[H_{11}] = \{1\}^T[g_1]^T[k_{cebb}],
\tag{8.30a}
$$

$$
[H_{12}] = \{1\}^T[g_1]^T[k_{cebt}],
\tag{8.30b}
$$

$$
[H_{21}] = \{1\}^T[g_2]^T[k_{cetb}],
\tag{8.30c}
$$

$$
[H_{22}] = \{1\}^T[g_2]^T[k_{cett}].
\tag{8.30d}
$$

8.3.1 Both Ends Yielding

Solutions of λ_1 and λ_2 and further formulations of the elasto-plastic stiffness equation of the element are related to the deformation condition of the element ends, which are discussed as follows.

When both ends of the element yield, $\{d\delta_{cp1}\} \neq 0$ and $\{d\delta_{cp2}\} \neq 0$, then $\lambda_1 \neq 0$ and $\lambda_2 \neq 0$, with which Equation (8.28) becomes

$$\begin{bmatrix} k_{bb} & k_{bt} \\ k_{tb} & k_{tt} \end{bmatrix} \begin{Bmatrix} \lambda_1 \\ \lambda_2 \end{Bmatrix} = \begin{bmatrix} [H_{11}] & [H_{12}] \\ [H_{21}] & [H_{22}] \end{bmatrix} \begin{Bmatrix} d\delta_{c1} \\ d\delta_{c2} \end{Bmatrix}. \tag{8.31}$$

The solution of λ_1 and λ_2 with the above equation is

$$\begin{Bmatrix} \lambda_1 \\ \lambda_2 \end{Bmatrix} = \begin{bmatrix} k_{bb} & k_{bt} \\ k_{tb} & k_{tt} \end{bmatrix}^{-1} \begin{bmatrix} [H_{11}] & [H_{12}] \\ [H_{21}] & [H_{22}] \end{bmatrix} \begin{Bmatrix} d\delta_{c1} \\ d\delta_{c2} \end{Bmatrix}. \tag{8.32}$$

Substituting Equation (8.32) into Equation (8.17) leads to

$$\begin{Bmatrix} \{d\delta_{cp1}\} \\ \{d\delta_{cp1}\} \end{Bmatrix} = \begin{bmatrix} [g_1] & 0 \\ 0 & [g_2] \end{bmatrix} \begin{bmatrix} \{1\} & 0 \\ 0 & \{1\} \end{bmatrix} \begin{bmatrix} k_{bb} & k_{bt} \\ k_{tb} & k_{tt} \end{bmatrix}^{-1} \begin{bmatrix} [H_{11}] & [H_{12}] \\ [H_{21}] & [H_{22}] \end{bmatrix} \begin{Bmatrix} d\delta_{c1} \\ d\delta_{c2} \end{Bmatrix}. \tag{8.33}$$

Equation (8.30) can be rewritten in partitioned matrix form as

$$\begin{bmatrix} [H_{11}] & [H_{12}] \\ [H_{21}] & [H_{22}] \end{bmatrix} = \begin{bmatrix} \{1\} & 0 \\ 0 & \{1\} \end{bmatrix}^T \begin{bmatrix} [g_1] & 0 \\ 0 & [g_2] \end{bmatrix}^T \begin{bmatrix} k_{bb} & k_{bt} \\ k_{tb} & k_{tt} \end{bmatrix}^{-1}. \tag{8.34}$$

Let

$$[G] = \begin{bmatrix} [g_1] & 0 \\ 0 & [g_2] \end{bmatrix}, \tag{8.35}$$

$$[E] = \begin{bmatrix} \{1\} & 0 \\ 0 & \{1\} \end{bmatrix}, \tag{8.36}$$

$$[L] = \begin{bmatrix} [k_{bb}] & [k_{bt}] \\ [k_{tb}] & [k_{tt}] \end{bmatrix}^{-1}. \tag{8.37}$$

Equation (8.33) can then be rewritten as

$$\{d\delta_{cp}\} = [G][E][L][E]^T[G]^T[k_{ce}]\{d\delta_c\}. \tag{8.38}$$

Noting Equation (8.7), one has

$$\{d\delta_{ce}\} = \{d\delta_c\} - \{d\delta_{cp}\} = ([I] - [G][E][L][E]^T[G]^T[k_{ce}])\{d\delta_c\}, \tag{8.39}$$

where $[I]$ is the unit matrix.

Substituting Equation (8.39) into Equation (8.20) yields

$$\{df_c\} = ([k_{ce}] - [k_{ce}][G][E][L][E]^T[G]^T[k_{ce}])\{d\delta_c\}. \tag{8.40}$$

The above is the elemental stiffness equation for the column element whose both ends are in plastic state.

8.3.2 Only End 1 Yielding

When only end 1 of the element yields, $\{d\delta_{cp1}\} \neq 0$ and $\{d\delta_{cp2}\} = 0$, then $\lambda_1 \neq 0$ and $\lambda_2 = 0$, with which Equation (8.28a) becomes

$$k_{bb}\lambda_1 = [H_{11}]\{d\delta_{c1}\} + [H_{12}]\{d\delta_{c2}\}. \tag{8.41}$$

Solving λ_1 with the above equation, one has

$$\lambda_1 = (1/k_{bb})([H_{11}]\{d\delta_{c1}\} + [H_{12}]\{d\delta_{c2}\}). \tag{8.42}$$

Equation (8.42) can also be written in matrix form as

$$\begin{Bmatrix} \lambda_1 \\ \lambda_2 \end{Bmatrix} = \begin{bmatrix} 1/k_{bb} & 0 \\ 0 & 0 \end{bmatrix} \begin{bmatrix} [H_{11}] & [H_{12}] \\ [H_{21}] & [H_{22}] \end{bmatrix} \begin{Bmatrix} d\delta_{c1} \\ d\delta_{c2} \end{Bmatrix}. \tag{8.43}$$

Combining Equations (8.34) and (8.17) with the above equation, one has

$$\{d\delta_{cp}\} = [G][E][L_b][E]^T[G]^T[k_{ce}]\{d\delta_c\}, \tag{8.44}$$

where

$$[L_b] = \begin{bmatrix} 1/k_{bb} & 0 \\ 0 & 0 \end{bmatrix}. \tag{8.45}$$

Substituting Equation (8.44) into Equation (8.7) yields $\{d\delta_{ce}\}$ and then substituting $\{d\delta_{ce}\}$ into Equation (8.20) yields the stiffness equation of the element with end 1 yielding as

$$\{df_c\} = ([k_{ce}] - [k_{ce}][G][E][L_b][E]^T[G]^T[k_{ce}])\{d\delta_c\}. \tag{8.46}$$

8.3.3 Only End 2 Yielding

When only end 2 of the element yields, $\{d\delta_{cp1}\} = 0$ and $\{d\delta_{cp2}\} \neq 0$, then $\lambda_1 = 0$ and $\lambda_2 \neq 0$, with which Equation (8.28b) becomes

$$k_{tt}\lambda_2 = [H_{21}]\{d\delta_{c1}\} + [H_{22}]\{d\delta_{c2}\}. \tag{8.47}$$

With a similar procedure for end 1 yielding, the stiffness equation of the element when plasticity yields at end 2 can be derived as

$$\{df_c\} = ([k_{ce}] - [k_{ce}][G][E][L_t][E]^T[G]^T[k_{ce}])\{d\delta_c\}, \tag{8.48}$$

where

$$[L_t] = \begin{bmatrix} 0 & 0 \\ 0 & 1/k_{tt} \end{bmatrix}. \tag{8.49}$$

8.3.4 Summary

In summary, the stiffness equation of the column element based on the elasto-plastic hinge model can be expressed in the uniform equation as

$$\{df_c\} = [k_{cp}]\{d\delta_c\}, \tag{8.50}$$

where $[k_{cp}]$ is the tangent elasto-plastic stiffness matrix of the element, given by

$$[k_{cp}] = [k_{ce}] - [k_{ce}][G][E][L][E]^T[G]^T[k_{ce}], \tag{8.51}$$

where matrix $[L]$ can be determined according to the following four cases:

(a) no yielding occurring at both ends of the element:

$$[L] = \begin{bmatrix} 0 & 0 \\ 0 & 0 \end{bmatrix};$$
(8.52a)

(b) yielding occurring at only end 1 of the element:

$$[L] = \begin{bmatrix} 1/k_{bb} & 0 \\ 0 & 0 \end{bmatrix};$$
(8.52b)

(c) yielding occurring at only end 2 of the element:

$$[L] = \begin{bmatrix} 0 & 0 \\ 0 & 1/k_{tt} \end{bmatrix};$$
(8.52c)

(d) yielding occurring at both ends of the element:

$$[L] = \begin{bmatrix} k_{bb} & k_{bt} \\ k_{tb} & k_{tt} \end{bmatrix}^{-1}.$$
(8.52d)

Noting Equations (8.29), (8.35) and (8.36), $[L]$ for the above four cases can also be uniformly expressed as

$$[L] = ([E]^T[G]^T([k_{ce}] + [k_n])[G][E])^{-1}.$$
(8.52e)

8.4   ELASTIC AND ELASTO-PLASTIC STIFFNESS EQUATIONS OF COLUMN ELEMENT SUBJECTED TO UNIAXIAL BENDING

The force and deformation components for the column element subjected to uniaxial bending (about y-axis assumed) are

$$\{f_c\} = [N_{z1}, \quad Q_{x1}, \quad M_{y1}, \quad N_{z2}, \quad Q_{x2}, \quad M_{y2}]^T,$$
(8.53)

$$\{\delta_c\} = [\delta_{z1}, \quad \delta_{x1}, \quad \theta_{y1}, \quad \delta_{z2}, \quad \delta_{x2}, \quad \theta_{y2}]^T.$$
(8.54)

The elastic stiffness equation of the element has the same form as Equation (8.5), but the stiffness matrix becomes

$$[k_{ce}] = \begin{bmatrix} \dfrac{EA}{l} & 0 & 0 & -\dfrac{EA}{l} & 0 & 0 \\[2mm] 0 & \dfrac{12EI_y}{l^3}\psi_{1y} & \dfrac{6EI_y}{l^2}\psi_{2y} & 0 & -\dfrac{12EI_y}{l^3}\psi_{1y} & \dfrac{6EI_y}{l^2}\psi_{2y} \\[2mm] 0 & \dfrac{6EI_y}{l^2}\psi_{2y} & \dfrac{4EI_y}{l}\psi_{3y} & 0 & -\dfrac{6EI_y}{l^2}\psi_{2y} & \dfrac{2EI_y}{l}\psi_{4y} \\[2mm] -\dfrac{EA}{l} & 0 & 0 & \dfrac{EA}{l} & 0 & 0 \\[2mm] 0 & -\dfrac{12EI_y}{l^3}\psi_{1y} & -\dfrac{6EI_y}{l^2}\psi_{2y} & 0 & \dfrac{12EI_y}{l^3}\psi_{1y} & -\dfrac{6EI_y}{l^2}\psi_{2y} \\[2mm] 0 & \dfrac{6EI_y}{l^2}\psi_{2y} & \dfrac{2EI_y}{l}\psi_{4y} & 0 & -\dfrac{6EI_y}{l^2}\psi_{2y} & \dfrac{4EI_y}{l}\psi_{3y} \end{bmatrix}.$$
(8.55)

Similarly, the elasto-plastic stiffness equation of the element has the same form as Equation (8.50), and the stiffness matrix of the element has the same form as Equation (8.51), but the matrices in the expression of the stiffness matrix become

$$[G] = \text{diag} \left[\frac{\partial \Gamma_1}{\partial N_{z1}}, \quad 0, \quad \frac{\partial \Gamma_1}{\partial M_{y1}}, \quad \frac{\partial \Gamma_2}{\partial N_{z2}}, \quad 0, \quad \frac{\partial \Gamma_{21}}{\partial M_{y2}} \right], \tag{8.56}$$

$$[E] = \begin{bmatrix} 1 & 0 \\ 1 & 0 \\ 1 & 0 \\ 0 & 1 \\ 0 & 1 \\ 0 & 1 \end{bmatrix}, \tag{8.57}$$

$$[k_n] = \text{diag}[\alpha_1 k_{ce11}, \quad \alpha_1 k_{ce22}, \quad \alpha_1 k_{ce33}, \quad \alpha_2 k_{ce44}, \quad \alpha_2 k_{ce55}, \quad \alpha_2 k_{ce66}], \tag{8.58}$$

where, α_1 and α_2 are the elasto-plastic hinge factors for the two ends of the element; k_{ceii} is the ith row and the ith column element in the stiffness matrix $[k_{ce}]$.

8.5 AXIAL STIFFNESS OF TAPERED COLUMN ELEMENT

The discussions in Sections 8.2–8.4 are for prismatic column elements. The axial stiffness in the stiffness equations presented hereinabove for prismatic column elements is not suitable to tapered column elements. The axial stiffness of tapered column elements can be determined with the following method.

8.5.1 Elastic Stiffness

For a tapered column element, axial force is given by $N = EA(\mathrm{d}u/\mathrm{d}x)$, where the sectional area A is a function of the axial coordinate (see Figure 3.1). The differential equilibrium equation of the element in the axial direction is

$$\frac{\mathrm{d}N}{\mathrm{d}z} = E \left[A(z) \frac{\mathrm{d}^2 u}{\mathrm{d}z^2} + \left(\frac{\mathrm{d}A}{\mathrm{d}z} \right) \frac{\mathrm{d}u}{\mathrm{d}z} \right] = 0, \tag{8.59}$$

where $A(z)$ is the sectional area varying with the axial coordinate z and $u(z)$ is the corresponding axial deformation. Solving the above equation leads to the axial stiffness

$$k_z = \frac{E}{\displaystyle\int_0^L \frac{1}{A(z)} \, \mathrm{d}z}. \tag{8.60}$$

8.5.2 Elasto-Plastic Stiffness

If the effects of residual stresses and initial deflection are considered, the axial force–deformation relationship of the tapered column element can be obtained from Equations (7.81) and (7.82). For CRC tangent modulus (Galambos, 1988), the axial deformation of the element can be expressed as

$$u = \frac{0.5 N_p L}{EA} + \int_{0.5 N_p}^{N} \frac{L}{AE_t} \, \mathrm{d}N, \quad \text{for} \quad N > 0.5 N_p. \tag{8.61}$$

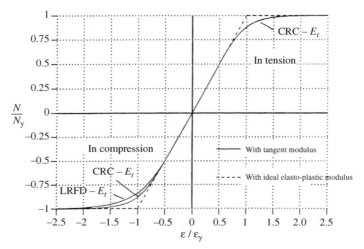

Figure 8.2 Nondimensionalized axial force–strain relationship of the tapered column element

Substituting Equation (7.81) into Equation (8.61) and integrating it gives the expression for the nondimensionalized axial force–deformation relationship of the element as

$$\frac{N}{N_p} = \frac{1}{1 + \exp(2 - 4\varepsilon/\varepsilon_y)}, \qquad \text{for} \quad N > 0.5N_p, \tag{8.62}$$

where ε and ε_y are the axial strain and the yielding strain, respectively. Then $\varepsilon/\varepsilon_y = Eu/(\sigma_y L)$, where σ_y is the yielding strength of material and L is the element length.

With the same approach, the similar axial force–deformation relationships of the tapered column element based on LRFD tangent modulus (AISC, 1994) can be obtained as

$$u = \frac{0.39N_p L}{EA} + \int_{0.39N_p}^{N} \frac{L}{AE_t}\,dN, \qquad \text{for} \quad N > 0.39N_p, \tag{8.63}$$

$$\frac{N}{N_p} = \exp\left[-0.9416 \exp\left(2.7243\left[0.39 - \frac{\varepsilon}{\varepsilon_y}\right]\right)\right], \qquad \text{for} \quad N > 0.39N_p. \tag{8.64}$$

The curves corresponding to Equations (8.62) and (8.64) are given in Figure 8.2. The axial stiffness of the tapered column element can be obtained from the differential operation of dN/du.

8.6 EXPERIMENT VERIFICATION

To verify the elastic and elasto-plastic stiffness equations of column elements, and the hysteretic model proposed in Chapter 5 for the column section subjected to biaxial bending, five box-section columns and four H-section columns were tested in Tongji University (Li, Huang and Shen , 1993; Li, Shen and Huang, 1999). The cyclic behaviour of these columns was investigated under a constant vertical load and repeated and reversed horizontal loads in two principal directions, by which the stiffness equations for column elements proposed have been validated.

8.6.1 Experiment Specimen

The box-section columns and H-section columns tested are shown in Figures 8.3 and 8.4, respectively. The steel material for the specimens is Chinese Q235 and material properties are as follows:

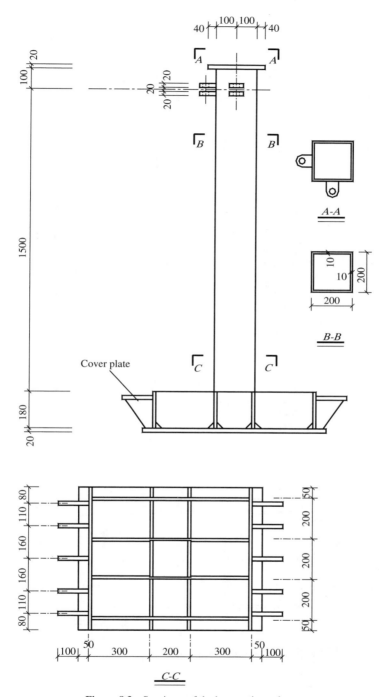

Figure 8.3   Specimen of the box-section column

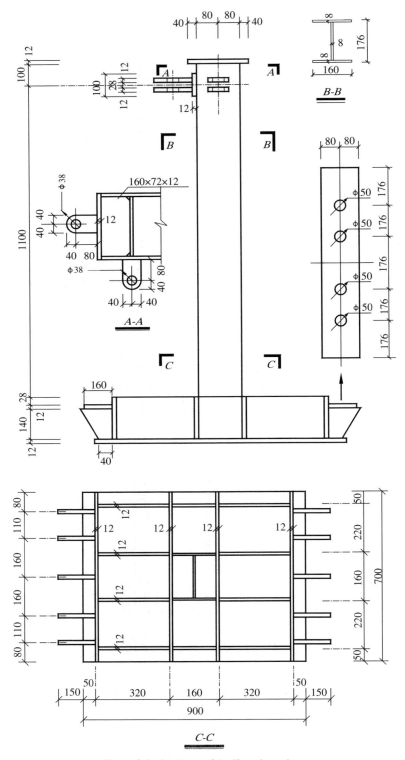

Figure 8.4   Specimen of the H-section column

(1) For box-section columns:

  – yielding strength $\sigma_s = 261.97$ MPa,

  – ultimate tensile strength $\sigma_b = 415.61$ MPa,

  – elastic modulus $E = 1.915$ MPa.

(2) For H-section columns:

  – yielding strength $\sigma_s = 290.08$ MPa,

  – ultimate tensile strength $\sigma_b = 440.37$ MPa,

  – elastic modulus $E = 1.972$ MPa.

The specimens are designed as cantilever columns to model the half of the frame columns with consideration that the point of contraflexure is at mid-height.

8.6.2   Set-Up and Instrumentation

The test set-up is shown in Figure 8.5. Because horizontal displacements at the top of the column specimen in two directions will be produced during test, the two horizontal orthotropic push–pull jacks should have some degree of freedom to certain extent in the horizontal plane. To achieve it, the jack base is connected with a hinge to its support frame. The sequence of loading devices in the horizontal direction from column top is column, horizontal hinge, loading sensor, jack, hinge and support frame (see Figure 8.5(a)).

The difficulty of vertical loading is that a large axial force should be applied and kept constant at the top of the column, whereas the nature of the free end of the cantilever column should be ensured. For the purpose of meeting the above requirements, two rows of roll shafts orthotropic to each other, separated by a steel plate, are placed against the vertical support frame. Beneath the roll shafts, a ball hinge is used to connect to the column specimen. The vertical support frame is a steel beam supported by two portal frames. The sequence of loading devices in the vertical direction from column top is column, jack, ball hinge, two rows of roll shafts and support frame (see Figure 8.5(b)).

One cable from the two horizontal force sensors is linked to the monitor for convenience of test control, and the other is linked to the computer for data acquisition.

Six displacement transducers are placed in two horizontal directions (each three), as shown in Figure 8.5(a). A glass piece is glued on the interface of the column and transducer to reduce friction. The central

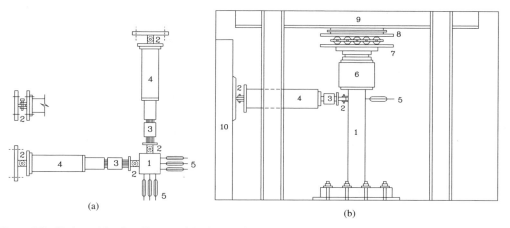

Figure 8.5   Horizontal loading diagram of the box-section column: (a) plan ; (b) elevation. *Notes*: (1) specimen; (2) hinge; (3) loading sensor; (4) horizontal jack; (5) displacement transducer; (6) vertical jack; (7) ball hinge; (8) roll shafts; (9) support frame; (10) support wall

transducer in each direction is linked to the monitor to control test, and the others are linked to the auto-acquisition computer.

8.6.3   Horizontal Loading Scheme

8.6.3.1   For box-section column

First, a vertical load of 600 kN is applied, and the ratio of axial force to squash load of the column is $n = 0.3$. Then, cyclic horizontal loads are applied at the column top. Five horizontal loading schemes are selected for five specimens, four of which are as follows:

- Scheme 1: Two horizontal jacks keep the same pace to completely simulate earthquake forces in two orthotropic directions in the same frequency and the same phase (see Figure 8.6(a)).

- Scheme 2: One jack keeps constant force, whereas the other produces cyclic force. Then, gradually increase the constant force and the cyclic force until the specimen fails (see Figure 8.6(b)).

- Scheme 3: Always keep force of one jack constant, whereas increase that of the other and make the jacks in two directions change the nature of loading alternatively until the specimen fails (see Figure 8.6(c)).

- Scheme 4: Simulate earthquake forces in two directions in the same period but with a phase difference of $\pi/2$ and gradually increase the forces of two jacks until the specimen fails (see Figure 8.6(d)).

The loading history for the four loading schemes measured is given in Figure 8.7.

8.6.3.2   For H-section column

First the vertical load is applied and then cyclic horizontal loads are applied in two horizontal orthotropic directions. Four horizontal loading schemes are selected for four specimens, three of which are as follows:

- Scheme 1: The jack in the sectional weak axis keeps constant force, whereas the other exerts cyclic force. Then gradually increase the constant force and the cyclic force until the specimen fails

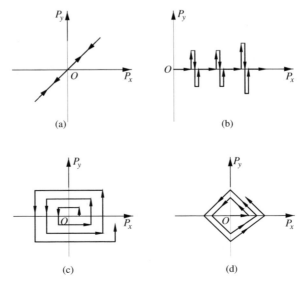

Figure 8.6   Horizontal loading schemes for box-section column specimens: (a) scheme 1; (b) scheme 2; (c) scheme 3; (d) scheme 4

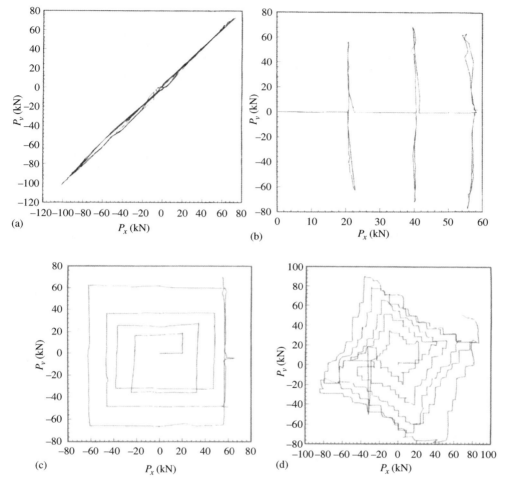

Figure 8.7 Loading history measured for box-section column specimens: (a) scheme 1; (b) scheme 2; (c) scheme 3; (d) scheme 4

(see Figure 8.8(a)). The axial force applied for this plan is 300 kN and the ratio of the axial force to squash load of the column is $n = 0.265$.

- Scheme 2: This plan is same as plan 3 for the box-section column specimen except that no axial force is applied , namely $n = 0$ (see Figure 8.8(b)).

- Scheme 3: This plan is same as plan 3 for the box-section column specimen except that the axial force applied is 300 kN, namely $n = 0.265$ (see Figure 8.8(c)).

The loading history for the three loading schemes measured is given in Figure 8.9.

8.6.4   Theoretical Predictions of Experiments

The column specimen tested can be regarded as a structure with only one column. Representing the column with one column element results in the global stiffness equation of the structure as

$$[k_c]\{d\delta_c\} = \{df_c\}, \tag{8.65}$$

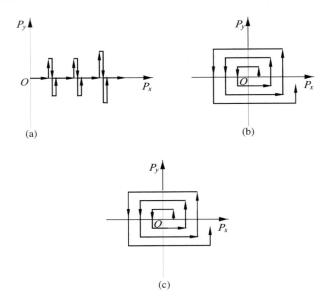

Figure 8.8   Horizontal loading schemes for H-section column specimens: (a) scheme 1 ($n = 0.265$); (b) scheme 2 ($n = 0$); (c) scheme 3 ($n = 0.265$)

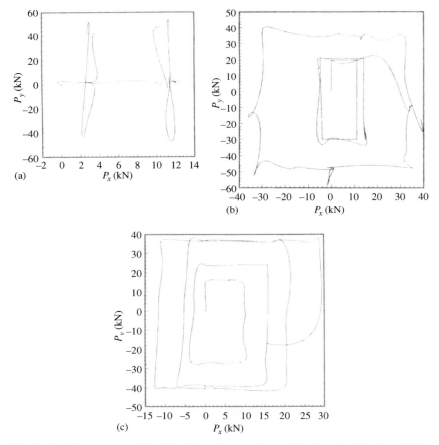

Figure 8.9   Loading history measured for H-section column specimens: (a) scheme 1; (b) scheme 2; (c) scheme 3

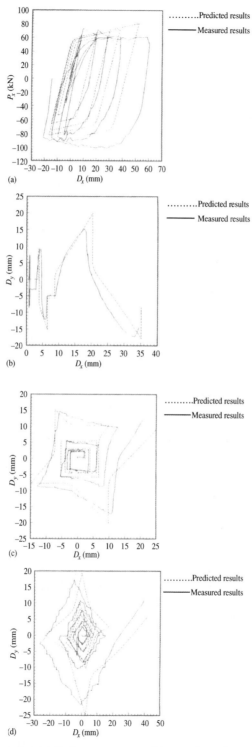

Figure 8.10 Predicted versus measured displacement curves of box-section column specimens: (a) scheme 1; (b) scheme 2; (c) scheme 3; (d) scheme 4

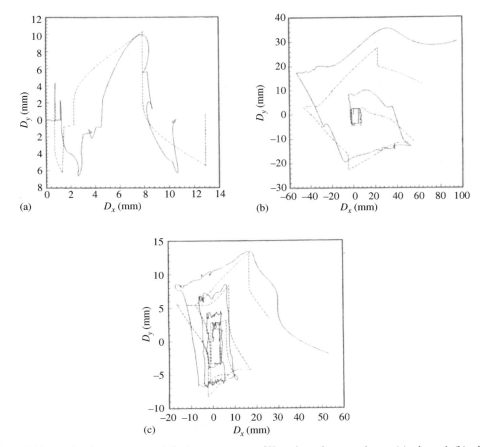

Figure 8.11   Predicted versus measured displacement curves of H-section column specimens: (a) scheme 1; (b) scheme 2; (c) scheme 3

where $[k_c]$ is the elemental stiffness matrix of the column element. If the column is in elastic state, $[k_c]$ is determined by Equation (8.6), whereas if it is elasto-plastic, $[k_c]$ can be determined by Equation (8.52).

Equation (8.65) can be partitioned as

$$\begin{bmatrix} [k_{cbb}] & [k_{cbt}] \\ [k_{ctb}] & [k_{ctt}] \end{bmatrix} \begin{Bmatrix} \{d\delta_{c1}\} \\ \{d\delta_{c2}\} \end{Bmatrix} = \begin{Bmatrix} \{df_{c1}\} \\ \{df_{c2}\} \end{Bmatrix}. \tag{8.66}$$

Introducing boundary conditions to the above equation yields the equation for the lateral displacements at the top of the column as

$$\{d\delta_{c2}\} = [k_{ctt}]^{-1}\{df_{c2}\}. \tag{8.67}$$

8.6.5   Comparison of Analytical and Tested Results

The predicted and measured horizontal displacements for box-section and H-section column specimens are compared in Figures 8.10 and 8.11, respectively, where D_x and D_y are the displacements at the top of the columns corresponding to the horizontal forces P_x and P_y. Good agreement can be found between analytical and experimental results.

Both analytical and experimental results indicate that interaction effects exist in steel columns subjected to biaxial bending. The interaction is that after yielding in columns under biaxial bending, incremental force in one horizontal direction produces the displacement not only in the same direction but also in the other horizontal orthotropic direction. It also includes that the displacement in the direction with smaller horizontal force is more evidently influenced by the larger horizontal force in the other direction. In addition, for the H-section column, the effect of horizontal force in the strong axis on the displacement in the weak axis is much more evident than that of the force in the weak axis on the displacement in the strong axis.

9 Effects of Joint Panel and Beam–Column Connection

9.1 BEHAVIOUR OF JOINT PANEL

9.1.1 Elastic Stiffness of Joint Panel

The four edges of a beam–column joint panel are subjected to the reaction forces from beams and columns connected to the joint panel, as shown in Figure 9.1, in which M_{gL}, Q_{gL}, M_{gR} M_{gR} and Q_{gR} are, respectively, the moments and shears from the left and right beam ends, whereas M_{cT}, Q_{cT}, N_{cT}, M_{cB}, Q_{cB} and N_{cB} are, respectively, the moments, shears and thrusts from the top and bottom column ends. All the directions of actions shown in Figure 9.1 are positive. All these actions make the joint panel in a shear state, which can be equalized to that indicated in Figure 9.2, in which Q_H and Q_V are, respectively, the equivalent horizontal and vertical shear forces and can be written as

$$Q_H = -\frac{M_{gL} + M_{gR}}{h_g} + \frac{1}{2}(Q_{cB} - Q_{cT}),$$ (9.1a)

$$Q_V = -\frac{M_{cT} + M_{cB}}{h_c} + \frac{1}{2}(Q_{gR} - Q_{gL}).$$ (9.1b)

The horizontal and vertical shear stresses in the joint panel can then be obtained by

$$\tau_H = \frac{Q_H}{h_c t_p},$$ (9.2a)

$$\tau_V = \frac{Q_V}{h_g t_p},$$ (9.2b)

where h_c and h_g are sectional heights of columns and beams, respectively, and t_p is the thickness of the joint panel.

The determination of the thickness of various joint panels is given in Figure 9.3.

According to the symmetry of shear stresses, the horizontal and vertical shear stresses in the joint panel are equal, i.e.

$$\frac{Q_H}{h_c t_p} = \frac{Q_V}{h_g t_p}$$ (9.3a)

or

$$Q_H h_g = Q_V h_c.$$ (9.3b)

Advanced Analysis and Design of Steel Frames Guo-Qiang Li and Jin-Jun Li
© 2007 John Wiley & Sons, Ltd

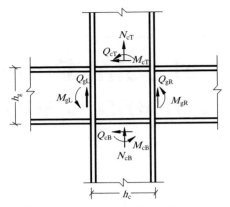

Figure 9.1   Forces applied on a joint panel

The shear moment of the joint panel can be defined as

$$M_\gamma = Q_H h_g = Q_V h_c = \frac{1}{2}(Q_H h_g + Q_V h_c). \tag{9.4}$$

Substituting Equation (9.1) into Equation (9.4) yields

$$M_\gamma = \frac{1}{2}\left[M_{cT} + M_{cB} - M_{gL} - M_{gR} + \frac{h_g}{2}(Q_{cB} - Q_{cT}) + \frac{h_c}{2}(Q_{gR} - Q_{gL})\right]. \tag{9.5}$$

Then, the shear stress of the joint panel may be expressed as

$$\tau = \frac{M_\gamma}{h_g h_c t_p}, \tag{9.6}$$

and the shear strain of the joint panel may be expressed as

$$\gamma = \frac{M_\gamma}{G h_g h_c t_p}, \tag{9.7}$$

where G is the shear elastic modulus.

The elastic stiffness of the joint panel is therefore defined as

$$k_{\gamma e} = G h_g h_c t_p. \tag{9.8}$$

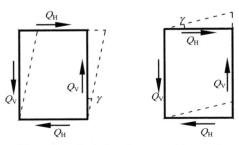

Figure 9.2   Equivalent shears on a joint panel

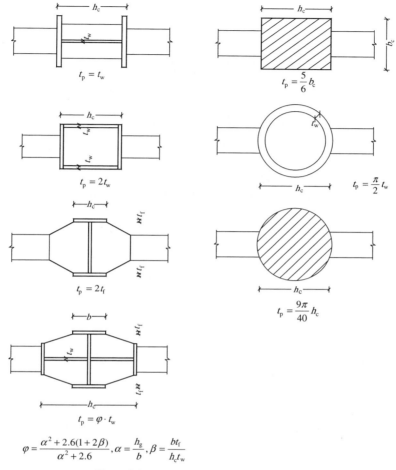

$$\varphi = \frac{\alpha^2 + 2.6(1 + 2\beta)}{\alpha^2 + 2.6}, \alpha = \frac{h_g}{b}, \beta = \frac{bt_f}{h_c t_w}$$

Figure 9.3   Values of t_p for various joint panels

9.1.2   Elasto-Plastic Stiffness of Joint Panel

An experimental curve for the relationship between shear deformation and shear moment of a joint panel is shown in Figure 9.4, which indicates that the bilinear model can be used to simulate the hysteretic and elasto-plastic behaviour of joint panels, as shown in Figure 9.5. In Figure 9.5, q is the hardening factor of joint panels and generally $q = 0.015$, and γ_p and $M_{\gamma p}$ are the shear yielding strain and shear yielding moment of joint panels, respectively. In view of the existence of axial forces in columns and with application of the Mises yielding rule, γ_p and $M_{\gamma p}$ can be determined with

$$\gamma_p = \frac{\sigma_s}{\sqrt{3}G}\sqrt{1 - \left(\frac{\bar{N}}{P_s}\right)^2}, \tag{9.9}$$

$$M_{\gamma p} = k_{\gamma e}\gamma_p, \tag{9.10}$$

in which

$$\gamma_p = \frac{\sigma_s}{\sqrt{3}G}\sqrt{1 - \left(\frac{\bar{N}}{P_s}\right)^2}, \tag{9.11}$$

$$P_s = A_c\sigma_s, \tag{9.12}$$

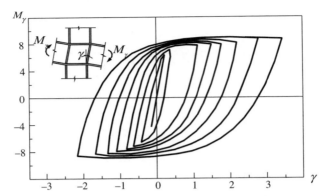

Figure 9.4   Hysteretic M–γ curves of the joint panel from test

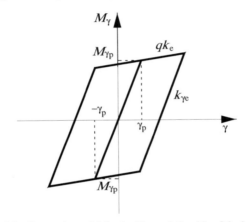

Figure 9.5   Hysteretic model for the M_γ–γ relationship of the joint panel

where σ_s is the yielding stress of joint panels and A_c is the cross-sectional area of the column at the position of joint panels.

9.2   EFFECT OF SHEAR DEFORMATION OF JOINT PANEL ON BEAM/COLUMN STIFFNESS

Inequality occurs on the end deformations of the beam and the column connected to the same joint due to shear deformation of the joint panel even if the connection is rigid. This discontinuity of deformation leads to difficulty in the assembly of global stiffness for the structure of frames using beam and column elements. For this purpose, the deformations of joint panels can be adopted as basic variables for frame structures, which include the horizontal displacement u, vertical displacement v, rotation θ and shear deformation γ, as shown in Figure 9.6. Based on these variables, the stiffness equation for beam or column elements with joint panels can be established.

9.2.1   Stiffness Equation of Beam Element with Joint Panel

As illustrated in Figure 9.7, the relationship between the deformations of a beam element and those of the adjacent joint panels is

$$\delta_1 = \delta_i + \frac{h_{cj}}{2}\theta_i - \frac{h_{ci}}{4}\gamma_i, \tag{9.13a}$$

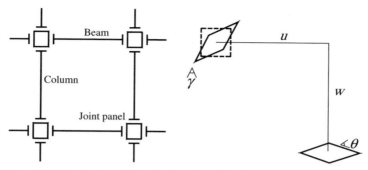

Figure 9.6 The basic variables of a frame involving joint-panel shear deformation

$$\theta_1 = \theta_i + \frac{1}{2}\gamma_i, \tag{9.13b}$$

$$\delta_2 = \delta_j - \frac{h_{cj}}{2}\theta_j + \frac{h_{cj}}{4}\gamma_j, \tag{9.13c}$$

$$\theta_2 = \theta_j + \frac{1}{2}\gamma_j. \tag{9.13d}$$

Let the deformation vector of the beam element and the deformation vector of the two adjacent joint panels be

$$\{\delta_g\} = \{\delta_1, \quad \theta_1, \quad \delta_2, \quad \theta_2\}^T, \tag{9.14a}$$

$$\{\delta_{g\gamma}\} = \{\delta_i, \quad \theta_i, \quad \gamma_i, \quad \delta_j, \quad \theta_j, \quad \gamma_j\}^T. \tag{9.14b}$$

Equation (9.13) can then be rewritten in matrix form as

$$\{\delta_g\} = [A_g]\{\delta_{g\gamma}\}, \tag{9.15}$$

where

$$[A_g] = \begin{bmatrix} 1 & \dfrac{h_{ci}}{2} & -\dfrac{h_{ci}}{4} & 0 & 0 & 0 \\ 0 & 1 & \dfrac{1}{2} & 0 & 0 & 0 \\ 0 & 0 & 0 & 1 & -\dfrac{h_{cj}}{2} & \dfrac{h_{cj}}{2} \\ 0 & 0 & 0 & 0 & 1 & \dfrac{1}{2} \end{bmatrix}. \tag{9.16}$$

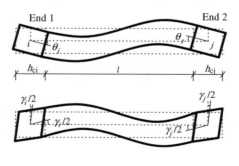

Figure 9.7 Deformation of the frame beam

Transition of the actions at the beam ends to those at the centres of the adjacent joint panels results in the following equations :

$$Q_i = Q_1, \tag{9.17a}$$

$$M_i = \frac{1}{2}h_{ci}Q_1 + M_1, \tag{9.17b}$$

$$Q_j = Q_2, \tag{9.17c}$$

$$M_j = -\frac{1}{2}h_{cj}Q_2 + M_2. \tag{9.17d}$$

Rewrite Equation (9.5) as

$$-M_\gamma = \left(-\frac{1}{2}M_{cT} + \frac{1}{4}h_g Q_{cT}\right) + \left(-\frac{1}{2}M_{cB} - \frac{1}{4}h_g Q_{cB}\right) + \left(\frac{1}{2}M_{gL} + \frac{1}{4}h_c Q_{gL}\right) + \left(\frac{1}{2}M_{gR} + \frac{1}{4}h_c Q_{gR}\right). \tag{9.18}$$

The terms in each parentheses on the right-hand side of the above equation are the contributions of the actions at the beam end or column end connected with the joint panel to the minus of shear moment, $-M_\gamma$. Note that M_{gR}, Q_{gR}, M_{gL} M_{gL} and Q_{gL} applied on the joint panels correspond to M_1, Q_1, M_2 and Q_2 applied on the two beam ends connected to the joint panels and define

$$M_{\gamma i} = \frac{1}{2}M_1 - \frac{1}{4}h_{ci}Q_1, \tag{9.19a}$$

$$M_{\gamma j} = \frac{1}{2}M_2 + \frac{1}{4}h_{cj}Q_2, \tag{9.19b}$$

where $M_{\gamma i}$ and $M_{\gamma j}$ are the contributions of the beam end actions to the minus of shear moments of the left and right adjacent joint panels.
Let

$$\{f_g\} = \{Q_1, \quad M_1, \quad Q_2, \quad M_2\}^T, \tag{9.20a}$$

$$\{f_{g\gamma}\} = \{Q_i, \quad M_i, \quad M_i, \quad Q_j, \quad M_j, \quad M_j\}^T. \tag{9.20b}$$

Combining Equations (9.17) and (9.19) results in

$$\{f_{g\gamma}\} = [A_g]^T\{f_g\}. \tag{9.21}$$

Express the incremental stiffness relation of the beam element as

$$\{df_g\} = [k_g]\{d\delta_g\}, \tag{9.22}$$

where $[k_g]$ is the stiffness matrix of the pure beam element. If the beam is in elastic state, $[k_g]$ is determined according to the method presented in Chapter 2 and $[k_g] = [k_{ge}]$, whereas if it is in elasto-plastic state, $[k_g]$ is determined according to the method presented in Chapter 5 and $[k_g] = [k_{gp}]$.
By Equations (9.15), (9.21) and (9.22), the incremental stiffness relation between $\{df_{g\gamma}\}$ and $\{d\delta_{g\gamma}\}$ can be derived as

$$\{df_{g\gamma}\} = [k_{g\gamma}]\{d\delta_{g\gamma}\}. \tag{9.23}$$

The above is the incremental stiffness equation of the beam element with joint panels, where $[k_{g\gamma}]$ is the stiffness matrix of the beam element including shear deformation of the joint panel given by

$$[k_{g\gamma}] = [A_g]^T[k_g][A_g]. \tag{9.24}$$

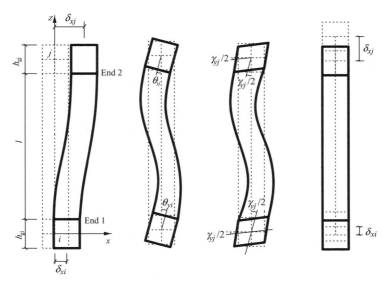

Figure 9.8   Deformation of the frame column

9.2.2   Stiffness Equation of Column Element with Joint Panel Subjected to Uniaxial Bending

The deformation of a frame column subjected to uniaxial bending is given in Figure 9.8. Let

$$\{\delta_c\} = \{\delta_{z1}, \quad \delta_{x1}, \quad \theta_{y1}, \quad \delta_{z2}, \quad \delta_{x2}, \quad \theta_{y2}\}^T, \tag{9.25a}$$

$$\{\delta_{c\gamma}\} = \{\delta_{zi}, \quad \delta_{xi}, \quad \theta_{yi}, \quad \gamma_{yi}, \quad \delta_{zj}, \quad \delta_{xj}, \quad \theta_{yj}, \quad \gamma_{yj}\}^T, \tag{9.25b}$$

$$\{f_c\} = \{N_{z1}, \quad Q_{x1}, \quad M_{y1}, \quad N_{z2}, \quad Q_{x2}, \quad M_{y2}\}^T, \tag{9.26a}$$

$$\{f_{c\gamma}\} = \{N_{zi}, \quad Q_{xi}, \quad M_{yi}, \quad M_{yi}, \quad N_{zj}, \quad Q_{xj}, \quad M_{yj}, \quad M_{yj}\}^T, \tag{9.26b}$$

where $\{\delta_c\}$ is the vector of deformations at the column ends, $\{\delta_{c\gamma}\}$ is the deformation vector of the two joint panels adjacent to the column, $\{f_c\}$ is the vector of forces at the column ends and $\{f_{c\gamma}\}$ is the force vector of the two joint panels adjacent to the column.

Let M_{yyi} and M_{yyj} represent, respectively, the contributions of the actions at the column ends to the minus of shear moment in the adjacent top and bottom joint panels as

$$M_{yyi} = -\frac{1}{2}M_{y1} + \frac{1}{4}h_{gi}Q_{x1}, \tag{9.27a}$$

$$M_{yyj} = -\frac{1}{2}M_{y2} - \frac{1}{4}h_{gj}Q_{x2}. \tag{9.27b}$$

Similar to that discussed for the beam element with joint panels, through the geometry relationship the deformations of the column ends are related to those of the adjacent joint panels with

$$\{\delta_c\} = [A_c]\{\delta_{c\gamma}\}. \tag{9.28}$$

Similarly, by equilibrium relationship of forces, the forces of the joint panels can be expressed with those of the adjacent column ends as

$$\{f_{c\gamma}\} = [A_c]^T\{f_c\}, \tag{9.29}$$

where

$$
[A_c] =
\begin{bmatrix}
1 & 0 & 0 & 0 & 0 & 0 & 0 & 0 \\
0 & 1 & \dfrac{1}{2}h_{gi} & \dfrac{1}{4}h_{gi} & 0 & 0 & 0 & 0 \\
0 & 0 & 1 & -\dfrac{1}{2} & 0 & 0 & 0 & 0 \\
0 & 0 & 0 & 0 & 1 & 0 & 0 & 0 \\
0 & 0 & 0 & 0 & 0 & 1 & -\dfrac{1}{2}h_{gj} & -\dfrac{1}{4}h_{gj} \\
0 & 0 & 0 & 0 & 0 & 0 & 1 & -\dfrac{1}{2}
\end{bmatrix}.
\tag{9.30}
$$

Assume the incremental stiffness equation of the column element be expressed as

$$
\{df_c\} = [k_c]\{d\delta_c\},
\tag{9.31}
$$

where $[k_c]$ is the stiffness matrix of the pure column element and can be determined according to the deformation state of the element by the method presented in Chapter 6. If the column is in elastic state, $[k_c] = [k_{ce}]$, whereas if it is in elasto-plastic state, $[k_c] = [k_{cp}]$.

By Equations (9.28), (9.29) and (9.31), the incremental stiffness equation of the column element with joint panels subjected to uniaxial bending can be derived as

$$
\{df_{c\gamma}\} = [k_{c\gamma}]\{d\delta_{c\gamma}\},
\tag{9.32}
$$

where $[k_{c\gamma}]$ is the stiffness matrix of the column element subjected to uniaxial bending including shear deformation of the joint panel given by

$$
[k_{c\gamma}] = [A_c]^T[k_c][A_c].
\tag{9.33}
$$

9.2.3 Stiffness Equation of Column Element with Joint Panel Subjected to Biaxial Bending

The deformation and force vector of the two joint panels connected to the column subjected to biaxial bending may be expressed as

$$
\{\delta_{c\gamma}\} = [\{\delta_{c\gamma i}\}^T, \quad \{\delta_{c\gamma j}\}^T]^T,
\tag{9.34a}
$$

$$
\{\delta_{c\gamma i}\} = \{\delta_{zi}, \quad \delta_{xi}, \quad \theta_{yi}, \quad \gamma_{yi}, \quad \delta_{yi}, \quad \theta_{xi}, \quad \gamma_{xi}, \quad \theta_{zi}\}^T,
\tag{9.35a}
$$

$$
\{\delta_{c\gamma j}\} = \{\delta_{zj}, \quad \delta_{xj}, \quad \theta_{yj}, \quad \gamma_{yj}, \quad \delta_{yj}, \quad \theta_{xj}, \quad \gamma_{xj}, \quad \theta_{zj}\}^T,
\tag{9.35b}
$$

$$
\{f_{c\gamma}\} = [\{f_{c\gamma i}\}^T, \quad \{f_{c\gamma j}\}^T]^T,
\tag{9.36}
$$

$$
\{f_{c\gamma i}\} = \{N_{zi}, \quad Q_{xi}, \quad M_{yi}, \quad M_{\gamma yi}, \quad Q_{yi}, \quad M_{xi}, \quad M_{\gamma xi}, \quad M_{zi}\}^T,
\tag{9.37a}
$$

$$
\{f_{c\gamma j}\} = \{N_{zj}, \quad Q_{xj}, \quad M_{yj}, \quad M_{\gamma yj}, \quad Q_{yj}, \quad M_{xj}, \quad M_{\gamma xj}, \quad M_{zj}\}^T,
\tag{9.37b}
$$

where $\gamma_{xi}, \gamma_{xj}, \gamma_{yi}$ and γ_{yj} are, respectively, the shear deformations of the two joint panels about the x-axis and y-axis of the column section, M_{yi} and M_{yj} are the contributions of the actions at the column ends to the minus of shear moment in the joint panels about the y-axis of the column section and determined by Equation (9.27), and M_{xi} and M_{xj} are the contributions of the actions at the column ends to the minus of shear moment in the joint panels about the x- axis of the column section and determined by

$$
M_{\gamma xi} = -\frac{1}{2}M_{x1} + \frac{1}{4}h_{gi}Q_{y1},
\tag{9.38a}
$$

$$
M_{\gamma xj} = -\frac{1}{2}M_{x2} - \frac{1}{4}h_{gj}Q_{y2}.
\tag{9.38b}
$$

Noting that the deformation and force vectors of the column element subjected to biaxial bending are those expressed in Equations (8.2), (8.4) and (8.1), (8.3), the stiffness equation of the column element in biaxial bending with joint panels can be derived in the similar way as that for the column element in uniaxial bending with joint panels. The incremental stiffness equation obtained has the same form as Equation (9.32) and the expression of the corresponding stiffness matrix for the column element in biaxial bending with joint panels is the same as Equation (9.33), but

$$[A_c] = \begin{bmatrix} [A_{ci}] & 0 \\ 0 & [A_{cj}] \end{bmatrix}, \tag{9.39}$$

where

$$[A_{ci}] = \begin{bmatrix} 1 & 0 & 0 & 0 & 0 & 0 & 0 & 0 \\ 0 & 1 & \frac{1}{2}h_{gi} & \frac{1}{4}h_{gi} & 0 & 0 & 0 & 0 \\ 0 & 0 & 1 & -\frac{1}{2} & 0 & 0 & 0 & 0 \\ 0 & 0 & 0 & 0 & 1 & \frac{1}{2}h_{gi} & \frac{1}{4}h_{gi} & 0 \\ 0 & 0 & 0 & 0 & 0 & 1 & -\frac{1}{2} & 0 \\ 0 & 0 & 0 & 0 & 0 & 0 & 0 & 1 \end{bmatrix} \tag{9.40a}$$

and

$$[A_{cj}] = \begin{bmatrix} 1 & 0 & 0 & 0 & 0 & 0 & 0 & 0 \\ 0 & 1 & \frac{1}{2}h_{gj} & \frac{1}{4}h_{gj} & 0 & 0 & 0 & 0 \\ 0 & 0 & 1 & -\frac{1}{2} & 0 & 0 & 0 & 0 \\ 0 & 0 & 0 & 0 & 1 & \frac{1}{2}h_{gj} & \frac{1}{4}h_{gj} & 0 \\ 0 & 0 & 0 & 0 & 0 & 1 & -\frac{1}{2} & 0 \\ 0 & 0 & 0 & 0 & 0 & 0 & 0 & 1 \end{bmatrix}. \tag{9.40b}$$

9.3 BEHAVIOUR OF BEAM–COLUMN CONNECTIONS

Beams and columns in steel frames are generally connected with welding or bolting (usually using high-strength bolts). Subjected to the bending moments at beam ends, deformation of connections occurs more or less and is dominant by relative rotations as shown in Figure 9.9. For the connection with the top and bottom flanges of the beam welded to the column while the web of the beam connected to the column with high-strength bolts or also welding (see Figure 9.9(e)), rigid connection assumption with no relative rotation between the beam and the column can be adopted in frame analysis because of the sufficient rotation stiffness of the connection. For the connection where only the web of the beam is connected to the column (especially with bolts, see Figure 9.9(a)), pinned connection assumption allowing for arbitrary relative rotation between the beam and the column can be used in frame analysis because of lack of enough rotation stiffness.

Rigid beam–column connections have good moment–resistance behaviour, but the difficulty in fabrication and erection for such connections is high. Configuration of pinned connections is relatively simple, but their stiffness and energy-absorption capacity are low, which is disadvantageous to aseismic design. Semi-rigid beam–column connections, however, are the compromise of rigid and pinned connections, and when used in steel frames, good comprehensive technique-economy index is hoped to achieve. As the behaviour of semi-rigid connections evidently affects the stiffness and load-carrying capacity of steel frames, clarifying the moment–rotation relationship and hysteretic behaviour of semi-rigid beam–

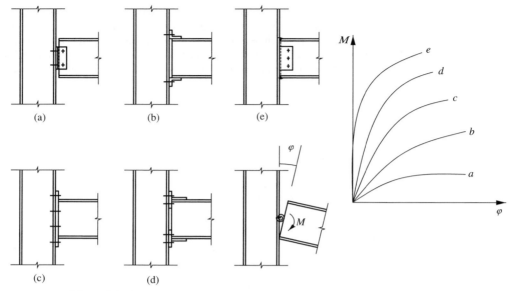

Figure 9.9   Connection configurations and the corresponding moment–rotation curves

column connections is a key point to static and dynamic analysis and design of steel frames involving such connections.

This section is to summarize, based on previous investigations, the moment–curvature relationship and hysteretic behaviour of some types of semi-rigid connections.

9.3.1   Moment–Rotation Relationship

9.3.1.1   Top- and seat-angle connection

The connection configuration is shown in Figure 9.10 and its moment–rotation relationship was given by Maxwell, Jenkins and Howlett (1981) as

$$M = k\varphi, \qquad \text{when} \quad \varphi < \varphi_p, \tag{9.41a}$$

$$M = M_p, \qquad \text{when} \quad \varphi \geq \varphi_p, \tag{9.41b}$$

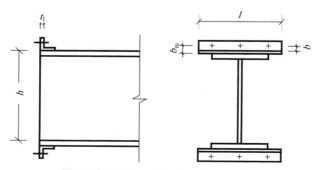

Figure 9.10   Top- and seat-angle connection

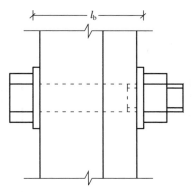

Figure 9.11 Effective length of a bolt

in which

$$k = \frac{Eh^2}{10} \left[\frac{(b_m - d/2)^3}{lt_1^3} + \frac{l_b}{3A_b n_b} \right],$$

$$M_p = \frac{\sigma_s lt_1^2}{2} \left(\frac{h + b_m}{b} - 1 \right),$$

$$\varphi_p = \frac{M_p}{k},$$

where A_b is the effective cross-sectional area of a bolt, l_b is the effective length of the bolt, as shown in Figure 9.11, n_b is the number of bolts in the first row at the vertical leg of the angle in tension, l and t_1 are the length and thickness of the angle, respectively, b_m and b are the distances from the bolt centre to the back of the angle and to the arc edge of the angle, respectively, d is the diameter of the bolt, h is the height of the beam, E is the elastic modulus and σ_s is the yielding strength of the angle steel.

A similar moment–rotation relationship was also given by Frye and Morris (1975) as

$$\varphi = 7.49 \times 10^{-6}(kM) + 7.00 \times 10^{-11}(kM)^3 + 6.37 \times 10^{-19}(kM)^5, \tag{9.42}$$

in which

$$k = 2.14 \times 10^{-4} t_1^{0.5} h^{-1.5} d^{-1.1} l^{-0.7},$$

where φ is the relative rotation between the beam and the column of the connection. The units of the variables in Equation (9.42) are as follows: φ is in rad, M is in N m and all units of dimension in the calculation of k are in m.

9.3.1.2 Top- and seat-angle connection with double web angles

The connection configuration is shown in Figure 9.12 and its moment–rotation relationship was given by Frye and Morris (1975) as

$$\varphi = 1.976 \times 10^{-7}(kM) + 1.283 \times 10^{-14}(kM)^3 + 1.732 \times 10^{-22}(kM)^5, \tag{9.43}$$

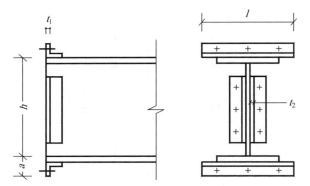

Figure 9.12   Top- and seat-angle connection with double web angles

in which

$$k = 3.41 \times 10^{-4} t_1^{-1.128} d^{-1.287} t_2^{-0.4145} l^{-0.6941} \left(a - \frac{d}{2} \right)^{1.35},$$

where t_1 and t_2 are the thicknesses of the angles at the flange and web of the beam and a is the length of the vertical leg of top and seat angles. The units of the variables in Equation (9.43) are same as those in Equation (9.42).

9.3.1.3 Flange- and web-plate connection

The configuration of the connection is shown in Figure 9.13 and its moment–rotation relationship was given by Ackroyd (1987) as

$$M = \frac{k\varphi}{[1 + (k\varphi/M_{\mathrm{p}})^{15.5}]^{1/15.5}}, \tag{9.44}$$

in which

$$k = 0.5 E b t h^2 / l,$$
$$M_{\mathrm{p}} = b t h \sigma_{\mathrm{s}},$$

where l, b and t are the length, width and thickness of the flange-connection plate, respectively, h is the height of the beam and σ_{s} is the yielding strength of the connection plate.

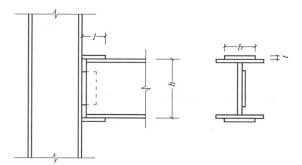

Figure 9.13   Flange- and web-plate connection

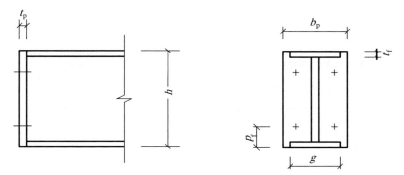

Figure 9.14 Flush end-plate connection

9.3.1.4 Flush end-plate connection

The connection configuration is shown in Figure 9.14 and its moment–rotation relationship was given by Kukreti, Murray and Abolmaali (1987) as

$$\varphi = 5.695 \times 10^{-5} (cM)^{1.356}, \tag{9.45}$$

in which

$$c = 1.230 \times 10^{-10} p_{\mathrm{f}}^{2.227} h^{-2.616} t_{\mathrm{w}}^{-0.501} t_{\mathrm{f}}^{-0.038} d^{-0.849} g_{\mathrm{b}}^{-0.519} b_{\mathrm{p}}^{-0.218} t_{\mathrm{p}}^{-1.539},$$

$$g_{\mathrm{b}} = \frac{1}{3} \frac{\sigma_{\mathrm{bs}}}{\sigma_{\mathrm{s}}} \frac{A_{\mathrm{b}}}{d},$$

A similar moment–rotation relationship was given by Kishi and Chen (1986) as

$$\varphi = 1.62 \times 10^{-5} (kM) + 7.21 \times 10^{-11} (kM)^3 + 3.47 \times 10^{-16} (kM)^5, \tag{9.46}$$

in which

$$k = 1.383 \times 10^{-7} h^{-2.4} t_{\mathrm{p}}^{-0.4} t_{\mathrm{fc}}^{-1.5},$$

where t_{f} and t_{w} are the thicknesses of the flange and web of the beam, respectively, t_{fc} is the thickness of the column flange, b_{p} and t_{p} are the width and thickness of the end plate, respectively, A_{b} is the cross-sectional area of a bolt, p_{f} is the minimum distance from the bolt centre to the outer edge of the beam flange, and σ_{bs} and σ_{s} are the yielding strengths of bolts and end plate, respectively. The units of length, moment and rotation in above equations are same as those in Equation (9.42), and the unit of strength is Pa.

9.3.1.5 Extended end-plate connection

The connection configuration is shown in Figure 9.15 and its moment–rotation relationship was given by Krishnamurthy *et al.* (1979) as

$$\varphi = 7.076 \times 10^{-8} \frac{cM^{1.58}}{t_{\mathrm{p}}^{1.38}}, \tag{9.47}$$

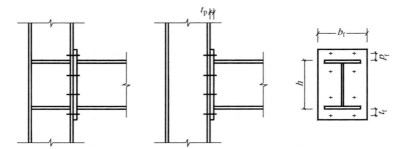

Figure 9.15   Extended end-plate connection

in which

$$c = 172.1 \beta \mu p_f^{2.03} A_{bl}^{-0.36},$$

$$\beta = 2.267 \times 10^{-5} b_f^{0.61} t_f^{1.03} d^{-1.30} t_w^{-0.26} w^{-1.58},$$

$$\mu = 6.38 \times 10^{10} \sigma_{bs}^{-1.20}.$$

A similar moment–rotation relationship was given by Frey and Morris (1975) as

- for unstiffened connection:

$$\varphi = 1.62 \times 10^{-5}(kM) + 7.21 \times 10^{-11}(kM)^3 + 3.47 \times 10^{-16}(kM)^5, \tag{9.48}$$

in which

$$k = 1.383 \times 10^{-7}(h + 2p_f)^{-2.4} t_p^{-0.4} t_{fc}^{-1.5};$$

- for stiffened connection:

$$\varphi = 1.58 \times 10^{-5}(kM) + 1.21 \times 10^{-10}(kM)^3 + 1.11 \times 10^{-14}(kM)^5, \tag{9.49}$$

in which

$$k = 1.639 \times 10^{-5}(h + 2p_f)^{-2.4} t_p^{-0.6},$$

where t_p is the thickness of the end plate, b_f and t_f are the width and thickness of the beam flange, respectively, t_w is the thickness of the beam web, t_{fc} is the thickness of the column flange, w is the bending resistance moment of the beam, A_{bl} is the total cross-sectional area of the connection bolts in one row and p_f is the distance from the centre of outside bolts to the edge of the beam flange. All units of above quantities are same with those used in Equation (9.42).

Yee and Melchers (1986) and Boswell and O'Conner (1988) proposed another moment–rotation relationship (as shown in Figure 9.16) for the extended end-plate connection as

$$M = M_p \left\{ 1 - \exp\left[\frac{-(k_e - k_p + c\varphi)\varphi}{M_p} \right] \right\} + k_p \varphi, \tag{9.50}$$

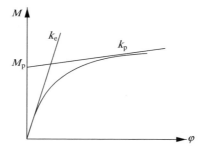

Figure 9.16   Moment–rotation curve expressed in Equation (9.50)

where M_p is the plastic moment of the connection, k_e is the initial stiffness of the connection, k_p is the hardening stiffness of the connection and c is a factor with a value of 3.5 for the stiffened connection and 1.5 for the unstiffened connection.

A method to calculate M_p, k_e and k_p was proposed in detail by Yee and Melchers (1986).

9.3.2 Hysteretic Behaviour

The hysteretic behaviour of semi-rigid connections is actually the moment–rotation relationship of connections in the process of loading–unloading–reloading. It has been verified from experiments that the connection stiffness in unloading is equal to the initial stiffness (see Figure 9.17), by which a hysteretic moment–rotation relationship of semi-rigid connections was proposed as shown in Figure 9.18 (Ackroyd

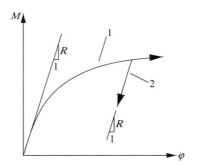

Figure 9.17   Unloading stiffness of a semi-rigid connection

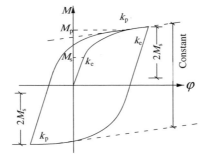

Figure 9.18   Hysteretic M–φ relationship of a semi-rigid connection

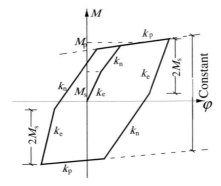

Figure 9.19   Trilinear hysteretic M–φ model of a semi-rigid connection

and Gerstle, 1982; Azizinamini, Bradburn and Radziminski, 1987), in which the definitions of the variables are same as those in Equation (9.50) except that M_s is defined as the initial yielding moment of the connection.

A triple-linear model was used to approximately predict the hysteretic behaviour of semi-rigid connections by Moncarz and Gerstle (1981) and Stelmack, Marley and Gerstle (1986), as shown in Figure 9.19. In the triple-linear model, k_n is the average stiffness between the initial stiffness k_e and the hardening stiffness k_p of connections.

For sake of application convenience, a hysteretic model as shown in Figure 9.20 was proposed by Li and Shen (1990) with consideration of nonlinearity of the moment–rotation relationship of semi-rigid connections. In this model, φ_p is the rotation corresponding to the plastic moment of connections. The tangent stiffness of connections under arbitrary moment M can be determined as

- for loading:

$$k = \frac{k_e}{1 + \left(\dfrac{k_e}{k_p - 1}\right)\left(\dfrac{|M|}{M_p}\right)^n}, \qquad |M| < M_p, \tag{9.51a}$$

$$k = k_p, \qquad |M| \geq M_p; \tag{9.51b}$$

- for unloading:

$$k = k_e, \tag{9.52}$$

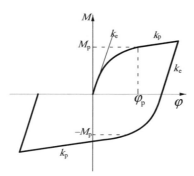

Figure 9.20   Curvilinear hysteretic M–φ model of a semi-rigid connection governed by Equations (9.51) and (9.52)

where

$$n = \frac{\dfrac{k_e}{k_p} - 1}{\dfrac{\varphi_e}{\varphi_p} - 1} - 1, \qquad \varphi_e = \frac{M_p}{k_e}.$$

The model illustrated in Figure 9.20 uses four parameters, namely initial stiffness k_e, hardening stiffness k_p, plastic moment M_p and the corresponding φ_p, to determine the hysteretic moment–rotation relationship of semi-rigid connections. If the M–φ relationship is

$$M = f(\varphi), \tag{9.53}$$

then

$$k_e = \left.\frac{\mathrm{d}M}{\mathrm{d}\varphi}\right|_{\varphi=0} = f'(\varphi), \tag{9.54}$$

$$k_p = \left.\frac{\mathrm{d}M}{\mathrm{d}\varphi}\right|_{\varphi=\varphi_p} = f'(\varphi_p), \tag{9.55}$$

or

$$\varphi = g(M), \tag{9.56}$$

then

$$k_e = \frac{1}{\left.\dfrac{\mathrm{d}\varphi}{\mathrm{d}M}\right|_{M=0}} = \frac{1}{g'(0)}, \tag{9.57}$$

$$k_p = \frac{1}{\left.\dfrac{\mathrm{d}\varphi}{\mathrm{d}M}\right|_{M=M_p}} = \frac{1}{g'(M_p)}. \tag{9.58}$$

If parameters φ_p and M_p cannot be determined, the value of φ_p can be determined approximately by

$$\varphi_p = 4\varphi_e = \frac{4M_p}{k_e} \tag{9.59}$$

and M_p can be solved from

$$g(M_p) = 4\frac{M_p}{k_e}. \tag{9.60}$$

9.4 EFFECT OF DEFORMATION OF BEAM–COLUMN CONNECTION ON BEAM STIFFNESS

Due to the deformation of the beam–column connection, the rotation of the beam end is not compatible with that of the edge of the beam–column joint panel. To overcome this incompatibility, the stiffness equation of the beam element with beam–column connections can be developed by regarding the vertical translation and rotation of the edge of the two adjacent joint panels as unknown variables.

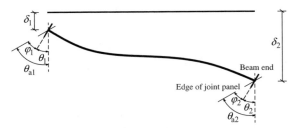

Figure 9.21   Deformation of a beam with connection

9.4.1   Stiffness Equation of Beam Element with Beam–Column Connections

The deformations of the beam element with beam–column connections are given in Figure 9.21, where δ_1 and δ_2 are vertical displacements at both element ends (namely the displacements of the edges of the joint panels adjacent to the two ends of the beam), θ_{a1} and θ_{a2} are rotations of the edges of the joint panels, φ_1 and φ_2 are relative rotations due to flexible connections between the beam ends and the corresponding edges of the joint panels, and θ_1 and θ_2 are the real rotations of the beam ends.

Let

$$\{\delta_g\} = \{\delta_1, \quad \theta_1, \quad \delta_2, \quad \theta_2\}^T, \tag{9.61}$$

$$\{\delta_{ga}\} = \{\delta_1, \quad \theta_{a1}, \quad \delta_2, \quad \theta_{a2}\}^T, \tag{9.62}$$

$$\{\delta_a\} = \{0, \quad \varphi_1, \quad 0, \quad \varphi_2\}^T. \tag{9.63}$$

Then

$$\{\delta_{ga}\} = \{\delta_g\} + \{\delta_a\}. \tag{9.64}$$

Assuming that the shear forces at both beam ends are, respectively, Q_1 and Q_2, and the moments are M_1 and M_2, let

$$\{f_g\} = \{Q_1, \quad M_1, \quad Q_2, \quad M_2\}^T, \tag{9.65}$$

$$\{f_{gt}\} = \{Q_1, \quad 0, \quad Q_2, \quad 0\}^T, \tag{9.66}$$

$$\{f_{gn}\} = \{0, \quad M_1, \quad 0, \quad M_2\}^T, \tag{9.67}$$

and then

$$\{f_g\} = \{f_{gt}\} + \{f_{gn}\}. \tag{9.68}$$

By the incremental stiffness equation of the pure beam element, the increments of $\{f_g\}$ and $\{\delta_g\}$ are related by Equation (9.22). The incremental equation for the moment at the end of the beam and the relative rotation between the beam and the column of the connection adjacent to the beam end can be obtained with the tangent stiffness of the beam-to-column connection as

$$\{df_{gn}\} = [k_a]\{d\delta_a\}, \tag{9.69}$$

in which

$$[k_a] = \text{diag}\,[0, \quad k_1, \quad 0, \quad k_2], \tag{9.70}$$

where k_1 and k_2 are tangent stiffnesses of the beam-to-column connection corresponding to the moments at the two ends of the beam, M_1 and M_2, respectively, which can be determined with the moment–rotation relationship or hysteretic model of the connections, such as given by Equations (9.51) and (9.52).

With Equation (9.68), one has

$$\{df_{gt}\} = \{df_g\} - \{df_{gn}\}. \tag{9.71}$$

Substituting Equations (9.22) and (9.69) into Equation (9.71) leads to

$$
\begin{aligned}
\{df_{gt}\} &= [k_g]\{d\delta_g\} - [k_a]\{d\delta_a\} = [k_g](\{d\delta_{ga}\} - \{d\delta_g\}) - [k_a]\{d\delta_a\} \\
&= [k_g]\{d\delta_{ga}\} - ([k_g] + [k_a])\{d\delta_a\}.
\end{aligned}
\tag{9.72}
$$

Express $\{d\delta_a\}$ as

$$\{d\delta_a\} = [H]^{\mathrm{T}}\{d\varphi\}, \tag{9.73}$$

where

$$\{d\varphi\} = [d\varphi_1, \quad d\varphi_2]^{\mathrm{T}}, \tag{9.74}$$

$$[H] = \begin{bmatrix} 0 & 1 & 0 & 0 \\ 0 & 0 & 0 & 1 \end{bmatrix}. \tag{9.75}$$

It is known from Equations (9.63) and (9.66) that $\{d\delta_a\}$ is orthotropic with $\{df_{gt}\}$, namely

$$\{d\delta_a\}^{\mathrm{T}}\{df_{gt}\} = \{d\varphi\}^{\mathrm{T}}[H]\{df_{gt}\} = 0. \tag{9.76}$$

Due to the connection flexibility, $\{d\varphi\} \neq \{0\}$ and then

$$[H]\{df_{gt}\} = 0. \tag{9.77}$$

Substituting Equation (9.73) into Equation (9.72) and then into Equation (9.77) results in

$$[H][k_g]\{d\delta_{ga}\} - [H]([k_g] + [k_a])[H]^{\mathrm{T}}\{d\varphi\} = 0. \tag{9.78}$$

Solution of $\{d\varphi\}$ from the above equation is obtained as

$$\{d\varphi\} = ([H]([k_g] + [k_a])[H]^{\mathrm{T}})^{-1}[H][k_g]\{d\delta_{ga}\}. \tag{9.79}$$

Substituting Equation (9.79) into Equation (9.73) leads to

$$\{d\delta_a\} = [H]^{\mathrm{T}}([H]([k_g] + [k_a])[H]^{\mathrm{T}})^{-1}[H][k_g]\{d\delta_{ga}\}, \tag{9.80}$$

and then

$$
\begin{aligned}
\{d\delta_g\} &= \{d\delta_{ga}\} - \{d\delta_a\} \\
&= ([I] - [H]^{\mathrm{T}}([H]([k_g] + [k_a])[H]^{\mathrm{T}})^{-1}[H][k_g])\{d\delta_{ga}\}.
\end{aligned}
\tag{9.81}
$$

Finally, substituting Equation (9.81) into Equation (9.22) yields

$$\{df_g\} = [k_{ga}]\{d\delta_{ga}\}, \tag{9.82}$$

in which

$$[k_{\text{ga}}] = [k_{\text{g}}] - [k_{\text{g}}][H]^{\mathrm{T}}([H]([k_{\text{g}}] + [k_{\text{a}}])[H]^{\mathrm{T}})^{-1}[H][k_{\text{g}}], \tag{9.83}$$

where $[k_{\text{ga}}]$ is the stiffness matrix of the beam element including flexibility of beam-to-column connections.

9.4.2 Stiffness Equation of Beam Element with Connections and Joint Panels

When effects of connection flexibility and joint-panel shear deformation are considered simultaneously, the stiffness equation of the beam element with connections and joint panels can be established. In this case, the basic variables include vertical displacement of the centres of joint panels adjacent to the beam, and rotation and shear deformations of the joint panels. With the similar derivation as in Sections 9.2.1 and 9.4.1, the expression of the stiffness equation desired for the beam element with connections and joint panels can be obtained as

$$\{\mathrm{d}f_{\text{g}\gamma}\} = [A_{\text{g}}]^{\mathrm{T}}\{\mathrm{d}f_{\text{g}}\} = [A_{\text{g}}]^{\mathrm{T}}[k_{\text{ga}}]\{\mathrm{d}\delta_{\text{ga}}\} = [A_{\text{g}}]^{\mathrm{T}}[k_{\text{ga}}][A_{\text{g}}]\{\mathrm{d}\delta_{\text{g}\gamma}\} \tag{9.84a}$$

or

$$\{\mathrm{d}f_{\text{g}\gamma}\} = [k_{\text{ga}\gamma}]\{\mathrm{d}\delta_{\text{g}\gamma}\}, \tag{9.84b}$$

in which

$$[k_{\text{ga}\gamma}] = [A_{\text{g}}]^{\mathrm{T}}[k_{\text{ga}}][A_{\text{g}}], \tag{9.85}$$

where $[k_{\text{ga}\gamma}]$ is the stiffness matrix of the beam element including flexibility of beam-to-column connections and joint-panel shear deformations.

9.5 EXAMPLES

9.5.1 Effect of Joint Panel

To examine the effect of joint panels on the lateral displacement of steel frames, consider a simple frame as shown in Figure 9.22. This frame is actually a substructure of realistic frames (see Figure 9.23) and can

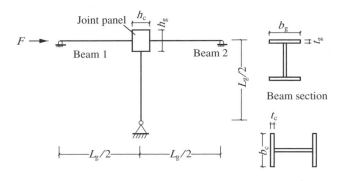

Figure 9.22 A subframe structure with the joint panel

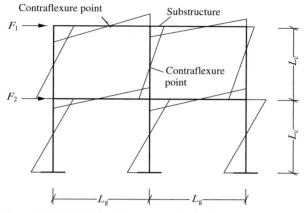

Figure 9.23 Typical bending moment distribution of the sway frame

reflect the characteristics of the lateral displacement behaviour of realistic frames. The geometries of the structure size are as follows:

$$L_g = 6\,\text{m}, L_c = 3.5\,\text{m}, h_g = 0.6\,\text{m}, h_c = 0.5\,\text{m},$$
$$b_g = 0.3\,\text{m}, t_g = 0.02\,\text{m}, b_c = 0.4\,\text{m}, t_c = 0.02\,\text{m}.$$

The column and beam of the example frame are both symmetric H sections. For sake of simplification, omit the web plate in the calculation of inertial moments of the beam, I_g, and the column, I_c. Then

$$I_g = 2b_g t_g \left(\frac{h_g}{2}\right)^2 = 1.08 \times 10^{-3}\,\text{m}^4,$$

$$I_c = 2b_c t_c \left(\frac{h_c}{2}\right)^2 = 1.00 \times 10^{-3}\,\text{m}^4.$$

Assume the thickness of the joint panel to be

$$t_p = c t_c = 0.02c,$$

where c is a factor greater than zero, on the value of which the thickness and stiffness of the joint panel depend. The relationship between the elastic lateral displacement of the frame and the value of c is discussed in the following.

To stand out the effect of the joint panel, effects of axial force and shear deformation are ignored in the elemental stiffness, and axial deformations of the structural components are also excluded. The basic variables of the structure are horizontal displacement u, rotation θ and shear deformation γ of the joint panel.

Subjected to the above assumption, each of the beams and columns is represented with a clamped-hinged beam element (see Figure 9.24), and the elemental lengths are

$$l_g = \frac{L_g}{2} - \frac{h_c}{2} = 2.75\,\text{m}, \qquad l_c = \frac{L_c}{2} - \frac{h_g}{2} = 1.45\,\text{m}.$$

By introducing boundary conditions, elemental stiffness equations for each member of the structure are as follows:

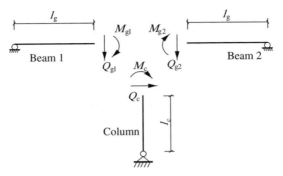

Figure 9.24 Element division

- for beam 1:

$$\left\{\begin{matrix} Q_{g1} \\ M_{g1} \end{matrix}\right\} = \frac{EI_g}{l_g}\begin{bmatrix} 3/l_g^2 & -3/l_g \\ -3/l_g & 3 \end{bmatrix}\left\{\begin{matrix} \delta_{g1} \\ \theta_{g1} \end{matrix}\right\} = \begin{bmatrix} 0.1558 & -0.4284 \\ -0.4284 & 1.1782 \end{bmatrix}\times 10^{-3}E\left\{\begin{matrix} \delta_{g1} \\ \theta_{g1} \end{matrix}\right\};$$

- for beam 2:

$$\left\{\begin{matrix} Q_{g2} \\ M_{g2} \end{matrix}\right\} = \frac{EI_g}{l_g}\begin{bmatrix} 3/l_g^2 & 3/l_g \\ 3/l_g & 3 \end{bmatrix}\left\{\begin{matrix} \delta_{g2} \\ \theta_{g2} \end{matrix}\right\} = \begin{bmatrix} 0.1558 & 0.4284 \\ 0.4284 & 1.1782 \end{bmatrix}\times 10^{-3}E\left\{\begin{matrix} \delta_{g2} \\ \theta_{g2} \end{matrix}\right\};$$

- for column:

$$\left\{\begin{matrix} Q_c \\ M_c \end{matrix}\right\} = \frac{EI_c}{l_c}\begin{bmatrix} 3/l_c^2 & -3/l_c \\ -3/l_c & 3 \end{bmatrix}\left\{\begin{matrix} \delta_c \\ \theta_c \end{matrix}\right\} = \begin{bmatrix} 0.9841 & -1.4269 \\ -1.4269 & 2.0690 \end{bmatrix}\times 10^{-3}E\left\{\begin{matrix} \delta_c \\ \theta_c \end{matrix}\right\},$$

where δ_{g1}, θ_{g1}, δ_{g2}, θ_{g2} δ_c and θ_c are the displacement components corresponding to the force components Q_{g1}, M_{g1}, Q_{g2}, M_{g2}, Q_c and M_c, respectively, and E is the elastic modulus.
Considering the effect of the joint panel, the stiffness equations of the beam and the column with the joint panel are (referring to Sections 9.2.1 and 9.2.2)

- for beam 1:

$$\left\{\begin{matrix} Q_{gv1} \\ M_{g\theta1} \\ M_{g\gamma1} \end{matrix}\right\} = \begin{bmatrix} 1 & 0 \\ -h_c/2 & 1 \\ h_c/4 & 1/2 \end{bmatrix}\begin{bmatrix} 0.1558 & -0.4284 \\ -0.4284 & 1.1782 \end{bmatrix}\times 10^{-3}E\begin{bmatrix} 1 & -h_c/2 & h_c/4 \\ 0 & 1 & 1/2 \end{bmatrix}\left\{\begin{matrix} v \\ \theta \\ \gamma \end{matrix}\right\}$$

$$= \begin{bmatrix} 0.1558 & -0.4674 & -0.1947 \\ -0.4674 & 1.4021 & 0.5842 \\ -0.1947 & 0.5842 & 0.2435 \end{bmatrix}\times 10^{-3}E\left\{\begin{matrix} v \\ \theta \\ \gamma \end{matrix}\right\},$$

- for beam 2:

$$\left\{\begin{matrix} Q_{gv2} \\ M_{g\theta2} \\ M_{g\gamma2} \end{matrix}\right\} = \begin{bmatrix} 1 & 0 \\ -h_c/2 & 1 \\ h_c/4 & 1/2 \end{bmatrix}\begin{bmatrix} 0.1558 & 0.4284 \\ 0.4284 & 1.1782 \end{bmatrix}\times 10^{-3}E\begin{bmatrix} 1 & -h_c/2 & h_c/4 \\ 0 & 1 & 1/2 \end{bmatrix}\left\{\begin{matrix} v \\ \theta \\ \gamma \end{matrix}\right\}$$

$$= \begin{bmatrix} 0.1558 & 0.4674 & 0.1947 \\ 0.4674 & 1.4021 & 0.5842 \\ 0.1947 & 0.5842 & 0.2435 \end{bmatrix}\times 10^{-3}E\left\{\begin{matrix} v \\ \theta \\ \gamma \end{matrix}\right\};$$

- for column:

$$\left\{\begin{array}{c} Q_{c1} \\ M_{c1} \\ M_{c1} \end{array}\right\} = \left[\begin{array}{cc} 1 & 0 \\ -h_g/2 & 1 \\ h_g/4 & 1/2 \end{array}\right] \left[\begin{array}{cc} 0.9841 & -1.4296 \\ -1.4296 & 2.0690 \end{array}\right] \times 10^{-3} E \left[\begin{array}{ccc} 1 & -h_g/2 & h_g/4 \\ 0 & 1 & 1/2 \end{array}\right] \left\{\begin{array}{c} u \\ \theta \\ \gamma \end{array}\right\}$$

$$= \left[\begin{array}{ccc} 0.9841 & -1.7221 & 0.5658 \\ -1.7221 & 3.0137 & -0.9902 \\ 0.5658 & -0.9902 & 0.3254 \end{array}\right] \times 10^{-3} E \left\{\begin{array}{c} u \\ \theta \\ \gamma \end{array}\right\},$$

where Q_{gv1}, $M_{g\theta1}$, $M_{g\gamma1}$, Q_{gv2}, $M_{g\theta2}$ and $M_{g\gamma2}$ are the vertical shear, bending moment and shear moment (negative) at the joint panels adjacent to beams 1 and 2, respectively, Q_{cv}, $M_{c\theta}$ and $M_{c\gamma}$ are the horizontal shear, bending moment and shear moment (negative) at the joint panels adjacent to the column, respectively, and u, v, θ and γ are the horizontal displacement, vertical displacement, rotation and shear deformation of the panel of the example frame, respectively.

The elemental stiffness equation of the joint panel is

$$M_\gamma = G h_g h_c t_p \gamma = (E/2.6) \times 0.6 \times 0.5 \times 0.02 c \gamma = 2.3077 c \gamma.$$

The relationship $G = E/2.6$ is used in the above equation.

Force equilibrium of the structure includes

$$Q_{cu} = F,$$
$$M_{g\theta1} + M_{g\theta2} + M_{c\theta} = 0,$$
$$M_{g\gamma1} + M_{g\gamma2} + M_{c\gamma} = -M_\gamma.$$

Ignoring the axial deformation of the column, namely $v = 0$, results in the global stiffness equation of the structure as

$$\left\{\begin{array}{c} F \\ 0 \\ 0 \end{array}\right\} = \left[\begin{array}{ccc} 0.9841 & -1.7221 & 0.5658 \\ -1.7221 & 3.0137 + 2 \times 1.4021 & -0.9902 + 2 \times 0.5842 \\ 0.5658 & -0.9902 + 2 \times 0.5842 & 0.3254 + 2 \times 0.2435 + 2.3077c \end{array}\right] \times 10^{-3} E \left\{\begin{array}{c} u \\ \theta \\ \gamma \end{array}\right\}$$

$$= \left[\begin{array}{ccc} 0.9841 & -1.7221 & 0.5658 \\ -1.7221 & 5.8179 & 0.1782 \\ 0.5658 & 0.1782 & 0.8124 + 2.3077c \end{array}\right] \times 10^{-3} E \left\{\begin{array}{c} u \\ \theta \\ \gamma \end{array}\right\}.$$

The solution of u from the above equation is

$$u = \left(2.108 + \frac{0.737}{c}\right) \times 10^3 \frac{F}{E}.$$

To specify the effect of the joint panel on the lateral displacement of the frame, two simplified approaches neglecting the effect of joint-panel shear deformation are used to analyse the structure additionally. They are

- *approach I*: neglecting the shear deformation of joint panels and treating joint panels as rigid bodies;

- *approach II*: neglecting the size of joint panels and extending the lengths of frame beams and columns to the distances between the central lines of the frame members. For this example, let $l_g = L_g/2 = 3.00$ m and $l_c = L_c/2 = 1.75$ m.

Table 9.1   Variation of u/u_I and u/u_{II} with c

c	0.25	0.5	0.75	1	2	4	10
u/u_I	2.400	1.700	1.467	1.350	1.175	1.088	1.035
u/u_{II}	1.578	1.118	0.965	0.888	0.773	0.716	0.681

The lateral displacements by the above two simplified approaches are

$$u_I = 2.108 \times 10^3 \frac{F}{E},$$

$$u_{II} = 3.204 \times 10^3 \frac{F}{E},$$

and then

$$\frac{u}{u_I} = 1 + \frac{0.350}{c},$$

$$\frac{u}{u_{II}} = 0.658 + \frac{0.230}{c}.$$

The relationship of u/u_I and u/u_{II} with c is listed in Table 9.1, from which the following conclusions can be drawn:

(1) The value of u/u_I is always more than 1, which indicates that the neglect of joint-panel shear deformation overestimates the frame stiffness.

(2) In approach II, the extension of beam and column lengths to the distances between the central lines of the frame members underestimates frame stiffness, which partially counteracts the overestimation from the neglect of joint-panel shear deformation. The value of u/u_{II} depends on the strength of the above two factors. When the effect of the former is evident, $u/u_{II} < 1$, and when the effect of the latter is dominant, $u/u_{II} > 1$. If $u/u_{II} = 1$, these two effects kill each other.

(3) For general H-section columns, the range of c is 0.5–1.0. Obviously, good estimation can be obtained in elastic analysis of steel frames with approach II strategy if frame columns are H sections.

(4) Generally $c = 2$ for box-section columns, which underestimates frame stiffness in approach II. As approach I will overestimate frame stiffness, a trade-off method is needed to approximately analyse the frame with box-section columns. This trade-off method can be that neglect shear deformation of joint panels and treat joint panels as rigid bodies, but the half size of a real panel is assigned to the corresponding rigid body.

(5) Number of degrees of freedom reduces if joint-panel shear deformation is neglected in frame analysis. However, if yielding of joint panels is possible in elasto-plastic analysis, the effect of joint-panel shear deformation should be included, otherwise the analysis cannot catch the frame stiffness change due to panel yielding.

9.5.2   Effect of Beam–Column Connection

To examine the effect of beam-to-column connection flexibility on lateral displacement of steel frames, consider a single-storey frame with semi-rigid connections as shown in Figure 9.25, where I_g and I_c are the inertial moments of beams and columns, respectively, l_g and l_c are the lengths of beams and columns, respectively and k_e is the elastic stiffness of the beam-to-column connection. The relationship between the elastic lateral displacement of the frame and the k_e value is discussed as follows.

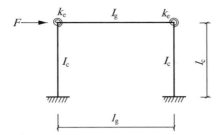

Figure 9.25   A single-storey frame with semi-rigid connections

To stand out the effect of connection flexibility, effects of axial force and shear deformation are ignored in the elemental stiffness, and axial deformations of the beams and column and shear deformation of the joint panels are also excluded. For sake of simplification, let

$$\frac{I_g}{l_g} = \frac{I_c}{l_c}, \qquad k_e = \frac{EI_c}{l_c} c, k_e = \frac{EI_c}{l_c} c,$$

where c is a non-negative factor and $c = 0$ indicates a hinged beam-to-column connection, whereas $c = \infty$ is a fixed connection. The elemental stiffness equation of the beam is

$$\left\{ \begin{array}{c} M_{g1} \\ M_{g2} \end{array} \right\} = \frac{EI_g}{l_g} \begin{bmatrix} 4 & 2 \\ 2 & 4 \end{bmatrix} \left\{ \begin{array}{c} \theta_{g1} \\ \theta_{g2} \end{array} \right\} = \frac{EI_c}{l_c} \begin{bmatrix} 4 & 2 \\ 2 & 4 \end{bmatrix} \left\{ \begin{array}{c} \theta_{g1} \\ \theta_{g2} \end{array} \right\},$$

and those of the two columns are

$$\left\{ \begin{array}{c} Q_{c1} \\ M_{c1} \end{array} \right\} = \frac{EI_c}{l_c} \begin{bmatrix} 12/l_c^2 & -6/l_c \\ -6/l_c & 4 \end{bmatrix} \left\{ \begin{array}{c} u \\ \theta_{c1} \end{array} \right\},$$

$$\left\{ \begin{array}{c} Q_{c2} \\ M_{c2} \end{array} \right\} = \frac{EI_c}{l_c} \begin{bmatrix} 12/l_c^2 & -6/l_c \\ -6/l_c & 4 \end{bmatrix} \left\{ \begin{array}{c} u \\ \theta_{c2} \end{array} \right\},$$

where M_{g1} and M_{g2} are the bending moments at the ends of the beam, Q_{c1}, M_{c1}, Q_{c2} and M_{c2} are the shears and bending moments at the top of the two columns, as shown in Figure 9.26, θ_{g1} and θ_{g2} are the rotations of the beam at the ends, and θ_{c1} and θ_{c2} are the rotations of the two columns at the top.

As the connection is flexible, inequality occurs between the rotation of the beam end and that of the adjacent column top, namely $\theta_{g1} \neq \theta_{c1}$ and $\theta_{g2} \neq \theta_{c2}$. The elemental stiffness equation of the column with connection is

$$\left\{ \begin{array}{c} M_{g1} \\ M_{g2} \end{array} \right\} = [k_{ga}] \left\{ \begin{array}{c} \theta_{g1} \\ \theta_{g2} \end{array} \right\},$$

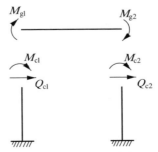

Figure 9.26   Element division and elemental forces

where

$$[k_{ga}] = [k_g] - [k_g][H]^T([H]([k_g] + [k_a])[H]^T)^{-1}[H][k_e].$$

Knowing

$$[k_g] = \frac{EI_c}{l_c}\begin{bmatrix} 4 & 2 \\ 2 & 4 \end{bmatrix}, \qquad [k_a] = \begin{bmatrix} k_e & 0 \\ 0 & k_e \end{bmatrix} = \frac{EI_c}{l}\begin{bmatrix} c & 0 \\ 0 & c \end{bmatrix}$$

and

$$[H] = \begin{bmatrix} 1 & 0 \\ 0 & 1 \end{bmatrix}$$

gives

$$[k_{ga}] = \frac{2c}{12 + 8c + c^2}\begin{bmatrix} 6 + 2c & c \\ c & 6 + 2c \end{bmatrix}.$$

The force equilibrium of the frame includes

$$M_{g1} + M_{c1} = 0,$$
$$M_{g2} + M_{c2} = 0,$$
$$Q_{c1} + Q_{c2} = F.$$

Then the global stiffness equation of the frame may be established as

$$\begin{Bmatrix} F \\ 0 \\ 0 \end{Bmatrix} = \frac{EI}{l_c}\begin{bmatrix} \dfrac{24}{l_c^2} & -\dfrac{6}{l_c} & -\dfrac{6}{l_c} \\ -\dfrac{6}{l_c} & 4 + \dfrac{2c(16 + 2c)}{12 + 8c + c^2} & \dfrac{2c^2}{12 + 8c + c^2} \\ -\dfrac{6}{l_c} & \dfrac{2c^2}{12 + 8c + c^2} & 4 + \dfrac{2c(16 + 2c)}{12 + 8c + c^2} \end{bmatrix}\begin{Bmatrix} u \\ \theta_{c1} \\ \theta_{c2} \end{Bmatrix}.$$

The solution of u from the above is

$$u = \frac{24 + 22c + 5c^2}{12(12 + 20c + 7c^2)}\frac{Fl_c^3}{EI_c}.$$

Table 9.2   Variation of u/u_0 and u/u_∞ with R

R	0	0.05	0.1	0.2	0.5	0.8	0.9	0.95	1.0
c	0	0.053	0.111	0.25	1.0	4.0	9.0	19.0	∞
u/u_0	1.000	0.963	0.926	0.855	0.654	0.471	0.413	0.385	0.357
u/u_∞	2.800	2.696	2.593	2.394	1.831	1.318	1.157	1.078	1.000

If the beam-to-column connection is hinged, i.e. $c = 0$, one has

$$u_0 = \frac{Fl_c^3}{6EI_c}.$$

If the beam-to-column connection is fixed, i.e. $c = \infty$, one has

$$u_\infty = \frac{5Fl_c^3}{84EI_c}.$$

Let

$$c = \frac{R}{1-R} \qquad \text{or} \qquad R = \frac{c}{1+c},$$

then $R = 0$ indicates a hinged connection, whereas $R = 1$ represents a fixed connection.

The relationship of u/u_0 and u/u_∞ with R is listed in Table 9.2, from which the following conclusions can be drawn:

(1) Stiffness of the beam-to-column connection significantly affects the behaviour of steel frames.

(2) When the stiffness factor $R < 0.1$, the connection acts as a hinged one, whereas when $R > 0.9$, it acts as a fixed one.

10 Brace Element and its Elastic and Elasto-Plastic Stiffness Equations

10.1 HYSTERETIC BEHAVIOUR OF BRACES

Braces are important members in steel frames to resist lateral loads. A brace is dominantly subjected to axial force and then can be represented with a truss element (see Figure 10.1). The force in braces is simple, but they are possibly buckled in compression, and elastic or elasto-plastic bending deformations take place (Figure 10.2), which makes the relationship between the axial force and the axial deformation of braces complex (Higginbotham and Hanson, 1976; Kahn and Hanson, 1976). A hysteretic curve of force versus deformation of a brace obtained by test is illustrated in Figure 10.3.

The following hysteretic characteristics of braces are found by tests:

(1) The critical load of the second time buckling of a brace is evidently lower than that of the first time buckling. And the buckling load thereafter is gradually reduced, but this trend is finally convergent.

(2) Bending deflection occurs in the brace buckled so that the axial stiffness of the buckled brace is small till it becomes straight again.

10.2 THEORETICAL ANALYSIS OF ELASTIC AND ELASTO-PLASTIC STIFFNESSES OF BRACE ELEMENT

A brace subjected to repeated and reversed loads (see Figure 10.4) can be categorized into seven states (Jain et al., 1980):

- State I: The brace is straight in elastic state (see 0–1 and 2–3 segments in Figure 10.4).

- State II: The brace buckles elastically in compression and bending deflection occurs (see 3–4 segment in Figure 10.4).

- State III: The plastic hinge is formed at the mid-span of the brace under the actions of compression and moment (see 4–5 segment in Figure 10.4).

- State IV: The brace is unloaded elastically but still in compression (see 5–6 segment in Figure 10.4).

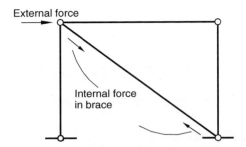

Figure 10.1 Force in a brace

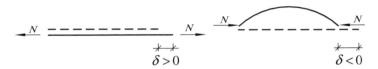

Figure 10.2 Deformation of a brace

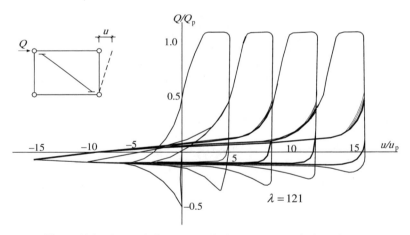

Figure 10.3 Hysteretic force versus displacement curve of a brace by test

- State V: The brace is loaded reversibly in tension and in elastic state (see 6–7 segment in Figure 10.4).

- State VI: The plastic hinge is formed at the mid-span of the brace under the actions of tension and moment (see 7–2 segment in Figure 10.4).

- State VII: The brace yields in tension and is straight (see 1–2 segment in Figure 10.4).

To obtain the axial stiffness of the brace in each of the above states, the following assumptions are made:

(1) the brace is subjected to only axial force, and the two ends are pinned;

(2) the stress–strain relationship of the brace is perfectly elasto-plastic;

(3) the brace is straight before loading and after yielding in tension;

(4) when the axial compression force of the brace reaches critical load N_{cr}, global lateral buckling occurs, but local buckling is precluded;

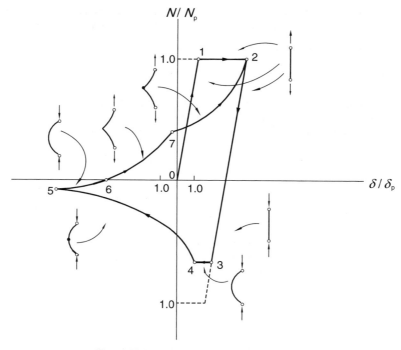

Figure 10.4   Force–deformation states of a brace

(5) if the axial force N (positive in tension) and bending moment $M = Ny_m$ (y_m is the lateral deflection at the mid-span of the brace, see Figure 10.5) of the brace satisfy the following yielding function, a concentrated plastic hinge is assumed to form at the mid-span of the brace:

$$\Gamma(M,N) = 1, \tag{10.1}$$

in which

$$\Gamma(M,N) = \left|\frac{M}{M_p}\right| + \left(\frac{N}{N_p}\right)^2, \tag{10.2}$$

where M_p is the plastic bending moment of the brace about the axis of buckling and N_p is the axial yielding load of the brace;

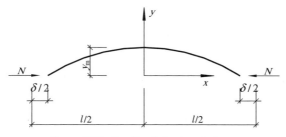

Figure 10.5   Bending deflection of a brace

(6) the rotation, stretch and contraction of the concentrated plastic hinge are assumed to be rigid plastic, and plastic axial deformation and plastic rotation of the plastic hinge are associated with plastic flow rules;

(7) the axial deformation of the brace is composed of three parts:

$$\delta = \delta_e + \delta_b + \delta_p, \tag{10.3}$$

where δ_e is the elastic axial deformation of the brace, δ_b is the axial deformation due to lateral deflection of the brace and δ_p is the plastic axial deformation of the brace.

For the sake of derivation convenience, define the following nondimensional parameters:

$$n = \frac{N}{N_p}, \quad n_{cr} = \frac{N_{cr}}{N_p}, \quad m = \frac{M}{M_p}, \quad \Delta = \frac{\delta\,EA}{l\,N_p}, \quad \xi = \frac{2x}{l},$$

$$n_e = \frac{N_e}{N_p}, \quad \eta = \frac{N_p}{M_p}y, \quad a = \frac{A}{I}\left(\frac{M_p}{N_p}\right)^2, \quad \omega = \frac{\pi}{2}\sqrt{\frac{|n|}{n_e}}, \quad \Lambda = \frac{\lambda}{\pi}\sqrt{\frac{\sigma_s}{E}}, \tag{10.4}$$

where A and I are the cross-sectional area and moment of inertia of the brace, respectively, λ is the slenderness of the brace, l is the length of the brace, E and σ_s are the elastic modulus and yielding strength of the brace material, respectively, y is the lateral deflection of the brace and N_e is the Euler critical load of the brace, given by

$$N_e = \frac{\pi^2 EI}{l^2}. \tag{10.5}$$

In view of the equilibrium of the brace after buckling, the differential equation is

$$\eta'' + \omega^2\eta = 0. \tag{10.6}$$

Introducing the boundary conditions $\eta|_{\xi=0} = \eta_m$ and $\eta|_{\xi=1} = 0$, one can solve η from the above equation as

$$\eta = \eta_m[\cos(\omega\xi) - \text{ctg}\omega\,\sin(\omega\xi)]. \tag{10.7}$$

Then the axial deformation due to the lateral deflection is

$$\Delta_b = -\frac{2a}{\pi^2\Lambda^2}\int_0^1 (\eta')^2 d\xi = \frac{-a\eta_m^2\omega^2}{\pi^2\Lambda^2}\left(\frac{1}{\sin^2\omega} + \frac{\text{ctgh}\omega}{\omega}\right). \tag{10.8}$$

When the axial force of the brace is in tension (see 6–7 and 7–2 segments in Figure 10.4), the differential equilibrium equation of the brace becomes

$$\eta'' - \omega^2\eta = 0, \tag{10.9}$$

and the contraction of the brace due to lateral deflection is

$$\Delta_b = \frac{-a\eta_m^2\omega^2}{\pi^2\Lambda^2}\left(\frac{1}{\sinh^2\omega} + \frac{\text{ctgh}\omega}{\omega}\right). \tag{10.10}$$

In view of the parameters defined in Equation (10.4), the yielding function expressed in Equation (10.2) can be rewritten as

$$\Gamma = |m| + n^2 = |\eta_m n| + n^2. \tag{10.11}$$

The axial stiffness of the brace k_b can be expressed with the stiffness parameter $\bar{k}_b$ as

$$k_b = \frac{dN}{d\delta} = \frac{EA}{l}\bar{k}_b, \tag{10.12}$$

where

$$\bar{k}_b = \frac{dn}{d\Delta} = \frac{1}{\dfrac{d\Delta_e}{dn} + \dfrac{d\Delta_b}{dn} + \dfrac{d\Delta_p}{dn}}, \tag{10.13}$$

$$\frac{d\Delta_2}{dn} \equiv 1. \tag{10.14}$$

Summarizing the above assumptions and derivations, one can obtain the stiffness parameters of the brace in different states:

- State I:

  The identifying condition is $-n_{cr} \leq n \leq 1$, $|m| = 0$.
  In this state, $\Delta_b = \Delta_p = 0$, so that

  $$\bar{k}_b = 1. \tag{10.15}$$

- State II:

  The identifying condition is $n_{cr} = n = \text{constant}$, $\Gamma < 0$, $|m| > 0$.
  Then

  $$\bar{k}_b = 0. \tag{10.16}$$

- State III:

  The identifying condition is $\Gamma = 1$, $n < 0$.
  In this state, a plastic hinge is formed at the mid-span of the brace. According to the plastic flow rule, the plastic axial deformation and rotation components of the plastic hinge, δ_p and θ_p, respectively, are

  $$\delta_p = c\frac{\partial\Gamma}{\partial N}, \quad \theta_p = c\frac{\partial\Gamma}{\partial M}, \tag{10.17}$$

where c is the proportional factor and

$$\theta_p = -2y'|_{x=0}. \tag{10.18}$$

Equation (10.17) can be rewritten as

$$\Delta_p = c\frac{EA}{N_p^2 l}\frac{\partial\Gamma}{\partial n}, \quad \varphi_p = c\frac{2N_p}{M_p^2}\frac{\partial\Gamma}{\partial m}, \tag{10.19}$$

where

$$\varphi_p = -2\eta'|_{\xi=0} = -2\eta_m\omega\,\text{ctg}\,\omega. \tag{10.20}$$

Eliminating c from Equation (10.19) yields

$$\frac{d\Delta_p}{dn} = \frac{2a}{\pi^2\Lambda^2}\frac{\partial\Gamma/\partial n}{\partial\Gamma/\partial m}\frac{d\varphi_p}{dn} = \frac{2a}{\pi^2\Lambda^2}\frac{\partial\Gamma/\partial n}{\partial\Gamma/\partial m}\left(\frac{\partial\varphi_p}{\partial\eta_m}\frac{d\eta_m}{dn} + \frac{\partial\varphi_p}{\partial\omega}\frac{d\omega}{dn}\right). \tag{10.21}$$

It can be known from Equation (10.4) that

$$\frac{d\omega}{dn} = \frac{\omega}{2n}. \tag{10.22}$$

Noting that

$$\Gamma = \eta_m |n| + n^2 = 1, \tag{10.23}$$

one has

$$\frac{d\Gamma}{dn} = |n|\frac{d\eta_m}{dn} + \text{sgn}(n)\eta_m + 2n = 0, \tag{10.24}$$

where sgn $(\cdot)$ is the signal function.

Solving $d\eta_m/dn$ from Equations (10.23) and (10.24) results in

$$\frac{d\eta_m}{dn} = -\frac{1}{|n|}\left(n + \frac{1}{n}\right). \tag{10.25}$$

From Equation (10.11), one has

$$\frac{\partial\Gamma/\partial n}{\partial\Gamma/\partial m} = \frac{2n}{\text{sgn}(m)} = \frac{2n}{\text{sgn}(n)} = 2|n|. \tag{10.26}$$

Substituting Equations (10.20), (10.22), (10.25) and (10.26) into Equation (10.21) leads to

$$\frac{d\Delta_p}{dn} = \frac{8a|n|}{\pi^2\Lambda^2}\left[\omega\,\text{ctg}\,\omega\frac{1}{|n|}\left(n + \frac{1}{n}\right) - \eta_m\left(\text{ctg}\,\omega - \frac{\omega}{\sin^2\omega}\right)\frac{\omega}{2n}\right]. \tag{10.27}$$

From Equation (10.8), one has

$$\begin{aligned}\frac{d\Delta_b}{dn} &= \frac{\partial\Delta_b}{\partial\eta_m}\frac{d\eta_m}{dn} + \frac{\partial\Delta_b}{\partial\omega}\frac{d\omega}{dn} \\ &= \frac{a\eta_m^2\omega^2}{\pi^2\Lambda^2}\left[2\left(\frac{1}{\sin^2\omega} + \frac{\text{ctgh}\omega}{\omega}\right)\frac{1}{|n|}\left(n + \frac{1}{n}\right) - \eta_m\left(\frac{\text{ctg}\omega}{\omega^2} + \frac{1}{\omega\sin^2\omega} - \frac{2\cos\omega}{\sin^3\omega}\right)\frac{\omega}{2n}\right].\end{aligned} \tag{10.28}$$

The stiffness parameter $\bar{k}_b$ can then be determined by substituting Equations (10.14), (10.27) and (10.28) into Equation (10.13).

- State IV:

The identifying condition is $\Gamma < 0$, $n \le 0$, $|m| > 0$.

This state is elastic and the plastic rotation φ_p at the end of state III remains unchanged, which yields from Equation (10.21),

$$\frac{d\Delta_p}{dn} = 0.$$

Combining Equations (10.8) and (10.20), one has

$$\Delta_b = -\frac{a\varphi_p^2}{4\pi^2\Lambda^2}\left(\frac{1}{\cos^2\omega} + \frac{\text{tg}\omega}{\omega}\right) \tag{10.29}$$

and

$$\frac{d\Delta_b}{dn} = \frac{d\Delta_b}{d\omega}\frac{d\omega}{dn} = -\frac{a\varphi_p^2}{4\pi^2\Lambda^2}\left(\frac{2\sin\omega}{\cos^3\omega} - \frac{\text{tg}\omega}{\omega^2} + \frac{1}{\omega\cos^2\omega}\right)\frac{\omega}{2n}. \tag{10.30}$$

- State V:

The identifying condition is $\Gamma < 0, \quad n \geq 0, \quad |m| > 0$.
Compared to state IV, only the axial force changes its sign in this state. Note that φ_p can be expressed as

$$\varphi_p = -2\eta'|_{\xi=0} = -2\eta_m\omega\,\text{ctgh}\omega, \tag{10.31}$$

and from Equation (10.10), one has

$$\frac{d\Delta_b}{dn} = -\frac{a\varphi_p^2}{4\pi^2\Lambda^2}\left(-\frac{2\sinh\omega}{\cosh^3\omega} - \frac{\text{tgh}\omega}{\omega^2} + \frac{1}{\omega\cosh^2\omega}\right)\frac{\omega}{2n}. \tag{10.32}$$

In this state, it remains

$$\frac{d\Delta_p}{dn} = 0.$$

- State VI:

The identifying condition is $\Gamma = 1, \quad n > 0$.
Taking a similar derivation as in state III yields

$$\frac{d\Delta_p}{dn} = \frac{8a|n|}{\pi^2\Lambda^2}\left[\omega\,\text{ctgh}\omega\frac{1}{|n|}\left(n + \frac{1}{n}\right) - \eta_m\left(\text{ctgh}\omega - \frac{\omega}{\sinh\omega}\right)\frac{\omega}{2n}\right], \tag{10.33}$$

$$\frac{d\Delta_b}{dn} = \frac{a\eta_m^2\omega^2}{\pi^2\Lambda^2}\left[2\left(\frac{1}{\sinh^2\omega} + \frac{\text{ctgh}\omega}{\omega}\right)\frac{1}{|n|}\left(n + \frac{1}{n}\right) - \eta_m\left(\frac{\text{ctgh}\omega}{\omega^2} + \frac{1}{\omega\sin^2\omega} - \frac{2\cosh\omega}{\sinh^3\omega}\right)\frac{\omega}{2n}\right]. \tag{10.34}$$

- State VII:

The identifying condition is $n = 1$.
Then

$$\bar{k}_b = 0. \tag{10.35}$$

10.3 HYSTERETIC MODEL OF ORDINARY BRACES

The hysteretic model of braces is a mathematical approach for describing the hysteretic relationship between the axial force and the axial deformation of braces. The hysteretic model can be used to determine the axial stiffness of braces subjected to repeated and reversed loading.

It is found from experimental investigations and theoretical analysis that the hysteretic behaviour of braces is very complicated. For the purpose of engineering application, the hysteretic model of braces can be simplified by reserving the principle characteristics, as shown in Figure 10.6.

In Figure 10.6, point A is the yielding strength in tension, B is the positive (in tension) unloading point, C (C') is the first critical load in compression, C'' is the subsequent critical load in compression, D (D') is the load-carrying point after the first compressive instability, E is the negative (in compression) unloading point, G is the intersection point of line BE and the horizontal line, $N = N_{cr}/2$, H is the intersection point of line BC and the horizontal line, $N = N_{cr}/2$, and F mid-point of GH.

The relation of the axial force and deformation of braces is divided into five phases in this model : (1) elastic deformation phase, AOC', BHC; (2) yielding in tension, AB; (3) instability in compression, DE; (4) disappearing of bending deformation and stretching, EF; (5) recovery to straightness in tension, FB.

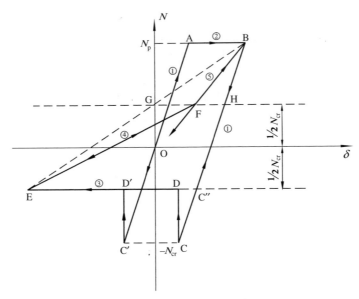

Figure 10.6   Simplified hysteretic model of ordinary braces

The hysteretic phase of braces depends on the cyclic state of the axial force and deformation of braces in the loading process. Let N be the axial force of braces and $d\delta$ the incremental of the axial deformation (stretch is positive). The identifications of the different phases of a brace are as follows:

(1) If the brace is in elastic state 1 and not subjected to first compressive instability:

   – when $-N_{cr} < N < N_p$, the brace retains elastic state (1);

   – when $N \geq N_p$, the brace enters yielding in tension (2);

   – when $N \leq N_{cr}$, the brace enters instability in compression (3).

(2) If the brace is in elastic state 1, but subjected to first compressive instability:

   – when $-\frac{1}{2}N_{cr} < N < N_p$, the brace retains elastic state (1);

   – when $N \geq N_p$, the brace enters yielding in tension (2);

   – when $N \leq -\frac{1}{2}N_{cr}$, the brace enters instability in compression (3).

(3) If the brace is in yielding in tension phase (2):

   – when $d\delta \geq 0$, the brace retains phase (2);

   – when $d\delta < 0$, the brace enters elastic state (1).

(4) If the brace is in compression instability phase (3):

   – when $d\delta \leq 0$, the brace retains phase (3);

   – when $d\delta > 0$, the brace enters stretching phase (4).

(5) If the brace is in stretching phase (4):

   – when $-\frac{1}{2}N_{cr} < N < \frac{1}{2}N_{cr}$, the brace retains phase (4);

   – when $N \geq \frac{1}{2}N_{cr}$, the brace enters recovery to straightness phase (5);

   – when $N \leq -\frac{1}{2}N_{cr}$, the brace enters instability in compression phase (3).

(6) If the brace is in recovery to straightness phase (5):

- when $-\frac{1}{2}N_{cr} < N < N_p$, the brace retains phase (5);
- when $N \geq N_p$, the brace enters yielding in tension phase (2);
- when $N \leq -\frac{1}{2}N_{cr}$, the brace enters instability in compression phase (3).

The axial stiffness of the brace in different phases is

- in phase (1):

$$k_b = \frac{EA}{l};$$ (10.36)

- in phases (2) and (3):

$$k_b = 0;$$ (10.37)

- in phase (4):

$$k_b = \frac{N_{cr}}{\delta_F - \delta_E};$$ (10.38)

- in phase (5):

$$k_b = \frac{N_p - N_{cr}/2}{\delta_B - \delta_F},$$ (10.39)

where δ_B, δ_E and δ_F are the axial deformations at points B, E and F, respectively.

10.4 HYSTERETIC CHARACTERISTICS AND MODEL OF BUCKLING-RESTRAINED BRACE

A buckling-restrained brace is based on an ordinary brace through coating a steel tube to restrain buckling of the brace to induce the strength and stiffness reduction. A free–slip interface is ensured between the inner steel brace and the outer steel tube without bonding (see Figure 10.7), or between the

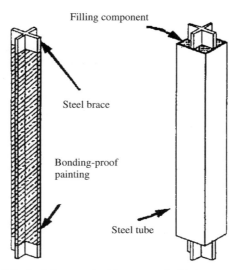

Filling component

Steel brace

Bonding-proof painting

Steel tube

Figure 10.7   Buckling-restrained brace coated with the outer steel tube

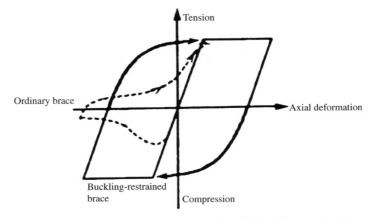

Figure 10.8   Axial force–deformation relationship of the buckling-restrained brace

inner steel brace and the outer reinforced concrete or plain concrete filled in steel tube with bonding-proof painting. Only the inner steel brace is connected with steel frame structures so that only the inner brace is loaded. The outer steel tube or concrete is to restrain the lateral deflection of the inner steel brace and to prevent it from local or global buckling. Therefore, the full strength and good ductility of the buckling-restrained brace can be achieved (Higgins and Newell, 2004; López, 2001; Wada *et al.*, 1998; Watanabe *et al.* 1988). The hysteretic behaviour of such buckling-restrained braces is clearly superior to that of ordinary braces (Figure 10.8).

The hysteretic model in Figure 10.9 can be used for buckling-restrained braces for practical convenience, where N_p is the axial yielding capacity of braces in tension and compression, k_b is the axial stiffness of braces and q is the hardening stiffness, normally $q = 0.015$. The following equations can be used to determine N_p and k_b:

$$N_P = A\sigma_s, \tag{10.40}$$

$$k_b = \frac{EA}{l}. \tag{10.41}$$

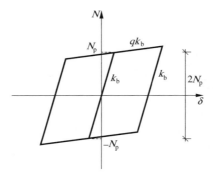

Figure 10.9   Hysteretic model of buckling-restrained bracing

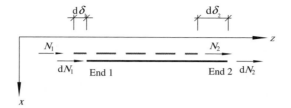

Figure 10.10   Force and deformation of the brace element

10.5 STIFFNESS EQUATION OF BRACE ELEMENT

With the stiffness k_b, the incremental stiffness equation of a brace in local coordinates as shown in Figure 10.10 can be expressed as

$$dN_1 = k_b(d\delta_1 - d\delta_2), \tag{10.42a}$$

$$dN_2 = k_b(d\delta_2 - d\delta_1). \tag{10.42b}$$

It can also be expressed in matrix form as

$$\begin{Bmatrix} dN_1 \\ dN_2 \end{Bmatrix} = k_b \begin{bmatrix} 1 & -1 \\ -1 & 1 \end{bmatrix} \begin{Bmatrix} d\delta_1 \\ d\delta_2 \end{Bmatrix}. \tag{10.43}$$

If the effect of joint panels needs to be considered in structural analysis, the displacement, rotation and shear deformation of joint panels should be included in the deformation vector, and the stiffness equation of the brace element with joint panels should be established (Figure 10.11).

By the geometric relation, the deformations of brace end 1 and end 2 due to the displacement of joint panels i and j are

$$d\delta_1 = [A_{bi}]\{d\delta_{b\lambda i}\}, \tag{10.44a}$$

$$d\delta_2 = [A_{bj}]\{d\delta_{b\lambda j}\}, \tag{10.44b}$$

in which

- for upwardly inclined brace:

$$[A_{bi}] = \left[\cos\alpha, \quad -\sin\alpha, \quad \frac{h_{gi}}{2}\cos\alpha - \frac{h_{ci}}{2}\sin\alpha, \quad \frac{h_{gi}}{4}\cos\alpha + \frac{h_{ci}}{4}\sin\alpha \right], \tag{10.45a}$$

$$[A_{bj}] = \left[\cos\alpha, \quad -\sin\alpha, \quad -\frac{h_{gj}}{2}\cos\alpha + \frac{h_{cj}}{2}\sin\alpha, \quad -\frac{h_{gj}}{4}\cos\alpha - \frac{h_{cj}}{4}\sin\alpha \right]; \tag{10.45b}$$

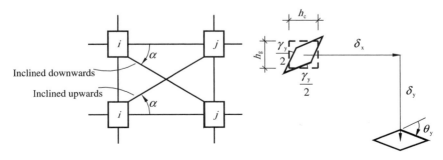

Figure 10.11   Braces in frame with joint-panel deformations

- for downwardly inclined brace:

$$[A_{bi}] = \left[\cos\alpha, \quad \sin\alpha, \quad -\frac{h_{gi}}{2}\cos\alpha + \frac{h_{ci}}{2}\sin\alpha, \quad -\frac{h_{gi}}{4}\cos\alpha - \frac{h_{ci}}{4}\sin\alpha \right], \tag{10.46a}$$

$$[A_{bj}] = \left[\cos\alpha, \quad \sin\alpha, \quad \frac{h_{gj}}{2}\cos\alpha - \frac{h_{cj}}{2}\sin\alpha, \quad \frac{h_{gj}}{4}\cos\alpha + \frac{h_{cj}}{4}\sin\alpha \right], \tag{10.46b}$$

where h_{gi}, h_{ci}, h_{gj} and h_{cj} are the height and width of joint panels i and j, respectively, and δ_{xi}, δ_{zi}, θ_{yi}, γ_{yi} and δ_{xj}, δ_{zj}, θ_{yj}, γ_{yj} are the horizontal displacement, vertical displacement, rotation and shear deformation of joint panels i and j, respectively.

The forces to joint panels i and j from the axial forces of brace end 1 and end 2 are

$$\{df_{byi}\} = [A_{bi}]^{T}dN_1, \tag{10.47a}$$

$$\{df_{byj}\} = [A_{bj}]^{T}dN_2, \tag{10.47b}$$

in which

$$\{df_{byi}\} = [dQ_{xi}, \quad dN_{zi}, \quad dM_{yi}, \quad dM_{\gamma yi}]^{T}, \tag{10.48a}$$

$$\{df_{byj}\} = [dQ_{xj}, \quad dN_{zj}, \quad dM_{yj}, \quad dM_{\gamma yj}]^{T}, \tag{10.48b}$$

where Q_{xi}, N_{zi}, M_{yi}, $M_{\gamma yi}$ and Q_{xj}, N_{zj}, M_{yj}, $M_{\gamma yj}$ are the horizontal force, vertical force, bending moment and shear moment at joint panels i and j due to the axial force of the brace.

Combining Equations (10.43), (10.44) and (10.47) leads to the incremental stiffness equation of the brace element with joint panels as

$$\{df_{by}\} = [k_{by}]\{d\delta_{by}\}, \tag{10.49}$$

in which

$$\{df_{by}\} = \left[\{df_{byi}\}^{T}, \quad \{df_{byj}\}^{T} \right]^{T}, \tag{10.50}$$

$$\{d\delta_{by}\} = \left[\{d\delta_{byi}\}^{T}, \quad \{d\delta_{byj}\}^{T} \right]^{T}, \tag{10.51}$$

$$[k_{by}] = k_b \begin{bmatrix} [A_{bi}]^{T}[A_{bi}] & -[A_{bi}]^{T}[A_{bj}] \\ -[A_{bj}]^{T}[A_{bi}] & [A_{bj}]^{T}[A_{bj}] \end{bmatrix}, \tag{10.52}$$

where $[k_{by}]$ is the stiffness matrix of the brace with joint panel.

11 Shear Beam and its Elastic and Elasto-Plastic Stiffness Equations

11.1 ECCENTRICALLY BRACED FRAME AND SHEAR BEAM

11.1.1 Eccentrically Braced Frame

Eccentrically braced frames (see Figure 1.4) were proposed based on the requirement of seismic resistance of structures. The building structures of good aseismic performance should have a balance among strength, stiffness and energy-absorption capacity. It is true that concentrically braced frames have good strength and stiffness behaviour, but their energy-absorption capacity is suspicious to satisfy aseismic requirement due to the possible buckling of braces in compression. Meanwhile, although pure steel frames have good elasto-plastic hysteretic behaviour and energy-absorption capacity, their stiffness is generally insufficient against lateral loads due to winds or earthquakes, otherwise the design with sufficient stiffness will sometimes lose economy. To satisfy the requirements of strength, stiffness and energy-absorption capacity simultaneously, eccentrically braced frames are proposed as a compromise of concentrically braced frames and pure steel frames for a better seismic resistance.

The working mechanism of eccentrically braced frames is different for different intensities of earthquakes. For small or moderate intensity of earthquakes, the contribution of the lateral stiffness of structures mainly comes from eccentric braces, which act as braces in concentrically braced frames. For large intensity of earthquakes, however, shear yielding occurs in eccentric beams to consume earthquake energy, and eccentric braces are prevented from compressive instability. Hence, eccentrically braced frames act similarly as pure frames for large intensity of earthquakes. So, the following two points should be noted in the design of eccentrically braced frames: (1) the brace should be strong enough to ensure the adjacent eccentric beam yielding taking place in advance; (2) the eccentric beam cannot be too long to ensure shear yielding rather than other failure modes to maximize the load-carrying capacity of the eccentric beam, while simultaneously keeping large lateral stiffness, and good ductility and energy-consumption capacity of the structure.

11.1.2 Condition of Shear Beam

The beam yielding mainly due to shear is termed as shear beam in this book.

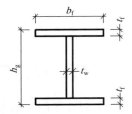

Figure 11.1   H-section beam

Generally, the beams of steel frames are of H sections (see Figure 11.1). It has three possible ways to form plastic hinges for H-sectional beams: (1) the hinge forms when the bending moment reaches the sectional plastic bending moment M_p, which is called the bending plastic hinge; (2) the hinge forms when the bending moment reaches the flange-yielding moment M_p^* ($M_p^* < M_p$) and the section of the beam is subjected to large shear force, which is called the bending-shear plastic hinge; (3) the hinge forms when the shear force reaches web-yielding shear of the section Q_p, although the bending moment is less than M_p^*, which is called the shear plastic hinge. M_p^* and Q_p can be determined as

$$M_p^* = (h_g - t_f)(b_f - t_w)t_f\sigma_s, \tag{11.1}$$

$$Q_p = (h_g - t_f)t_w\sigma_s/\sqrt{3}, \tag{11.2}$$

where σ_s is the yielding strength of the beam.

The curve of the moment–shear interaction and zones for the three types of plastic hinges are illustrated in Figure 11.2. The moment–shear interaction curve can be expressed as

$$\frac{(M - M_p^*)^2}{(M_p - M_p^*)^2} + \frac{Q}{Q_p} = 1, \quad \text{for} \quad M \geq M_p^*, \tag{11.3a}$$

$$Q \approx Q_p, \quad \text{for} \quad M < M_p^*. \tag{11.3b}$$

Obviously, the shear plastic hinge has the largest shear-carrying capacity.

The force applied on the eccentric beam segment of eccentrically brace frames is given in Figure 11.3, where the relationship of the moment and shear is

$$Ql = 2M. \tag{11.4}$$

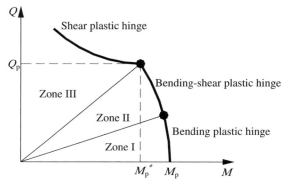

Figure 11.2   Conditions to form a plastic hinge in the beam section

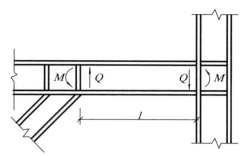

Figure 11.3 Forces in the eccentric beam

The critical length of the eccentric beam to form a shear plastic hinge can be derived from Equations (11.3b) and (11.4) as

$$l^* = \frac{2M_p^*}{Q_p}.$$

(11.5)

Substituting Equations (11.1) and (11.2) into Equation (11.5) leads to

$$l^* = \frac{3.5b_f t_f}{t_w}.$$

(11.6)

Then the condition of shear beam is

$$l \leq l^*.$$

(11.7)

11.2 HYSTERETIC MODEL OF SHEAR BEAM

Many specimens were tested to investigate the elasto-plastic hysteretic relationship between the shear and shear deformation of shear beams (Kasai and Popov, 1986; Roeder and Popov, 1978; Wyllie and Degenkolb, 1977; Yang, 1982, 1984). The cyclic test set-up and specimen of a shear beam is given in Figure 11.4(a), and the hysteretic shear versus shear deformation curves obtained are given in Figure 11.4(b). The following findings can be drawn from the test results:

(1) Shear beams have stable hysteretic performance and very good ductility. The ductility factor of shear beams (ratio of maximum shear strain to shear yielding strain) may be larger than 100.

(2) The dominant failure mode of shear beams is the local buckling of beam webs. This failure can be restrained or delayed by adding stiffeners to beam webs (see Figure 11.4(a)).

A bilinear model (see Figure 11.5) can be adopted to predict the hysteretic behaviour of a shear beam. In Figure 11.5, Q is the shear force of the beam, δ is the relative displacement at the two ends of the beam, and k_e and k_p are the elastic and elasto-plastic shear stiffnesses of the beam, respectively, given by

$$k_e = \frac{GA}{\mu l}, \quad k_p = qk_e,$$

(11.8)

where l is the length of the beam, A is the cross-sectional area, μ is the sectional shear shape factor and q is the hardening factor of the beam.

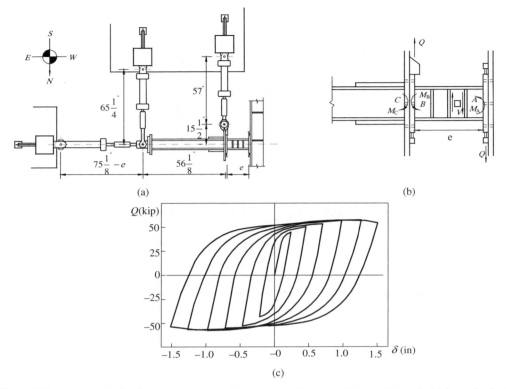

(a)   (b)

(c)

Figure 11.4 Experiment of a shear beam under cyclic loading: (a) test set-up (1 in. = 25.4 mm); (b) forces in shear beam; (c) hysteretic curves

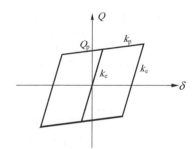

Figure 11.5 Hysteretic model of the shear beam

11.3 STIFFNESS EQUATION OF SHEAR BEAM

The stiffness equation of a shear beam can be derived from that of a general beam element involving the effect of shear deformation as (neglecting the effect of axial force)

$$\{df_g\} = [k_{gQ}] \{d\delta_g\}, \tag{11.9}$$

in which

$$\{df_g\} = [dQ_1, \quad dM_1, \quad dQ_2, \quad dM_2]^T,$$
$$\{d\delta_g\} = [d\delta_1, \quad d\theta_1, \quad d\delta_2, \quad d\theta_2]^T,$$

$$[k_{gQ}] = \frac{k}{(1+\bar{r})} \begin{bmatrix} 1 & \dfrac{l}{2} & -1 & \dfrac{l}{2} \\[2mm] \dfrac{l}{2} & (4\bar{r}+1)\dfrac{l^2}{12\bar{r}} & -\dfrac{l}{2} & (2\bar{r}-1)\dfrac{l^2}{12\bar{r}} \\[2mm] -1 & -\dfrac{l}{2} & 1 & -\dfrac{l}{2} \\[2mm] \dfrac{l}{2} & (2\bar{r}-1)\dfrac{l^2}{12\bar{r}} & -\dfrac{l}{2} & (4\bar{r}+1)\dfrac{l^2}{12\bar{r}} \end{bmatrix},$$

$$(11.10)$$

$$\bar{r} = \frac{kl^3}{12EI}, \tag{11.11}$$

$$k = k_e \quad \text{for the beam in elastic state,} \tag{11.12a}$$

or

$$k = k_p \quad \text{for the beam in elasto-plastic state,} \tag{11.12b}$$

where Q_1, M_1, Q_2, M_2, δ_1, θ_1, δ_2, are θ_1, are, respectively, the shear forces and moments at the two ends of the shear beam and the corresponding deformations.

According to the method presented in Section 7.2.1, the stiffness equation of the shear beam with joint panel (only one end of shear beam with joint panel) can be derived as

$$\{df_{g\gamma}\} = [k_{gQ\gamma}]\{d\delta_{g\gamma}\}, \tag{11.13}$$

in which

$$\{df_{g\gamma}\} = [dQ_1, \quad dM_1, \quad dQ_j, \quad dM_j, \quad dM_{\gamma j}]^T,$$
$$\{d\delta_g\} = [d\delta_1, \quad d\theta_1, \quad d\delta_j, \quad d\theta_j, \quad d\gamma_j]^T,$$
$$[k_{gQ\gamma}] = [A_{gQ}]^T [k_{gQ}][A_{gQ}], \tag{11.14}$$

$$[A_{gQ}] = \begin{bmatrix} 1 & 0 & 0 & 0 & 0 \\[1mm] 0 & 1 & 0 & 0 & 0 \\[1mm] 0 & 0 & 1 & \dfrac{h_{cj}}{2} & \dfrac{h_{cj}}{4} \\[2mm] 0 & 0 & 0 & 1 & \dfrac{1}{2} \end{bmatrix}, \tag{11.15}$$

where δ_j, θ_j, γ_j, Q_j, M_j and $M_{\gamma j}$ are, respectively, the deflection, rotation and shear deformation of the joint panel adjacent to the shear beam and the corresponding forces.

12 Elastic Stability Analysis of Planar Steel Frames

12.1 GENERAL ANALYTICAL METHOD

The elastic stability, which will be discussed in this chapter, is a classic eigenvalue (i.e. critical load) problem in structural analysis. The following conditions are required in elastic stability analysis (Figure 12.1):

(1) Only concentrated loads in the vertical direction are applied on the nodes (intersections of two or more frame members) of the steel frames considered.

(2) The frame is absolutely elastic.

(3) The axial deformation of frame columns is small and can be neglected.

(4) The frame is proportionally loaded.

The vertical concentrated loads applied to the frame can be expressed with vector $\{P\}$, and according to condition (4) it can be written as $\{P\} = \alpha\{P_0\}$ ($\{P_0\}$ is a reference load vector and α is the load factor). The global stiffness equation of the frame, an assembly of the elemental stiffness equations of all the beams, columns, bracings and joint panels of the steel frame, is

$$[k]\{D\} = \{F\}, \tag{12.1}$$

in which

$$\{D\} = \left\{ \begin{array}{c} \{u\} \\ \{\theta\} \\ \{\gamma\} \end{array} \right\}, \qquad \{F\} = \left\{ \begin{array}{c} \{F_u\} \\ \{F_\theta\} \\ \{F_\gamma\} \end{array} \right\}, \tag{12.2}$$

where $\{u\}$ is the horizontal deflection vector involving the deflections of all the floors of the frame, $\{\theta\}$ and $\{\gamma\}$ are, respectively, the rotation vector and shear deformation vector of the frame joint panels, and $\{F_u\}$, $\{F_\theta\}$ and $\{F_\gamma\}$ are the force vectors corresponding to the deformation vectors $\{u\}$, $\{\theta\}$ and $\{F_\gamma\}$, respectively. For the loading case described in Figure 12.1, $\{F\} = 0$ because $\{F_u\} = 0$, $\{F_\theta\} = 0$ and $\{F_\gamma\} = 0$.

As the elemental stiffness matrix of the frame columns depends on the axial forces in them due to the effect of geometrical nonlinearity and the axial forces are related to the vertical load $\{P\}$, the global stiffness

Advanced Analysis and Design of Steel Frames   Guo-Qiang Li and Jin-Jun Li
© 2007 John Wiley & Sons, Ltd

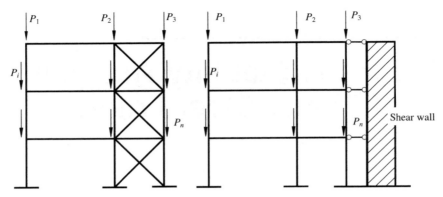

Figure 12.1 Frame subjected to vertical loads on nodes

matrix of the frame $[K]$ should be a function of $\{P\}$ and finally a function of the load factor α under the condition of proportional loading, namely

$$[k] = [k(\{P\})] = [k(\alpha\{P_0\})] = [k(\alpha)]. \tag{12.3}$$

Before the frame buckles, the global stiffness matrix of the frame is positive definite, and it can be solved from Equation (12.1) that $\{D\} = 0$. When structural instability occurs, $\{D\} \neq 0$, which means that Equation (12.1) has nonzero solution and the global stiffness matrix satisfies

$$|[k(\alpha)]| = 0. \tag{12.4}$$

Solving the above equation can yield a series of α and the minimum positive one, α_{cr}, is desired. And the critical load of the frame is

$$\{P_{\text{cr}}\} = \alpha_{\text{cr}}\{P_0\}. \tag{12.5}$$

12.2 EFFECTIVE LENGTH OF PRISMATIC FRAME COLUMN

12.2.1 Concept of Effective Length

Analytical investigations indicate that there are two possible buckling modes for steel frames under vertical loads applied at nodes: non-sway mode and sway mode (see Figure 12.2). Generally, sway instability of frames is more possible because the critical load is lower. However, if frames are restrained in the lateral direction, for example, with braces or concrete shear wall or tube (see Figure 12.1), the non-sway buckling mode will occur.

Economical and rational design of a steel frame should ensure all of the frame columns buckling simultaneously, which means that the stability of the frame is equivalent to the stability of an arbitrary column in the frame, and the frame stability analysis can then be transferred to the analysis of the elastic critical load of the arbitrary column. Express the critical load of the frame column, the axial force in the frame column when the frame buckles, as the Euler load of a column with the same section of the frame column as

$$N_{\text{cr}} = \frac{\pi^2 EI}{l_0^2}, \tag{12.6}$$

where l_0 is called as the effective length of the frame column, given by

$$l_0 = \pi\sqrt{\frac{EI}{N_{\text{cr}}}}, \tag{12.7}$$

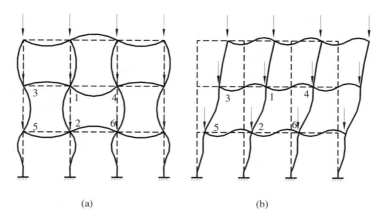

(a)   (b)

Figure 12.2   Buckling modes of steel frames: (a) non-sway buckling; (b) sway buckling

and define η as the effective length factor of the frame column, expressed by

$$\eta = \frac{l_0}{l} = \frac{\pi}{l}\sqrt{\frac{EI}{N_{\mathrm{cr}}}}, \qquad (12.8)$$

where l is the realistic length of the frame column.

A physical explanation of the effective length of a frame column is the distance between two antiflexural points of the frame column in the buckling mode, as shown in Figure 12.3.

It can be seen that if instability occurs in all of the columns in a frame, exactly or approximately simultaneously, the frame stability problem is actually that of the effective length determination of the frame columns.

12.2.2   Assumption and Analytical Model

To simplify the calculation of effective length of frame columns, the following assumptions are adopted:

(1) Ignore the size and shear deformation effect of joint panels; the lengths of frame beams and columns are treated as the distances between the intersections of the axes of beams and columns.

(2) Ignore the effect of shear deformation in frame beams and columns.

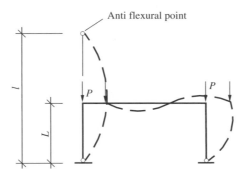

Figure 12.3   Instability of the portal frame

(3) When instability of a frame occurs, the restrain moment provided by the beam is distributed to the columns adjacent to the beam according to the linear stiffness ratio of the columns.

(4) The inter-storey sway-resistant stiffness provided by braces, shear walls and so on is distributed to the frame columns on the same storey according to the sway stiffness ratio of the columns.

(5) In non-sway instability of a frame, the rotations at the two ends of a frame beam are of the same magnitude in reverse directions, whereas in sway instability, the rotations at the two ends of the beam are not only of the same magnitude but also in the same direction.

To determine the effective length of one arbitrary column in the frame with sway-resisting stiffness, as shown in Figure 12.1, only the effect of the beams directly connected with the column is considered, whereas that of other beams is ignored. For example, in the analysis of the effective length of column 1–2 in Figure 12.2, only subframe 1–2–3–4–5–6 is considered. Furthermore, a simple model as shown in Figure 12.4 can be used to calculate the effective length of the column, where C_1 and C_2 are the rotational restrain stiffnesses to the top and bottom of the column from the adjacent beams, B is the sway-restrained stiffness to the column from the braces, shear walls and so on. Based on assumptions (3) and (4), C_1, C_2 and B can be determined by

$$B = \frac{D_c}{\sum D_c} B_b \tag{12.9}$$

- for non-sway instability:

$$C_1 = 2\frac{i_c}{\sum i_{c1}}\sum i_{1g}, \tag{12.10}$$

$$C_2 = 2\frac{i_c}{\sum i_{c2}}\sum i_{2g}; \tag{12.10}$$

- for sway instability:

$$C_1 = 6\frac{i_c}{\sum i_{c1}}\sum i_{1g}, \tag{12.10}$$

$$C_2 = 6\frac{i_c}{\sum i_{c2}}\sum i_{2g}, \tag{12.10}$$

where D_c is the sway stiffness of the column, which can be determined with the D-value method (Long and Bao, 1981; Yang, 1979), $\sum D_c$ is the sum of the sway stiffness of all the columns at the same storey as the column considered, B_b is the inter-storey sway stiffness to the column, provided by the braces or shear walls, which can be calculated as the inverse of the inter-storey drift of the braces or shear walls when a unit load is applied at the location of the top of the column considered to the braces or shear walls, i_c is the linear stiffness of the column considered, which is equal to the ratio of the inertial moment to the length of the column, $\sum i_{1c}$ and $\sum i_{2c}$ are the sum of the linear stiffnesses of the columns intersected at the top and the

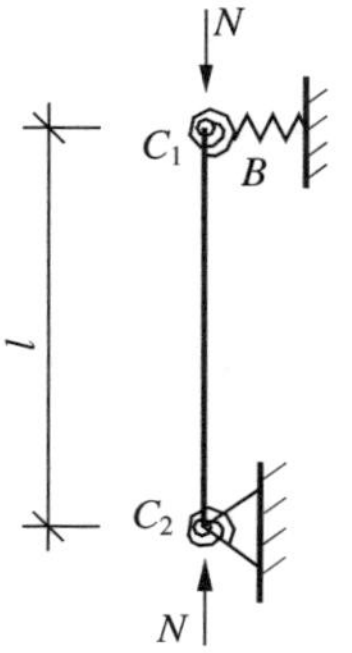

Figure 12.4   Analytical model of the effective length of the frame column

bottom of the column considered, respectively, and $\sum i_{1g}$ and $\sum i_{2g}$ are the sums of the linear stiffnesses of the beams intersected at the top and the bottom of the column considered, respectively.

12.2.3 Formulations of Effective Length

Based on the elemental stiffness equation of columns, the approach for determining the effective length of a frame column with arbitrary sway-restrain stiffness can be established. Let θ_1 and θ_2 be the rotations at the top and bottom ends of the frame column and u the relative drift between the two ends of the column; the equilibrium equation at buckling of the subframe shown in Figure 12.4 is

$$
\begin{aligned}
k_{11}\theta_1 + k_{12}\theta_2 + k_{13}u &= 0, \\
k_{21}\theta_1 + k_{22}\theta_2 + k_{23}u &= 0, \\
k_{31}\theta_1 + k_{32}\theta_2 + k_{33}u &= 0,
\end{aligned}
\tag{12.11}
$$

in which

$$
k_{11} = 4i_c\psi_3(\alpha l) + C_1,
$$

$$
k_{12} = k_{21} = 2i_c\psi_4(\alpha l),
$$

$$
k_{13} = k_{31} = \frac{6i_c}{l}\psi_2(\alpha l),
$$

$$
k_{22} = 4i_c\psi_3(\alpha l) + C_2,
$$

$$
k_{23} = k_{32} = \frac{6i_c}{l}\psi_2(\alpha l),
$$

$$
k_{33} = \frac{12i_c}{l^2}\psi_1(\alpha l) + B,
$$

where l is the length of the column and $\psi_1(\alpha l) - \psi_4(\alpha l)$ are the modification factors to stiffness of elements in compression. When the effects of shear deformation are neglected, $\psi_1(\alpha l) - \psi_4(\alpha l)$ are determined by Equation (2.44). The factor α is determined by Equations (2.10) and (12.8) as

$$
\alpha l = l\sqrt{\frac{N}{EI}} = \frac{\pi}{\eta},
\tag{12.12}
$$

where N is the axial force applied to the column and η is the effective length factor of the column.

In view of Equation (12.4), the instability condition of the subframe shown in Figure 12.4 is

$$
\begin{vmatrix}
k_{11} & k_{12} & k_{13} \\
k_{21} & k_{22} & k_{23} \\
k_{31} & k_{32} & k_{33}
\end{vmatrix} = 0.
\tag{12.13}
$$

Equation (12.13) is suitable not only to sway instability of frames, but also to non-sway instability of frames. For the case of non-sway instability, let $B = \infty$, whereas for the case of sway instability, B should be calculated with Equation (12.9). The effective length factor of the frame column can then be determined as

$$
\eta = \max(\eta_1, \quad \eta_2),
\tag{12.14}
$$

where η_1 is the effective length factor of the frame column in the non-sway instability mode and η_2 is that in the sway instability mode.

Solving η_1 and η_2 from Equation (12.13), respectively, leads to

$$\left[\left(\frac{\pi}{\eta_1}\right)^2 + 2(k_1 + k_2) - 4k_1 k_2\right]\frac{\pi}{\eta_1}\sin\frac{\pi}{\eta_1}$$
$$-2\left[(k_1 + k_2)\left(\frac{\pi}{\eta_1}\right)^2 + 4k_1 k_2\right]\cos\frac{\pi}{\eta_1} + 8k_1 k_2 = 0, \tag{12.15}$$

$$\left(\frac{\pi}{\eta_2}\right)^3\left\{6(k_1 + k_2)\frac{\pi}{\eta_2}\cos\frac{\pi}{\eta_2} + \left[36k_1 k_2 - \left(\frac{\pi}{\eta_2}\right)^2\right]\sin\frac{\pi}{\eta_2}\right\}$$
$$+ k\left\{\left[\left(\frac{\pi}{\eta_2}\right)^2 + 6(k_1 + k_2) - 36k_1 k_2\right]\frac{\pi}{\eta_2}\sin\frac{\pi}{\eta_2}\right. \tag{12.16}$$
$$\left. -6\left[(k_1 + k_2)\left(\frac{\pi}{\eta_2}\right)^2 + 12k_1 k_2\right]\cos\frac{\pi}{\eta_2} + 72k_1 k_2\right\} = 0,$$

in which

$$k_1 = \frac{\sum i_{1g}}{\sum i_{1c}}, \qquad k_2 = \frac{\sum i_{2g}}{\sum i_{2c}}, \qquad k = \frac{Bl^2}{i_c}. \tag{12.17}$$

From Equations (12.14)–(12.16), the effective length factor of the frame columns can be calculated. Let $k = 0$, $k = 1$ and $k = \infty$; the effective length factors η_0, η_1 and η_∞ for different values of k_1 and k_2 are calculated and tabulated in Tables 12.1–12.3, respectively, where η_0 is the effective length factor of the columns in pure frames under the condition of sway instability without any sway-restrain stiffness, whereas η_∞ is that in frames under the condition of non-sway instability with infinitely large sway-restrain stiffness.

When the beam-to-column connection is pinned, rigid or semi-rigid, k_1 and k_2 in Equations (12.15) and (12.16) and Tables 12.1–12.3 should be modified as follows:

(1) For the column of single-storey frames, or the column of the first storey in multi-storey frames, $k_2 = 0$ if the column base is pinned and $k_2 = \infty$ if the column base is fixed to the foundation.

(2) If the connection at the immediate end of the beam to the column considered is pinned, the linear stiffness of the beam is taken as $i_g = I_g/l_g = 0$ for calculating k_1 or k_2 in Equation (12.17).

(3) If the connection at the far end of the beam to the column considered is pinned, the linear stiffness of the beam i_g should be multiplied with the factors a as

- for non-sway instability: $a = 1.5$;

- for sway instability: $a = 0.5$.

These two factors are determined by the ratio of the moments induced by unit rotation at the immediate end of the beam. For non-sway instability (see Figure 12.5(a) and (c)), the ratio of the two moments is $a = 3/2 = 1.5$; for sway instability (see Figure 12.5(b) and (d)), the ratio of the two moments is $a = 3/6 = 0.5$.

(4) If the connection at the far end of the beam to the column considered is fixed, the linear stiffness of the beam i_g should be multiplied with the factors a as

Table 12.1 Effective length factor of columns in the sway instability frame $(k = 0)$

k_2	k_1														
	0	0.05	0.1	0.2	0.3	0.4	0.5	1	2	3	4	5	10	20	∞
0	∞	6.02	4.46	3.42	3.01	2.78	2.64	2.33	2.17	2.11	2.08	2.07	2.03	2.02	2.00
0.05	6.02	4.16	3.47	2.86	2.58	2.42	2.31	2.07	1.94	1.90	1.87	1.86	1.83	1.82	1.80
0.1	4.46	3.47	3.01	2.56	2.33	2.20	2.11	1.90	1.79	1.75	1.73	1.72	1.70	1.63	1.67
0.2	3.42	2.86	2.56	2.23	2.05	1.94	1.87	1.70	1.60	1.57	1.55	1.54	1.52	1.51	1.50
0.3	3.01	2.58	2.33	2.05	1.90	1.80	1.74	1.58	1.49	1.46	1.45	1.44	1.42	1.41	1.40
0.4	2.78	2.42	2.20	1.94	1.80	1.71	1.65	1.50	1.42	1.39	1.37	1.37	1.35	1.34	1.33
0.5	2.64	2.31	2.11	1.87	1.74	1.65	1.59	1.45	1.37	1.34	1.32	1.32	1.30	1.29	1.28
1	2.33	2.07	1.90	1.70	1.58	1.50	1.45	1.32	1.24	1.21	1.20	1.19	1.17	1.17	1.16
2	2.17	1.94	1.79	1.60	1.49	1.42	1.37	1.24	1.16	1.14	1.12	1.12	1.10	1.09	1.08
3	2.11	1.90	1.75	1.57	1.46	1.39	1.34	1.21	1.14	1.11	1.10	1.09	1.07	1.06	1.06
4	2.08	1.87	1.73	1.55	1.45	1.37	1.32	1.20	1.12	1.10	1.08	1.08	1.06	1.05	1.04
5	2.07	1.86	1.72	1.54	1.44	1.37	1.32	1.19	1.12	1.09	1.08	1.07	1.05	1.04	1.03
10	2.03	1.83	1.70	1.52	1.42	1.35	1.30	1.17	1.10	1.07	1.06	1.05	1.03	1.03	1.02
20	2.02	1.82	1.68	1.51	1.41	1.34	1.29	1.17	1.09	1.06	1.05	1.04	1.03	1.02	1.01
∞	2.00	1.80	1.67	1.50	1.40	1.33	1.28	1.16	1.08	1.06	1.04	1.03	1.02	1.01	1.00

Table 12.2 Effective length factor of columns in a frame with partial sway restrain ($k = 1$)

k_2	k_1														
	0	0.05	0.1	0.2	0.3	0.4	0.5	1	2	3	4	5	10	20	∞
0	3.142	2.767	2.573	2.327	2.190	2.104	2.044	1.905	1.825	1.796	1.781	1.773	1.755	1.746	1.737
0.05	2.767	2.507	2.331	2.122	2.003	1.928	1.875	1.752	1.679	1.654	1.641	1.633	1.617	1.609	1.601
0.1	2.573	2.331	2.174	1.986	1.878	1.808	1.760	1.645	1.577	1.553	1.541	1.534	1.519	1.511	1.504
0.2	2.327	2.122	1.986	1.819	1.721	1.657	1.612	1.505	1.443	1.421	1.409	1.403	1.389	1.382	1.375
0.3	2.190	2.003	1.878	1.721	1.627	1.566	1.523	1.420	1.359	1.337	1.326	1.320	1.306	1.300	1.293
0.4	2.104	1.928	1.808	1.657	1.566	1.506	1.464	1.362	1.302	1.281	1.270	1.263	1.250	1.243	1.237
0.5	2.044	1.875	1.760	1.612	1.523	1.464	1.422	1.321	1.261	1.240	1.229	1.222	1.209	1.203	1.196
1	1.905	1.752	1.645	1.505	1.420	1.362	1.321	1.220	1.160	1.138	1.127	1.121	1.107	1.101	1.094
2	1.825	1.679	1.577	1.443	1.359	1.302	1.261	1.160	1.098	1.076	1.065	1.058	1.045	1.038	1.031
3	1.796	1.654	1.553	1.421	1.337	1.281	1.240	1.138	1.076	1.054	1.043	1.036	1.022	1.015	1.008
4	1.781	1.641	1.541	1.409	1.326	1.270	1.229	1.127	1.065	1.043	1.031	1.024	1.011	1.004	0.997
5	1.773	1.633	1.534	1.403	1.320	1.263	1.222	1.121	1.058	1.036	1.024	1.017	1.004	0.997	0.990
10	1.755	1.617	1.519	1.389	1.306	1.250	1.209	1.107	1.045	1.022	1.011	1.004	0.990	0.983	0.976
20	1.746	1.609	1.511	1.382	1.300	1.243	1.203	1.101	1.038	1.015	1.004	0.997	0.983	0.976	0.969
∞	1.737	1.601	1.504	1.375	1.293	1.237	1.196	1.094	1.031	1.008	0.997	0.990	0.976	0.969	0.962

Table 12.3 Effective length factor of columns in the non-sway instability frame ($k = \infty$)

k_2	k_1														
	0	0.05	0.1	0.2	0.3	0.4	0.5	1	2	3	4	5	10	20	∞
0	1.000	0.990	0.981	0.964	0.949	0.935	0.922	0.875	0.820	0.791	0.773	0.760	0.732	0.716	0.699
0.05	0.990	0.981	0.971	0.955	0.940	0.926	0.914	0.867	0.814	0.784	0.766	0.754	0.726	0.711	0.694
0.1	0.981	0.971	0.962	0.946	0.931	0.918	0.908	0.860	0.0.807	0.778	0.760	0.748	0.721	0.705	0.689
0.2	0.964	0.955	0.946	0.930	0.916	0.903	0.891	0.846	0.795	0.767	0.749	0.737	0.711	0.696	0.679
0.3	0.949	0.940	0.931	0.916	0.902	0.889	0.878	0.834	0.784	0.756	0.739	0.728	0.701	0.687	0.671
0.4	0.935	0.926	0.918	0.903	0.889	0.877	0.866	0.823	0.774	0.747	0.730	0.719	0.693	0.678	0.663
0.5	0.922	0.914	0.906	0.891	0.878	0.866	0.855	0.813	0.765	0.738	0.721	0.710	0.685	0.671	0.656
1	0.875	0.867	0.860	0.846	0.834	0.823	0.813	0.774	0.729	0.704	0.688	0.677	0.654	0.640	0.626
2	0.820	0.814	0.807	0.795	0.784	0.774	0.765	0.729	0.686	0.663	0.648	0.638	0.615	0.603	0.590
3	0.791	0.784	0.778	0.767	0.756	0.747	0.738	0.704	0.633	0.640	0.625	0.616	0.593	0.581	0.568
4	0.773	0.766	0.760	0.749	0.739	0.730	0.721	0.688	0.648	0.625	0.611	0.601	0.580	0.568	0.555
5	0.760	0.754	0.748	0.737	0.728	0.719	0.710	0.677	0.638	0.616	0.601	0.592	0.570	0.558	0.546
10	0.732	0.726	0.721	0.711	0.701	0.693	0.685	0.654	0.615	0.593	0.580	0.570	0.549	0.537	0.524
20	0.716	0.711	0.705	0.696	0.687	0.678	0.671	0.640	0.603	0.581	0.568	0.553	0.537	0.525	0.512
∞	0.699	0.694	0.689	0.679	0.671	0.663	0.656	0.626	0.590	0.568	0.555	0.546	0.524	0.512	0.500

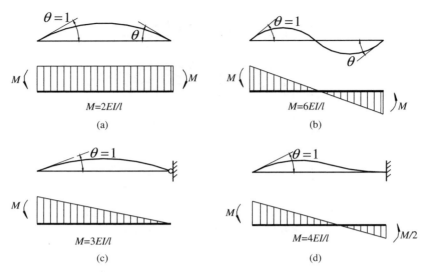

Figure 12.5   Moment of the beam with unit end rotation

- for non-sway instability: $a = 2$;
- for sway instability: $a = 2/3$.

These two factors are obtained by the ratio of the moments in Figure 12.5(d) to that in Figure 12.5(a) and to that Figure 12.5(b), respectively.

(5) If the connections at both ends of the beam are semi-rigid, i_g should be multiplied with the factor a as

- for non-sway instability: $a = 2b/(2 + b)$;
- for sway instability: $a = 6b/(6 + b)$,

where b is the factor of the rotational stiffness of the semi-rigid connections, k_e, i.e. $k_e = bi_g$ or $b = k_e/i_g$.

The calculation of a in this case is obtained according to the total rotational stiffness of the beam at both ends with semi-rigid connections depending on the serial relation of the rotational stiffness of the beam itself and that of the semi-rigid connections.

12.2.4   Simplified Formula of Effective Length

Equations (12.15) and (12.16) are transcendental equations, and it is not convenient to calculate the effective length factors of frame columns with them. Three special solutions of Equations (12.15) and (12.16) when $k = 0, k = 1$ and $k = \infty$ for the effective length of columns varying with k_1 and k_2 are listed in Tables 12.1– 12.3. For practical purpose, the values of the effective length of columns in Tables 12.1–12.3 can be fitted by the formula with variables k_1 and k_2.

It can be found by checking Equation (12.16) that if $k = Bl^2/i_c = 60$, namely $B = 60i_c/l^2$, the effective length factor of the column in the laterally restrained frame is nearly equal to that of the column in the non-sway instability frame. It can therefore be justified that the instability of the frame column with $B = 60i_c/l^2$ belongs to the non-sway instability type. Because the sway-restrain stiffness of the column itself is $S_c = 12i_c/l^2$, so when the external sway-restrained stiffness to the column is five times the sway-restrained stiffness of the column itself, instability of the frame column is generally non-sway.

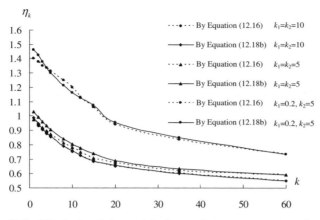

Figure 12.6   Effective length factor of the frame column versus sway-restrain stiffness

Through statistical study, the effective length factor of frame columns with arbitrary sway-restrain stiffness can be calculated with the following formula:

$$\eta_0 = \sqrt{\frac{1.6 + 4(k_1 + k_2) + 7.5k_1k_2}{k_1 + k_2 + 7.5k_1k_2}}, \qquad \text{for} \quad k = 0, \tag{12.18a}$$

$$\eta_k = \frac{\eta_0}{\sqrt{1 + \left(\frac{\eta_0^2}{\eta_\infty^2} - 1\right)\left(\frac{k}{60}\right)^{0.5}}}, \qquad \text{for} \quad 0 \le k \le 60, \tag{12.18b}$$

$$\eta_\infty = \frac{3 + 1.4(k_1 + k_2) + 0.64k_1k_2}{3 + 2(k_1 + k_2) + 1.28k_1k_2}, \qquad \text{for} \quad k > 60, \tag{12.18c}$$

where η_k is the effective length factor of the frame column with arbitrary sway-restrain stiffness, η_0 is that of the frame column with none of the restrain stiffness in sway instability and η_∞ is that of the frame column in non-sway instability.

The values of the effective length factor for frame columns with sway restrain obtained with Equation (12.16) through numerical analysis and with Equation (12.18b) are compared in Figure 12.6. Good agreement can be seen in that comparison.

12.2.5   Modification of Effective Length

The premise of using effective length of columns to the elastic stability analysis of a steel frame is that the instabilities of all of the frame columns occur simultaneously. Otherwise, the critical load by the effective length method will not agree the realistic one, and the effective length should be modified.

12.2.5.1   Sway instability frame

Analysis indicates that the total load supported by the columns on one storey of a frame does not vary much with various load distributions on the frame with sway instability, by which the effective length of the columns on the same storey can be modified.

Assume the forces and inertial moments of the columns on a certain storey of the frame are N_i and I_i ($i = 1, 2, \ldots, m$, where m is the sum of the columns in the storey), respectively, and the effective lengths and then the critical loads of these columns can be calculated according to the methods in Section 12.2.3 or 12.2.4 as

$$N_{\text{cri}} = \frac{\pi^2 EI_i}{(\eta_i l)^2}, \qquad i = 1, 2, \ldots, m, \tag{12.19}$$

where l is the height of the storey.

If

$$\frac{N_1}{N_{cr1}} = \frac{N_2}{N_{cr2}} = \cdots = \frac{N_m}{N_{crm}}, \tag{12.20}$$

the instabilities of all of the frame columns occur simultaneously, and thus the effective length in Equation (12.19) need not be modified.

If Equation (12.20) is not satisfied, modification of the effective length should be conducted. For this, let

$$\sum_{j=1}^{m} N'_{crj} = \sum_{j=1}^{m} (\alpha N_j) = \alpha \sum_{j=1}^{m} (N_j) = \sum_{j=1}^{m} N_{crj}. \tag{12.21}$$

Then, the critical load modified is

$$N'_{cri} = \alpha N_i = \frac{N_i}{\sum\limits_{j=1}^{m} (N_j)} \sum_{j=1}^{m} N_{crj}. \tag{12.22}$$

So, the modified effective length factor of the frame column becomes

$$\eta'_i = \frac{\pi}{l} \sqrt{\frac{EI_i}{N'_{cri}}}. \tag{12.23}$$

Combining Equations (12.19), (12.22) and (12.23), one has

$$\eta'_i = \sqrt{\frac{\sum\limits_{j=1}^{m} (N_j)}{N_i} \cdot \frac{I_i}{\sum\limits_{j=1}^{m} \left(\dfrac{I_j}{\eta_j^2}\right)}}. \tag{12.24}$$

12.2.5.2 Non-sway instability frame

Analysis also indicates that the total load supported by the columns in the same vertical line of a frame does not change with various load distributions on the frame with non-sway instability, by which the effective length of the columns at different storeys but in the same vertical line can be modified.

Assume the forces and inertial moments of the columns in the same vertical line of the frame are, respectively, N_i and I_i ($i = 1, 2, \ldots, n$, where n is the total storey number of the frame in the vertical line). The critical load represented by the effective length without modification is

$$N_{cri} = \frac{\pi^2 EI_i}{(\eta_i l_i)^2}, \qquad i = 1, 2, \ldots, n. \tag{12.25}$$

With the similar derivation as that for the sway instability frame, the modified effective length factor for the non-sway instability frame is obtained with

$$\eta'_i = \frac{1}{l_i} \sqrt{\frac{\sum\limits_{j=1}^{n} (N_j)}{N_j} \cdot \frac{I_i}{\sum\limits_{j=1}^{n} \left(\dfrac{I_j}{\eta_j^2 l_j^2}\right)}}. \tag{12.26}$$

12.2.6 Effect of Shear Deformation on Effective Length of Column

The effects of shear deformation of frame beams and columns are neglected in the above derivations. When the slenderness of the beams and the columns is small, it is necessary to include the shear deformation effect in the stability analysis of frames. In this section, a simplified method will be introduced.

The calculation of the effective length of the column remains as described above, but the linear stiffness i_g of the beam adjacent to the column is modified to involve the shear deformation effects as

- for non-sway instability:

$$i'_g = i_g; \tag{12.27a}$$

- for sway instability:

$$i'_g = \frac{1}{1+r} i_g, \tag{12.27b}$$

where i'_g is the modified linear stiffness of the beam with shear deformation and r is the impact factor of shear deformation, given by

$$r = \frac{12\mu_b EI_b}{GA_b l_b^2}, \tag{12.28}$$

where E and G are the stretch and shear elastic modulus, respectively, A_b and I_b are the sectional area and inertial moment of the beam, respectively, l_b is the length of the beam and μ_b is the shear shape factor of the beam as illustrated in Figure 2.3.

After the determination of the effective length factor η of the column involving shear deformation effects of all relevant beams, η can be modified further to involve the shear deformation effect of the column itself as

$$\eta' = \eta\sqrt{1 + \frac{\pi^2 \mu_c EI_c}{GA_c (\eta l_c)^2}}, \tag{12.29}$$

where η' is the modified effective length factor of the column with shear deformation, A_c and I_c are the sectional area and inertial moment of the column, respectively, l_c is the height of the column and μ_c is the shear shape factor of the column.

It should be noted that $I_b/A_b l_b^2$ and $I_c/A_c(\eta l_c)^2$ in Equations (12.28) and (12.29) are, respectively, the inverse of square of slenderness of beams and columns, which indicates that the smaller the slenderness of beams and columns, the larger the effects of shear deformation on the effective length of columns. Analytical investigations demonstrate that when the slenderness of frame beams and columns is larger than 35, the shear deformation effect can be ignored.

12.2.7 Examples

12.2.7.1 Effect of sway-restrain stiffness on effective length of columns

To examine the effect of sway-restrain stiffness on the effective length of frame columns, a symmetric single-storey frame as shown in Figure 12.7 is investigated using the general analysis method. In Figure 12.7, i represents the linear stiffness of the frame columns, and β and k are parameters with arbitrary values.

There are three degree of freedoms in this frame, namely two rotations at the beam ends, θ_1 and θ_2, and the lateral storey drift of the frame, u. As axial forces are applied to the frame columns, the stiffness equations of the two columns are established as beam elements in compression as

$$Ei \begin{bmatrix} \dfrac{12}{l^2}\psi_1(\alpha l) & -\dfrac{6}{l}\psi_2(\alpha l) \\ -\dfrac{6}{l}\psi_2(\alpha l) & 4\psi_3(\alpha l) \end{bmatrix} \begin{Bmatrix} u \\ \theta_1 \end{Bmatrix} = \begin{Bmatrix} Q_{c1} \\ M_{c1} \end{Bmatrix},$$

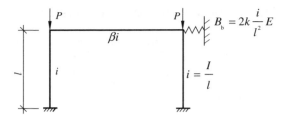

Figure 12.7   A symmetric single-storey frame

$$
Ei\begin{bmatrix} \dfrac{12}{l^2}\psi_1(\alpha l) & -\dfrac{6}{l}\psi_2(\alpha l) \\[2ex] -\dfrac{6}{l}\psi_2(\alpha l) & 4\psi_3(\alpha l) \end{bmatrix}\begin{Bmatrix} u \\ \theta_2 \end{Bmatrix} = \begin{Bmatrix} Q_{c2} \\ M_{c2} \end{Bmatrix},
$$

and the stiffness equation of the frame beam without the axial force is

$$
E\beta i\begin{bmatrix} 4 & 2 \\ 2 & 4 \end{bmatrix}\begin{Bmatrix} \theta_1 \\ \theta_2 \end{Bmatrix} = \begin{Bmatrix} M_{g1} \\ M_{g2} \end{Bmatrix},
$$

where Q_{c1}, M_{c1} and Q_{c2}, M_{c2} are the shears and bending moments at the top of the two columns, respectively, and M_{g1} and M_{g2} are the bending moments at the two beam ends.

The equilibrium equations of the frame include

$$
M_{c1} + M_{g1} = 0,
$$
$$
M_{c2} + M_{g2} = 0,
$$
$$
Q_{c1} + Q_{c2} = -B_b u = -2k\frac{i}{l^2}Eu.
$$

So, the global stiffness equation of the frame is obtained as

$$
Ei\begin{bmatrix} \dfrac{24}{l^2}\psi_1(\alpha l) + \dfrac{2k}{l^2} & -\dfrac{6}{l}\psi_2(\alpha l) & -\dfrac{6}{l}\psi_2(\alpha l) \\[2ex] -\dfrac{6}{l}\psi_2(\alpha l) & 4\psi_3(\alpha l) + 4\beta & 2\beta \\[2ex] -\dfrac{6}{l}\psi_2(\alpha l) & 2\beta & 4\psi_3(\alpha l) + 4\beta \end{bmatrix}\begin{Bmatrix} u \\ \theta_1 \\ \theta_2 \end{Bmatrix} = \begin{Bmatrix} 0 \\ 0 \\ 0 \end{Bmatrix}
$$

and the instability condition of the frame is

$$
\begin{vmatrix} \dfrac{24}{l^2}\psi_1(\alpha l) + \dfrac{2k}{l^2} & -\dfrac{6}{l}\psi_2(\alpha l) & -\dfrac{6}{l}\psi_2(\alpha l) \\[2ex] -\dfrac{6}{l}\psi_2(\alpha l) & 4\psi_3(\alpha l) + 4\beta & 2\beta \\[2ex] -\dfrac{6}{l}\psi_2(\alpha l) & 2\beta & 4\psi_3(\alpha l) + 4\beta \end{vmatrix} = 0.
$$

Substituting Equation (2.44) into the above equation one can obtain the positive minimum root of (αl), denoted as $(\alpha l)_{cr}$. And the effective length factor of the frame columns according to Equation (12.12) is obtained as

$$
\eta = \frac{\pi}{(\alpha l)_{cr}}.
$$

The ratio, a, of the sway-restrain stiffness B_b to the frame lateral stiffness and the effective length factor η of the frame column with various values of β and k are listed in Table 12.4, from which one can find that

Table 12.4   Variation of η and a with β and k

k	$\beta = 0.4$		$\beta = 1.0$		$\beta = 5.0$		$\beta = \infty$	
	a	η	a	η	a	η	a	η
0	0	1.329	0	1.157	0	1.033	0	1.000
0.5	0.098	1.281	0.073	1.124	0.049	1.011	0.042	0.980
1.0	0.197	0.237	0.146	1.094	0.099	0.990	0.083	0.962
2.0	0.394	1.162	0.292	1.041	0.198	0.951	0.167	0.927
3.0	0.591	1.099	0.438	0.995	0.297	0.917	0.250	0.896
5.0	0.985	1.000	0.729	0.929	0.495	0.859	0.417	0.843
8.0	1.576	0.893	1.167	0.832	0.792	0.790	0.667	0.779
10	1.970	0.840	1.458	0.787	0.990	0.752	0.833	0.744
20	3.939	0.683	2.917	0.641	1.979	0.624	1.667	0.621
30	5.909	0.663	4.375	0.626	2.969	0.548	2.500	0.548
50	9.848	0.663	7.292	0.626	4.948	0.546	4.167	0.500
∞	∞	0.663	∞	0.626	∞	0.546	∞	0.500

(1) the sway-restrain stiffness of frames affects the effective length of frame columns significantly;

(2) when ratio a of the sway-restrain stiffness to the corresponding frame lateral stiffness is more than 5, the effective length of the frame column is equal to that with $a = \infty$, which indicates that if $a \geq 5$, the frame buckles generally in the non-sway mode.

12.2.7.2   Calculation of effective length of frame columns

To demonstrate the determination of the effective length of frame columns, study a symmetric two-storey frame as shown in Figure 12.8.

In this frame, $k_1 = 3.0$ and $k_2 = 1.0$ for upper storey columns, and $k_1 = 1.0$ and $k_2 = \infty$ for lower storey columns. The ratio of the axial force applied to the inertial moment of all the frame columns is P/I, which implies that the critical loads of the upper storey columns are less than those of the lower storey ones. Evidently, the effective lengths of all the frame columns are of the same value, and the instability will occur

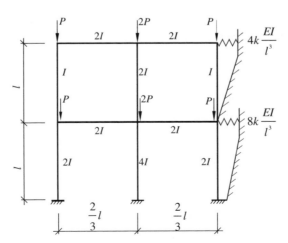

Figure 12.8   A symmetric two-storey frame

simultaneously to all the upper storey columns of the frame. Checking with Tables 12.1 and 12.3, the effective length factors of the frame columns for $k = 0$ and $k = \infty$ are

$$\eta_0 = 1.210, \qquad \eta_\infty = 0.704.$$

Or, the effective length factors can be obtained with Equations (12.18a) and (12.18c) as

$$\eta_0 = \sqrt{\frac{1.6 + 4 \times (3 + 1) + 7.5 \times 3 \times 1}{3 + 1 + 7.5 \times 3 \times 1}} = 1.230,$$

$$\eta_\infty = \frac{3 + 1.4 \times (3 + 1) + 0.64 \times 3 \times 1}{3 + 2 \times (3 + 1) + 1.28 \times 3 \times 1} = 0.709.$$

Good agreement between the table-listed and the formulated results indicates that the simplified formulae have satisfactory accuracy.

The sway-restrain stiffness factor of the upper storey column is k, and when $k = 1.0$, the effective length factor of the frame column can be calculated by Equation (12.18b) as

$$\eta_1 = \frac{1.21}{\sqrt{1 + \left(\dfrac{1.21^2}{0.704^2} - 1\right)\left(\dfrac{1}{60}\right)^{0.5}}} = 1.087.$$

When $k = 3.6$, the effective length factor of the column becomes

$$\eta_{3.6} = \frac{1.21}{\sqrt{1 + \left(\dfrac{1.21^2}{0.704^2} - 1\right)\left(\dfrac{3.6}{60}\right)^{0.5}}} = 0.995.$$

If the exact method, i.e. the general analysis method presented in Section 12.1, is used to analyse this frame, the effective length factors of the columns for $k = 1.0$ and $k = 3.6$ are, respectively, taken as

$$\eta_1 = 1.156, \qquad \eta_{3.6} = 0.964.$$

The relative errors between the above two methods are -6.5 and $3.2\,\%$, which indicates a good accuracy of the formulae for calculating the effective length of frame columns.

12.2.7.3 Modification of effective length of frame columns

Example 1

Figure 12.9 gives a single-storey frame with hinged column bases, the column parameters of which are

- for column A:

$$k_1 = \frac{30/2}{1} = 15 \quad \text{and} \quad k_2 = 0;$$

- for column B:

$$k_1 = \frac{\dfrac{30}{2} + \dfrac{30 \times 3}{10}}{4.5} = 5.333 \quad \text{and} \quad k_2 = 0;$$

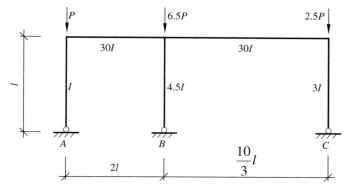

Figure 12.9 A single-storey frame

- for column C:

$$k_1 = \frac{30 \times 3}{10}/3 = 3 \quad \text{and} \quad k_2 = 0.$$

Checking with Table 12.1, the effective length factors of the above columns are

$$\eta_A = 2.04, \ \eta_B = 2.07, \ \eta_C = 2.13,$$

and the elastic critical loads of the columns are

- for column A, $N_A = \pi EI/(2.04l)^2 = 0.240 N_E$;
- for column B, $N_B = \pi E(4.5I)/(2.07l)^2 = 1.050 N_E$;
- for column C, $N_A = \pi E(3I)/(2.13l)^2 = 0.661 N_E$.

Although the actual axial forces applied on the columns are

$$P_A = P, \ P_B = 6.5P, \ P_C = 2.5P,$$

obviously

$$\frac{P_A}{N_A} \neq \frac{P_B}{N_B} \neq \frac{P_C}{N_C}.$$

As the precondition that instability occurs to all the columns of the frame simultaneously is not satisfied in this case, Equation (12.24) should be used to modify the effective length factor of the columns. The modified values are

$$\eta'_A = \sqrt{\frac{10P}{P} \cdot \frac{I}{\dfrac{I}{2.04^2} + \dfrac{4.5I}{2.07^2} + \dfrac{3I}{2.12^2}}} = 2.26,$$

$$\eta'_B = \sqrt{\frac{10P}{6.5P} \cdot \frac{4.5I}{\dfrac{I}{2.04^2} + \dfrac{4.5I}{2.07^2} + \dfrac{3I}{2.12^2}}} = 1.88,$$

$$\eta'_C = \sqrt{\frac{10P}{2.5P} \cdot \frac{3I}{\dfrac{I}{2.04^2} + \dfrac{4.5I}{2.07^2} + \dfrac{3I}{2.12^2}}} = 2.48.$$

The values modified coincide with those obtained by the exact analysis method.

Example 2

Figure 12.10 is a symmetric two-storey non-sway frame, the column information of which is as follows:

- for upper columns:

$$k_1 = 1, \, k_2 = 0.5 \text{ and then } \eta = 0.813;$$

- for lower columns:

$$k_1 = 0.5, \, k_2 = \infty \text{ and then } \eta = 0.656.$$

So, the critical load capacities of the frame columns are

- for upper columns: $N_T = \pi EI/(0.813l)^2 = 1.51N_E$;
- for lower columns: $N_B = \pi EI/(0.656l)^2 = 2.32N_E$.

The actual axial forces applied on the upper and lower columns are, respectively,

$$P_T = P, \, P_B = 2P.$$

Obviously,

$$\frac{P_T}{N_T} \neq \frac{P_B}{N_B}.$$

Then, the modification of effective length factors is necessary. For such purpose, let

$$P_T + P_B = N_T + N_B.$$

The critical value of P when instability of the frame occurs is then obtained as

$$P = \frac{1.51 + 2.32}{1 + 2} N_E = 1.28N_E.$$

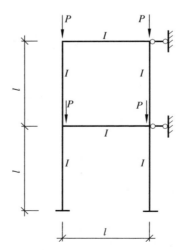

Figure 12.10   A symmetric two-storey non-sway frame

And the effective length factors modified are

- for upper columns: $\eta' = \sqrt{1/1.28} = 0.884$;
- for lower columns: $\eta' = \sqrt{1/(2 \times 1.28)} = 0.628$.

The values of the effective length factors modified above can also be calculated with Equation (12.26), which coincide with those obtained by the exact analysis method.

12.3 EFFECTIVE LENGTH OF TAPERED STEEL COLUMNS

12.3.1 Tapered Columns Under Different Boundary Conditions

Examine four types of common boundary conditions, as shown in Figure 12.11, for tapered columns. Define taper ratio as $\alpha = (d_1/d_2) - 1$, where d_1 and d_2 are the cross-sectional height at large and small ends of tapered columns, respectively. After obtaining the elastic critical load of a tapered column with a definite value of taper ratio, P_{cr}, one can obtain the effective length factor of the column with the reference of the parameter of the column at the small end as

$$\eta_\alpha = \sqrt{\frac{P_{E2}}{P_{cr}}} = \frac{\pi}{L}\sqrt{\frac{EI_2}{P_{cr}}}, \tag{12.30}$$

where I_2 is the inertial moment of the small end of the column, E is the elastic modulus and L is the length of the column.

If the effect of shear deformation is excluded, the effective length factor of tapered columns can be determined with the methods presented in Sections 12.2.2 and 12.2.3, which can be fitted by

$$\eta_\alpha = c + (K_0 - c) \cdot \exp\left(-\frac{\alpha}{1.56}\right), \tag{12.31}$$

where K_0 is the effective length factor of the corresponding prismatic columns with the same boundary conditions and c is a constant relevant to the boundary conditions. The values of K_0 and c are

$$K_0 = 2.0, \ c = 0.3800 \text{ for boundary condition a,}$$

$$K_0 = 1.0, \ c = 0.3155 \text{ for boundary condition b,}$$

$$K_0 = 0.7, \ c = 0.2218 \text{ for boundary condition c,}$$

$$K_0 = 0.5, \ c = 0.1656 \text{ for boundary condition d.}$$

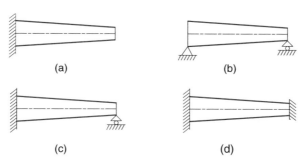

(a)   (b)

(c)   (d)

Figure 12.11   Four boundary conditions of tapered columns: (a) large end fixed and small end free; (b) both ends hinged; (c) large end fixed and small end hinged; (d) both ends fixed

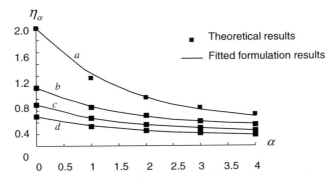

Figure 12.12   Effective length factor of tapered columns versus taper ratio under four boundary conditions (neglecting shear deformation effects)

The comparison of the values of η_α obtained with the exact analytical method and with Equation (12.31) is given in Figure 12.12.

Generally, the effect of shear deformation on column stiffness relates to the column slenderness. The slenderness of tapered columns can be defined using small end parameters as $\lambda = L/\sqrt{I_2/A_2}$. When the value of λ is small, the effect of shear deformation is large and otherwise it is small. For tapered columns, the effect of shear deformation increases with the taper ratio a. So, the effective length factor of tapered columns, η'_α, considering the effect of shear deformation can generally be expressed as

$$\eta'_\alpha = \eta_\alpha \cdot \beta(\alpha, \lambda), \tag{12.32}$$

where β is the magnification factor of the effective length factor due to reduction of column stiffness caused by shear deformation, which is a function of slenderness λ and taper ratio α of tapered columns simultaneously. The relationship of β with λ and α obtained by numerical study is given in Figure 12.13.

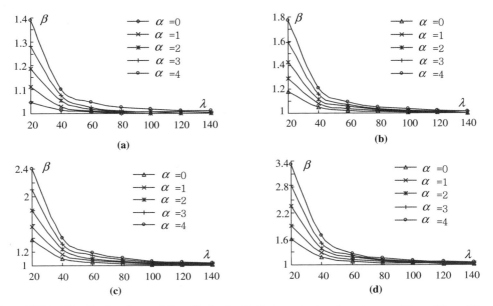

Figure 12.13   Magnification factor of the effective length factor under four boundary conditions: (a) condition a; (b) condition b; (c) condition c; (d) condition d

From Figure 12.13, it can be found that among the four boundary conditions shown in Figure 12.11, the effect of shear deformation is more evident on the value β under conditions c and d, and for the same boundary conditions, the greater the taper ratio with smaller slenderness, the greater the effect of shear deformation. So, when the taper ratio of tapered columns is small and at the same time the slenderness is relatively large, the effect of shear deformation on the effective length factor is negligible and this condition, to make $\beta \leq 1.05$, can be expressed as

$$\lambda \geq \lambda_0 + e_1\alpha + e_2\alpha^2 + e_3\alpha^3, \tag{12.33}$$

where the parameters are

$\lambda_0 = 20, \ e_1 = 21.30, \ e_2 = -7.50, \ e_3 = 1.17$ for boundary condition a,

$\lambda_0 = 39, \ e_1 = 12.17, \ e_2 = 1.00, \ e_3 = -0.167$ for boundary condition b,

$\lambda_0 = 65, \ e_1 = 32.85, \ e_2 = -14.36, \ e_3 = 2.58$ for boundary condition c,

$\lambda_0 = 80, \ e_1 = 24.76, \ e_2 = -7.86, \ e_3 = 1.67$ for boundary condition d.

If the condition in Equation (12.33) is not satisfied, the effect of shear deformation is proposed to be considered. Through numerical studies, the magnification factor β can be expressed as

$$\beta = f_0(\alpha) + f_1(\alpha) \cdot \exp\left(-\frac{\lambda - 20}{15}\right) \geq 1.0, \tag{12.34}$$

where $f_0(\alpha)$ and $f_1(\alpha)$ are given by

- for boundary condition a:

$$f_0(\alpha) = 1.0003 - 0.000\,9770\alpha + 0.0011\alpha^2,$$
$$f_1(\alpha) = 0.048\,37 + 0.056\,58\alpha + 0.006\,09\alpha^2;$$

- for boundary condition b:

$$f_0(\alpha) = 1.0058 - 0.002\,530\alpha + 0.000\,8571\alpha^2,$$
$$f_1(\alpha) = 0.1720 + 0.095\,73\alpha + 0.011\,86\alpha^2;$$

- for boundary condition c:

$$f_0(\alpha) = 1.0150 - 0.008\,160\alpha + 0.000\,6929\alpha^2$$
$$f_1(\alpha) = 0.3527 + 0.1735\alpha + 0.018\,89\alpha^2;$$

- for boundary condition d:

$$f_0(\alpha) = 1.0254 - 0.009\,220\alpha + 0.005\,410\alpha^2,$$
$$f_1(\alpha) = 0.5589 + 0.3153\alpha + 0.026\,82\alpha^2.$$

12.3.2 Tapered Column in Steel Portal Frame

The steel portal frame with tapered columns as shown in Figure 12.14 is widely used for industrial buildings. Effective length factors of such tapered columns are given in the Chinese specification (CECS, 2002) for simplifying safety check on the stability of tapered columns. But the specification values do not involve the

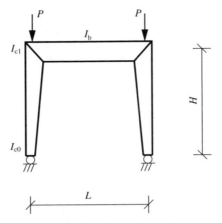

Figure 12.14   A flat steel portal frame

effect of shear deformation. In reality, tapered columns employ, in most cases, H sections with linearly varying web height and constant flange width.

12.3.2.1 *Effective length factor excluding effects of shear deformation*

Define the effective length factor of the tapered columns in a steel portal frame as

$$\eta_\alpha = \sqrt{\frac{\pi^2 E I_{c0}}{P_{cr} H^2}}, \tag{12.35}$$

where P_{cr} is the elastic critical load of the steel portal frame, I_{c0} is the inertial moment of the small end of the tapered column in the frame and E is the elastic modulus.

Introduce two parameters

$$n = \frac{I_{c1}}{I_{c0}}, \tag{12.36}$$

$$K = \frac{I_b H}{I_{c1} L}, \tag{12.37}$$

where I_{c1} is the inertial moment of the large end of the tapered column, I_b is the inertial moment of the frame girder, and H and L are the height and span of the portal frame, respectively.

After the numerical results of the elastic critical loads of steel portal frames are obtained, the effective length factors of tapered columns in steel portal frames can be calculated with Equation (12.35). These numerical values are compared with the specification values (CECS, 2002), as listed in Table 12.5. Good agreement between them can be noted.

12.3.2.2 *Effect of shear deformation*

Define the slenderness of tapered columns in steel portal frames as

$$\lambda_c = \frac{H}{\sqrt{I_{c0}/A_{c0}}}, \tag{12.38}$$

where A_{c0} is the sectional area of the small end of tapered columns.

Table 12.5   Values of η_α for tapered columns excluding effect of shear deformation

n		K					
		0.1	0.2	0.5	1.0	2.0	10.0
50	Given in CECS (2002)	0.706	0.591	0.518	0.494	0.484	0.475
	Calculated with Equation (12.35)	0.704	0.588	0.516	0.493	0.481	0.473
20	Given in CECS (2002)	1.095	0.889	0.758	0.713	0.693	0.682
	Calculated with Equation (12.35)	1.077	0.881	0.751	0.708	0.687	0.670
10	Given in CECS (2002)	1.473	1.208	1.008	0.942	0.929	0.869
	Calculated with Equation (12.35)	1.495	1.201	1.002	0.933	0.918	0.871
5	Given in CECS (2002)	2.053	1.641	1.341	1.229	1.176	1.140
	Calculated with Equation (12.35)	2.065	1.644	1.337	1.228	1.173	1.129
1	Given in CECS (2002)	—	3.420	2.630	2.330	2.170	2.000
	Calculated with Equation (12.35)	4.405	3.404	2.627	2.327	2.164	2.033

It is found that the effective length factor of tapered columns varies with only n and K if the effect of shear deformation is neglected (Li and Li, 2000). However, when the effect of shear deformation is considered, the slenderness of tapered columns λ_c has significant influence on the effective length factor η_α. In general, the larger the slenderness, the smaller the effect of shear deformation on the effective length factor of tapered columns in steel portal frames.

In a similar way with that in Section 12.3.1, define the magnification factor as

$$\beta = \frac{\eta'_\alpha}{\eta_\alpha}, \tag{12.39}$$

where η'_α and η_α are the effective length factors including and excluding the effect of shear deformation, respectively. It is very clear that the magnification factor β indicates the severity of the shear deformation effect on the stability of tapered columns in steel portal frames. If β is obtained, η'_α can easily be calculated with Equation (12.39) and the previously known values of η_α.

Table 12.6 gives the values of the magnification factor β varying with n and K when $\lambda_c = 23.67$. It can be found that the magnification factor reduces with n when K is constant. However, when n is constant, the magnification factor varies slightly with K. For the purpose of simplification, β_E can be used to represent all the values of β with K ranging from 0.1 to 10.0. The even value of the maximum and minimum values of β for each n can be used for β_E, as listed in Table 12.6.

12.3.2.3 Modification to specification values

By using β_E, the magnification factor becomes a function of λ_c and n. Figure 12.15 plots a group of curves for β_E versus λ_c under five values of n, by which one can quickly estimate the magnification factor of the effective length factors for tapered columns in practice.

Table 12.6   Values of β when $\lambda_c = 23.67$

n	K						β_E
	0.1	0.2	0.5	1.0	2.0	10.0	
50	1.278	1.330	1.365	1.372	1.360	1.358	1.325
20	1.177	1.215	1.227	1.226	1.208	1.182	1.202
10	1.161	1.159	1.168	1.158	1.119	1.142	1.144
5	1.124	1.126	1.128	1.125	1.115	1.091	1.110
1	1.064	1.064	1.071	1.067	1.059	1.066	1.065

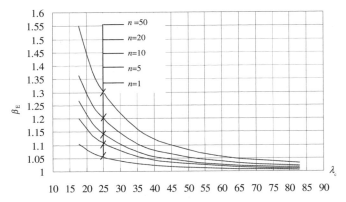

Figure 12.15 Magnification factor versus slenderness of tapered columns

After a careful examination of Figure 12.15, it can be found that when

$$\lambda_c = 36\sqrt{0.02n} + 26, \tag{12.40}$$

the magnification factor is less than 1.05, so that the effect of shear deformation on the stability of tapered columns in steel portal frames is negligible.

The values of β_E in Figure 12.15 can be further expressed with the following equation:

$$\beta_E = \beta_E(n, \lambda_c) = 1.0 + f_1(n) \cdot e^{-(\lambda_c - 17.75)/4} + f_2(n) \cdot e^{-(\lambda_c - 17.75)/22}, \tag{12.41}$$

where

$$f_1(n) = 0.029\,03 + 0.0099n - 0.000\,3416n^2 + 0.000\,004\,155n^3, \tag{12.42}$$

$$f_2(n) = 0.065\,23 + 0.011\,20n - 0.000\,1056n^2. \tag{12.43}$$

The fitted values in Equation (12.41) are compared with the original β_E values in Table 12.7.

Table 12.7 The analytical and fitted values of β_E

		n				
λ_c		50	20	10	5	1
17.750	Analytical value	1.550	1.364	1.267	1.200	1.105
	Fitted value	1.550	1.371	1.265	1.189	1.115
23.670	Analytical value	1.328	1.218	1.157	1.117	1.061
	Fitted value	1.319	1.217	1.150	1.107	1.067
35.505	Analytical value	1.157	1.102	1.072	1.054	1.028
	Fitted value	1.163	1.112	1.076	1.054	1.035
47.340	Analytical value	1.092	1.059	1.042	1.031	1.016
	Fitted value	1.094	1.064	1.044	1.031	1.020
59.175	Analytical value	1.056	1.038	1.027	1.020	1.010
	Fitted value	1.055	1.038	1.025	1.018	1.012
71.010	Analytical value	1.041	1.027	1.019	1.014	1.007
	Fitted value	1.032	1.022	1.015	1.011	1.007
82.845	Analytical value	1.031	1.020	1.015	1.010	1.005
	Fitted value	1.019	1.013	1.009	1.006	1.004

12.3.2.4 *Example*

To explain the modification process of the effective length factor of the tapered columns in steel portal frames due to shear deformation, take an example from the case shown in Table 12.6 for $\lambda_c = 23.67$. When $n = 50$ and $K = 0.1$, it can be checked out that $\beta = 1.278$ and $\beta_E = 1.325$ from Table 12.6 and that $\eta_\alpha = 0.706$ from Table 12.5. Hence, the accurate effective length factor considering the shear deformation effect is $\eta'_\alpha = \beta\eta_\alpha = 1.278 \times 0.706 = 0.902$, whereas the corresponding approximate value is $\eta''_\alpha = \beta_E\eta_\alpha = 1.325 \times 0.706 = 0.935$. The relative error between η''_α and η'_α is only 3.7 %, which indicates that β_E gives a good approximation of β.

13 Nonlinear Analysis of Planar Steel Frames

The structural nonlinearities include geometric and material nonlinearities. The geometric nonlinearity is due to the additional actions which result from the structural displacements and the position changes of loads applied on the structure considered. The geometric nonlinearity is generally associated with the second-order effect. For steel frames, the geometric nonlinearity is mainly due to the $P - \Delta$ effect, which on the resultant and deflection of a cantilevered column is illustrated in Figure 13.1.

The material nonlinearity is that the structural material exhibits nonlinear stress–strain relationship after the stresses of frame members exceed the linearly elastic limit. The structural resultants and displacements are influenced by the material nonlinearity.

The structural analysis methods can be divided into four types, according to whether the geometric or material nonlinearity is considered, as

(1) the first-order elastic analysis (neither nonlinearity considered);

(2) the second-order elastic analysis (geometric nonlinearity considered);

(3) the first-order elasto-plastic analysis (material nonlinearity considered);

(4) the second-order elasto-plastic analysis (both nonlinearities considered).

The tasks in structural nonlinear analysis include structural nonlinear response analysis (static and dynamic displacement and resultant analysis), ultimate load-carrying capacity analysis and structural dynamic stability (or collapse) analysis. The static nonlinear response analysis and ultimate load-carrying capacity analysis of planer steel frames will be discussed in this chapter.

13.1 GENERAL ANALYSIS METHOD

13.1.1 Loading Types

Two loading types can be categorized for the loading of steel frames used for buildings or other engineering structures. One loading type is that only vertical loads are applied (see Figure 13.2(a)), corresponding to the load combination of vertical dead load and vertical live load. The other is that both vertical and horizontal loads are applied (see Figure 13.2(b)), corresponding to the load combination of vertical dead load plus vertical live load plus horizontal loads induced by wind or earthquakes. Generally, the loading sequence of the second loading type is that first the vertical loads are applied and then the horizontal loads are applied.

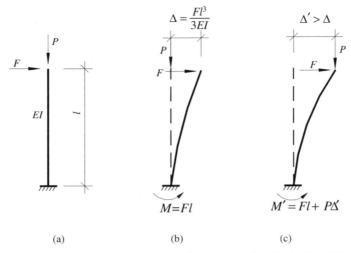

Figure 13.1   Effect of geometry nonlinearity: (a) loading condition; (b) excluding $P-\Delta$ effect; (c) including $P-\Delta$ effect

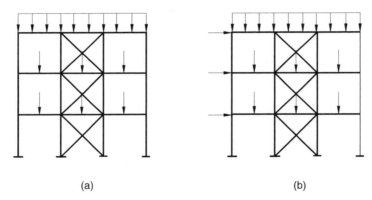

Figure 13.2   Loading of frame structures: (a) vertical loading; (b) vertical and horizontal loading

13.1.2   Criteria for the Limit State of Ultimate Load-Carrying Capacity

Due to the geometric and material nonlinearities, the structural stiffness relates to the structural resultants and displacements. Assume at one moment the vector of forces applied to the structure considered to be $\{F\}$ and the displacement vector to be $\{D\}$, so that the transient state of the structure can be defined as

$$\Omega = (\{F\}, \ \{D\}). \tag{13.1}$$

Based on the state Ω, let the incremental force vector be $\{dF\}$ and the corresponding incremental displacement vector be $\{dD\}$, which should satisfy the incremental stiffness equation as

$$\{dF\} = [K(\Omega)]\{dD\}, \tag{13.2}$$

where $[K(\Omega)]$ is the structural tangent stiffness matrix in the state Ω.

If the structure is in stability state (see Figure 13.3), the work done by $\{dF\}$ must be positive, namely $\{dD\}^{\mathrm{T}}\{dF\} = \{dD\}^{\mathrm{T}}[K(\Omega)]\{dD\} > 0$ and $[K(\Omega)]$ is a positive definite matrix. If the structure is in

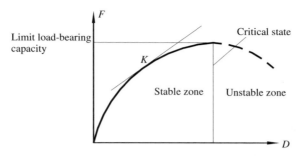

Figure 13.3   States of the frame loaded

instability state, the work done by $\{dF\}$ must not be positive, namely $\{dD\}^T\{dF\} = \{dD\}^T[K(\Omega)]\{dD\} \leq 0$ and $[K(\Omega)]$ should be a nonpositive definite matrix. It can be seen that whether $[K(\Omega)]$ is a positive definite matrix or not is a criterion for the limit state of the structural ultimate load-carrying capacity. Actually, the force vector $\{dF\}$ corresponding to the state Ω when $[K(\Omega)]$ becomes a nonpositive definite matrix is the structural load-carrying capacity.

According to theory of linear algebra, the necessary and sufficient condition of $[K]$ being positive definite is

$$k_{ii} > 0, \quad i = 1, 2, \ldots, n, \tag{13.3}$$

and

$$\begin{vmatrix} k_{11} & k_{12} & \cdots & k_{1i} \\ k_{21} & k_{22} & \cdots & k_{2i} \\ \cdots & \cdots & \cdots & \cdots \\ k_{i1} & k_{i2} & \cdots & k_{ii} \end{vmatrix} > 0, \quad i = 1, 2, \ldots, n, \tag{13.4}$$

where k_{ii} is the element of the ith row and the ith column in the matrix $[K]$.

If one of the above conditions is dissatisfied, $[K]$ is nonpositive definite.

13.1.3   Analysis Procedure

In the structural nonlinear analysis, the loading process can be divided into many load increments $\{dF_1\}$, $\{dF_2\}$ and so on. The loading process should meet the realistic loading condition, and it is not necessary to take a proportional loading. The calculation when the mth incremental loading is processed can be

$$\{F_{m-1}\} = \sum_{i=0}^{m-1} \{dF_i\}, \tag{13.5a}$$

$$\{D_{m-1}\} = \sum_{i=0}^{m-1} \{dD_i\}, \tag{13.5b}$$

$$\Omega_{m-1} = (\{F_{m-1}\}, \{D_{m-1}\}), \tag{13.5c}$$

$$\{dD_m\} = [K(\Omega_{m-1})]^{-1} \{dF_m\}, \tag{13.6}$$

where $\{F_0\} = \{dF_0\}$, $\{D_0\} = \{dD_0\}$ for $m = 1$ is the initial loading and displacement state. In the calculation of the mth incremental loading, if $[K(\Omega_{m-1})]$ is nonpositive definite, there can be surety that the ultimate load-bearing capacity of the structure is between $\{F_{m-2}\}$ and $\{F_{m-1}\}$. For the sake of conservation, $\{F_{m-2}\}$ can be taken as the ultimate load-bearing capacity of the structure.

To enhance the accuracy, the following can be referred in the calculation:

(1) Increase the number of loading increments, namely reduce the value of loading incremental, but this will also increase the computational cost.

(2) With the same number of loading increments, adopt larger loading incremental when the structure is in elastic state and smaller one when in elasto-plastic state.

(3) If $[K(\Omega_{m-1})]$ being nonpositive definite has been detected, more refined loading incrementals to represent $\{dF_{m-1}\}$ can be employed and the accuracy of the ultimate loads of the structures can be improved.

It should be noted that the geometric nonlinearity enlarges with the development of material nonlinearity in structures. Second-order elasto-plastic analysis, therefore, should be used for structural elasto-plastic response or ultimate load-carrying capacity analysis. Large error will be introduced if the first-order elasto-plastic analysis is used for that purpose.

13.1.4   Basic Elements and Unknown Variables

In the nonlinear analysis of planar steel frames, basic elements include beams, columns, braces and joint panels. Effects of shear deformation of beams and columns, axial deformation of columns, shear deformation of joint panels and relative rotations of beam-to-column connections need to be considered, whereas the effect of axial deformation of beams can be excluded. The basic unknown variables are the horizontal displacements on each frame storey, the vertical displacements at the centres of joint panels and the rotations and shear deformations of joint panels (see Figure 9.6).

When relatively large concentrated loads or distributed loads are applied within the span of frame beams, plastic hinges may occur within the beam spans, so that the beam elements with internal plastic hinges must be used, or the frame beam should be subdivided into two or more beam elements and the additional nodes are inserted at the possible positions to form plastic hinges. The vertical displacements and rotations of such nodes added are also the basic variables.

13.1.5   Structural Analysis of the First Loading Type

For the analysis of the planar steel frame, as shown in Figure 13.2(a), applied with only vertical loads, the actions of all the loads should be first equalized to vertical forces, moments and shear moments at the nodes or joint panels of the frame, and then the incremental global stiffness equation of the frame can be established by assembling the elemental stiffness equations of the beams, columns, braces and joint panels as

$$\begin{bmatrix} [K_{xx}] & [K_{xz}] \\ [K_{zx}] & [K_{zz}] \end{bmatrix} \begin{bmatrix} \{dD_x\} \\ \{dD_z\} \end{bmatrix} = \begin{bmatrix} \{dF_x\} \\ \{dF_z\} \end{bmatrix}, \tag{13.7}$$

where $\{dD_x\}$ is the incremental vector of horizontal displacements at each floor of the frame, $\{dD_z\}$ is the incremental vector of vertical displacements, rotations of joint panels and nodes added within beam spans, and shear deformations of joint panels, $\{dF_x\}$ is the incremental vector of horizontal forces at each floor of the frame and $\{dF_z\}$ is the incremental vector of vertical forces, moments at joint panels and nodes added within beam spans and shear moments at joint panels.

For the first loading type, $\{dF_x\} = 0$. By static condensation, Equation (13.7) becomes

$$[K_z]\{dD_z\} = \{dF_z\}, \tag{13.8}$$

where

$$[K_z] = [K_{zz}] - [K_{zx}][K_{xx}]^{-1}[K_{xz}]. \tag{13.9}$$

Based on Equation (13.8), using calculation procedures proposed above as in Equations (13.5) and (13.6), one can obtain the force–displacement relationship of the structure analyzed. Once $[K_z]$ is nonpositive definite, the structure reaches its limit state for supporting the vertical loads.

The difference between the structural elasto-plastic load-bearing capacity analysis of the first loading type mentioned above and the structural elastic critical load analysis given in Chapter 10 is that (1) loads applied within the beam spans and axial deformations of columns can be considered in the former, but not in the latter; (2) with the same vertical loads, the solution of the latter is the upper boundary of that of the former; (3) in mathematics, the former is the nonlinear simultaneous equations problem, whereas the latter is the eigenvalue problem.

13.1.6 Structural Analysis of the Second Loading Type

For the planar steel frame, as shown in Figure 13.2(b), applied with both vertical and horizontal loads, the analysis of the frame subjected to the vertical loads can be conducted at first as in Section 13.1.5. After that, the structure should be stable and then the horizontal loads are added in the following analysis.

In the analysis of the frame subjected to the horizontal loads, the structural state obtained in the analysis with only vertical loads is taken as its initial state. As $\{dF_z\} = 0$ in the analysis of applying horizontal loads, by static condensation, Equation (13.7) becomes

$$[K_x]\{dD_x\} = \{dF_x\}, \tag{13.10}$$

where

$$[K_x] = [K_{xx}] - [K_{xz}][K_{zz}]^{-1}[K_{zx}]. \tag{13.11}$$

Based on Equation (13.10), using calculation procedures proposed above as in Equations (13.5) and (13.6), one can obtain the force–displacement relationship of the structure analysed. Once $[K_x]$ is nonpositive definite, the structure reaches its limit state for supporting the horizontal loads with constant vertical loads.

13.1.7 Numerical Examples

Example 1

Four three-storey single-bay steel frames were tested and their elasto-plastic deformations and load-carrying capacities were obtained (Ding, 1987). Three of the frames tested are analysed here using the second-order elasto-plastic method with beam elements.

The geometric sizes of the frames tested are given in Figure 13.4, and the arrangement of the loads for the tests is shown in Figure 13.5. The loading sequence is that first the vertical loads are applied and kept

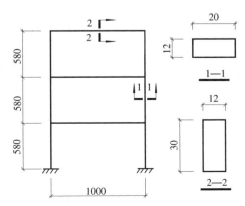

Figure 13.4   Geometrical sizes of the frame tested

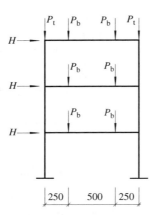

Figure 13.5   Loads on the frame tested

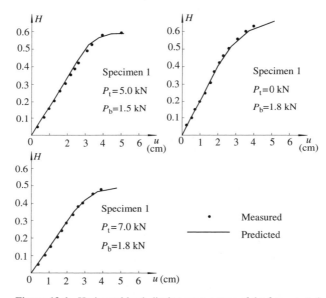

Figure 13.6   Horizontal load–displacement curves of the frame tested

constant, and then the horizontal loads are applied and increased until frame failure. The horizontal load versus horizontal displacement relationship results obtained through theoretical analysis and test measurements are both illustrated in Figure 13.6, and the values of the load-carrying capacities are compared in Table 13.1. Good agreement is found in those comparisons which indicates that the second-order elasto-plastic analysis can provide satisfactory prediction of the elasto-plastic behaviour and load-carrying capacity of steel frames.

Table 13.1   Horizontal load-carrying capacity of the frame tested (kN)

Number of specimens	H_t (tested)	H_c (analysed)	$\frac{H_c - H_t}{H_t} \times 100\%$
Specimen 1	0.570	0.579	1.58
Specimen 2	0.650	0.665	2.31
Specimen 3	0.490	0.485	−1.02

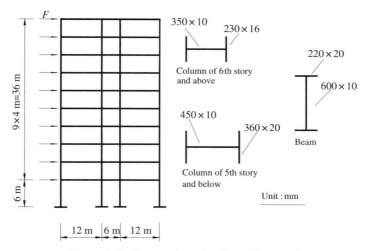

Figure 13.7   Frame and member sizes of Example 2

Example 2

A 10-storey three-bay semi-rigid steel frame is illustrated in Figure 13.7. The distributed vertical load on each storey (including roof) is 16 KN/m, the initial stiffness and plastic moment of the beam-to-column connections are, respectively, $k_e = 1.4 \times 10^5$ kN m/rad and $M_p = 9.5 \times 10^6$ kN m, and the moment–rotation relationship of the connections as shown in Figure 9.20 is assumed. Let F denote the horizontal loads on each floor of the frame and D denote the horizontal displacement at the top of the frame. The complete $F - D$ curves obtained with six different considerations are illustrated in Figure 13.8, where

- I represents considering F1–F5 effects;
- II represents considering F2–F5 effects;
- III represents considering F3–F5 effects;
- IV represents considering F4 and F5 effects;
- V represents considering F5 effect;
- VI represents considering F3–F5 effects and using the conventional structural analysis model.

The conventional structural analysis model is that the effect of joint panels is neglected and the lengths of frame beams and columns are calculated according to the distances between their central axes. Effects F1–F5 are defined as follows:

- F1: effect of beam-to-column connection flexibility;
- F2: effect of joint-panel shear deformation;
- F3: effect of geometric nonlinearity;
- F4: effect of beam and column shear deformation;
- F5: effect of material nonlinearity.

The results of analyses with considerations I–V show that the effects such as connection flexibility, joint-panel shear deformation, geometric nonlinearity, and beam and column shear deformations are significant for the performance of steel frames. A comparison between II and VI analyses indicates that

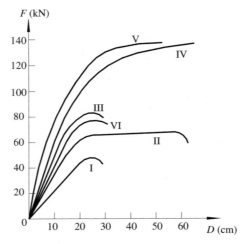

Figure 13.8   Horizontal load–displacement curves of Example 2 frame with various considerations

it is possible for the conventional structural analysis model to produce large error in elasto-plastic analysis of steel frames.

13.2   APPROXIMATE ANALYSIS CONSIDERING $P-\Delta$ EFFECT

13.2.1   Formulation

As the pure steel frame is relatively weak in lateral stiffness, the second-order effect ($P-\Delta$ effect) has large impact on the horizontal displacements and internal forces of the frame. It is therefore necessary to consider the $P-\Delta$ effect in the analysis of the frame of the second loading type. The numerical cost increases when the second-order elasto-plastic analysis is pursued for the $P-\Delta$ effect. In this section, an approximate approach for elastic analysis of steel frames considering the $P-\Delta$ effect is introduced for the purpose of engineering application.

Assume that the horizontal displacement at the ith floor of the frame by the first-order elastic analysis is u_i. The inter-storey drift Δ_i at the ith storey is then

$$\Delta_i = u_i - u_{i-1}. \tag{13.12}$$

As shown in Figure 13.9, the additional shear force at the ith storey due to the $P-\Delta$ effect is

$$dV_{i1} = \frac{P_i \Delta_i}{h_i}, \tag{13.13}$$

where V_i is the shear force at the ith storey, P_i is the total gravity load above the ith storey and h_i is the height of the ith storey.

The additional inter-storey drift due to the additional shear force dV_{i1} is

$$d\Delta_{i1} = \frac{dV_{i1}}{V_i} \Delta_i = \frac{P_i \Delta_i}{V_i h_i} \Delta_i. \tag{13.14}$$

In turn, $d\Delta_{i1}$ will lead to the additional shear force dV_{i2} as

$$dV_{i2} = \frac{P_i d\Delta_{i1}}{h_i} = \frac{P_i^2 \Delta_i^2}{V_i h_i^2}, \tag{13.15}$$

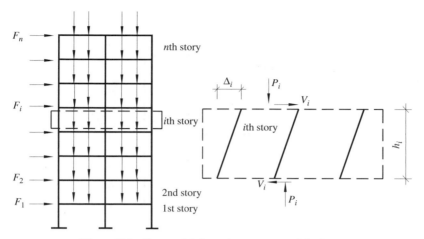

Figure 13.9   Force and deformation at a storey of the frame

and $\mathrm{d}V_{i2}$ will also further result in the additional inter-storey drift $\mathrm{d}\Delta_{i2}$ as

$$\mathrm{d}\Delta_{i2} = \frac{\mathrm{d}V_{i2}}{V_i}\Delta_i = \left(\frac{P_i\Delta_i}{V_ih_i}\right)^2 \Delta_i. \tag{13.16}$$

Repeating the above procedures, the final storey drift Δ'_i considering the $P-\Delta$ effect can be obtained as

$$\Delta'_i = \Delta_i + \mathrm{d}\Delta_{i1} + \mathrm{d}\Delta_{i2} + \mathrm{d}\Delta_{i3} + \cdots$$

$$= \Delta_i(1 + \alpha_i + \alpha_i^2 + \alpha_i^3 + \cdots) = \frac{\Delta_i}{1-\alpha_i}, \tag{13.17}$$

where

$$\alpha_i = \frac{P_i\Delta_i}{V_ih_i}. \tag{13.18}$$

So, to consider the $P-\Delta$ effect, the resultant of columns and braces at the ith storey of the frame by the first-order elastic analysis should be multiplied with the magnification factor $1/(1-\alpha_i)$, whereas the resultant of the beams with the magnification factor $1/(1-\bar{\alpha}_i)$, where $\bar{\alpha}_i$ is given by

$$\bar{\alpha}_i = \alpha_i, \quad \text{for the top storey,} \tag{13.19}$$

$$\bar{\alpha}_i = \frac{\alpha_i + \alpha_{i+1}}{2}, \quad \text{for other storeys.} \tag{13.20}$$

Obviously, presented above is the approximate approach for considering the $P-\Delta$ effect on steel frames, based on the results of the first-order elastic analysis.

13.2.2 Example

To specify the procedure and validity of the above simplified method considering the $P-\Delta$ effect, a five-storey frame is illustrated in Figure 13.10. The vertical loads and the horizontal loads on each floor of the frame are assumed to be the same. The horizontal displacements at each floor of the frame obtained with the

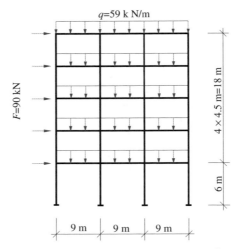

Figure 13.10   A five-storey frame. Bending stiffness of beams: $EI_g = 2.11 \times 10^5$ kN m; bending stiffness of columns: $EI_c = 1.22 \times 10^5$ kN m

Table 13.2   The analytical results of a five-storey frame with the $P-\Delta$ effect

Number of floor	First-order analysis				Second-order analysis			Relative error(%)		
	$u_i$cm	$\Delta_i = u_i - u_{i1}$	V_i(KN)	P_i(KN)	α_i	$\frac{1}{1-\alpha_i}$	Δ_i' (approximated)	Δ_i'' (exact)	$\frac{\Delta_i''-\Delta_i}{\Delta_i''}$	$\frac{\Delta_i''-\Delta_i'}{\Delta_i''}$
Fifth	3.909	0.261	50	1593	0.0185	1.019	0.266	0.271	3.69	1.85
Fourth	3.648	0.491	100	3186	0.0348	1.036	0.509	0.524	6.30	2.86
Third	3.157	0.736	150	4779	0.0521	1.055	0.776	0.804	8.46	3.48
Second	2.421	0.987	200	6372	0.0699	1.075	1.061	1.110	11.08	4.41
First	1.434	1.434	250	7965	0.0761	1.082	1.552	1.639	12.51	5.31

approximate and exact analysis methods are listed and compared in Table 13.2. By this comparison, it is shown that Equation (13.17) can be validly used to modify the results of the first-order elastic analysis.

13.3   SIMPLIFIED ANALYSIS MODEL CONSIDERING $P-\Delta$ EFFECT

13.3.1   Development of Simplified Model

To involve the $P-\Delta$ effect, a geometry stiffness matrix can be added to the elemental stiffness matrix of frame columns (see Equation (2.49)). As the geometry stiffness matrix relates to the axial force in columns and the axial force of columns is variable under the loads applied, the elastic stiffness matrix of column elements in frames is not constant, but is variable with horizontal loads in the second loading type. The global stiffness equation of a frame is no longer linear, but is nonlinear. To solve it, an iterative strategy is necessary. Although the iteration method is an exact one to consider the $P-\Delta$ effect, it is complicated in engineering application. For elastic analysis of steel frames, is it possible to calculate the $P-\Delta$ effect without iteration? If a positive answer is provided, it is of significance in practice.

As shown in Figure 13.11 for the second loading type, the axial forces of frame columns are produced not only by the vertical loads, but also by the horizontal loads. For columns at the same storey, some axial forces due to horizontal loads increase whereas others decrease. Actually, giving a section on all columns in one storey indicates by equilibrium condition that the axial force summation of all the columns on this storey is

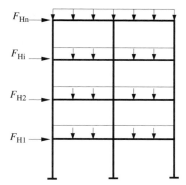

Figure 13.11   Loads on frame structures

dependent only on the summation of the vertical loads applied above this storey and independent of the horizontal loads applied. It can therefore be reasonably assumed from this point that P in the $P-\Delta$ effect of steel frames relates only to vertical loads, based on which a simplified analysis model considering the $P-\Delta$ effect is proposed as shown in Figure 13.12.

The global stiffness equation of a frame can be obtained by assembling the elemental stiffness equations of all the beams, columns, braces and joint panels in the frame while neglecting geometric nonlinearity as

$$\begin{bmatrix} [K_{uu}] & [K_{uw}] & [K_{u\theta}] & [K_{u\gamma}] \\ [K_{wu}] & [K_{ww}] & [K_{w\theta}] & [K_{w\gamma}] \\ [K_{\theta u}] & [K_{\theta w}] & [K_{\theta\theta}] & [K_{\theta\gamma}] \\ [K_{\gamma u}] & [K_{\gamma w}] & [K_{\gamma\theta}] & [K_{\gamma\gamma}] \end{bmatrix} \begin{Bmatrix} \{u\} \\ \{w\} \\ \{\theta\} \\ \{\gamma\} \end{Bmatrix} = \begin{Bmatrix} \{F_u\} \\ \{F_w\} \\ \{F_\theta\} \\ \{F_\gamma\} \end{Bmatrix}, \tag{13.21}$$

where $\{u\}$ is the vector of the horizontal displacements of all the floors, $\{w\}$, $\{\theta\}$ and $\{\gamma\}$ are the vectors of the vertical displacements, rotations and shear deformations of all joint panels, respectively, $\{F_u\}$ is the vector of the horizontal loads on all floors, and $\{F_w\}$, $\{F_\theta\}$ and $\{F_\gamma\}$ are the vectors of the vertical loads, bending moments and shear moments applied on all joint panels, respectively.

Assume the frame has n storeys and

$$\{u\} = [u_1, \quad u_2, \quad \ldots, \quad u_n]^{\mathrm{T}}, \tag{13.22}$$

$$\{F_u\} = [F_{u1}, \quad F_{u2}, \quad \ldots, \quad F_{un}]^{\mathrm{T}}. \tag{13.23}$$

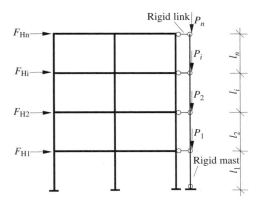

Figure 13.12   The simplified model to consider the $P-\Delta$ effect of the frame

In the second loading type, the vertical and horizontal loads are applied separately and then the results can be superimposed on each other. When only horizontal loads are applied, $\{F_w\} = \{F_\theta\} = \{F_\gamma\} = \{0\}$. Then, the lateral deflection stiffness equation of the frame can be obtained with static condensation as

$$[K_{uf}]\{u\} = \{F_u\}, \tag{13.24}$$

where

$$[K_{uf}] = [K_{uu}] - ([K_{uv}][K_{u\theta}][K_{u\gamma}]) \begin{bmatrix} [K_{vv}] & [K_{v\theta}] & [K_{v\gamma}] \\ [K_{\theta v}] & [K_{\theta\theta}] & [K_{\theta\gamma}] \\ [K_{\gamma v}] & [K_{\gamma\theta}] & [K_{\gamma\gamma}] \end{bmatrix}^{-1} \begin{bmatrix} [K_{vu}] \\ [K_{\theta u}] \\ [K_{\gamma u}] \end{bmatrix}. \tag{13.25}$$

The two parts are included in $\{F_u\}$ as

$$\{F_u\} = \{F_H\} + \{F_G\}, \tag{13.26}$$

$$\{F_H\} = [F_{H1}, \quad F_{H2}, \quad \ldots, \quad F_{Hn}]^T, \tag{13.27}$$

$$\{F_G\} = [F_{G1}, \quad F_{G2}, \quad \ldots, \quad F_{Gn}]^T, \tag{13.28}$$

where $\{F_{Hi}\}$ are the horizontal loads on the floor and $\{F_{Gi}\}$ are the additional horizontal forces on the ith floor due to the $P-\Delta$ effect.

By the model illustrated in Figure 13.12, one has

$$
\begin{aligned}
\{F_{Gi}\} &= N_i \frac{u_i - u_{i-1}}{l_i} - N_{i+1} \frac{u_{i+1} - u_i}{l_{i+1}} \\
&= \left(\frac{N_i}{l_i} + \frac{N_{i+1}}{l_{i+1}}\right) u_i - \left(\frac{N_i}{l_i} u_{i-1} + \frac{N_{i+1}}{l_{i+1}}\right) u_{i+1} \quad (i = 1, 2, \ldots, n),
\end{aligned} \tag{13.29}
$$

in which

$$N_i = \sum_{j=i}^{n} P_j, \tag{13.30}$$

where P_j is the summation of all the vertical loads applied on the ith floor, N_i is the summation of all the vertical loads applied on and above the ith floor and l_i is the height of the ith storey.

In Equation (13.29), $u_0 = 0$, $u_{n+1} = 0$ and $N_{n+1} = 0$. It can then be derived by Equation (13.29) that

$$\{F_G\} = [K_G]\{u\}, \tag{13.31}$$

where

$$[K_G] = \begin{bmatrix} \frac{N_1}{l_1} + \frac{N_2}{l_2} & -\frac{N_2}{l_2} & & & & \\ -\frac{N_2}{l_2} & \frac{N_2}{l_2} + \frac{N_3}{l_3} & -\frac{N_3}{l_3} & & & \\ & -\frac{N_3}{l_3} & \ddots & & & \\ & & & \frac{N_{n-1}}{l_{n-1}} + \frac{N_n}{l_n} & -\frac{N_n}{l_n} \\ & & & -\frac{N_n}{l_n} & \frac{N_n}{l_n} \end{bmatrix}. \tag{13.32}$$

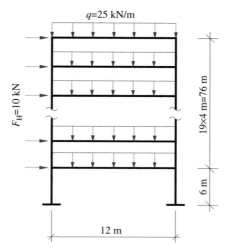

Figure 13.13 A 20-storey steel frame. Bending stiffness of beams: $EI_g = 1.22 \times 10^5$ kN m; bending stiffness of columns: $EI_c = 0.61 \times 10^5$ kN m (11th and upper storeys), $EI_c = 0.61 \times 10^5$ kN m (10th and lower storeys)

Combining Equations (13.24), (13.26) and (13.31) leads to

$$([K_{uf}] - [K_G])\{u\} = \{F_H\}. \tag{13.33}$$

As the matrix $[K_G]$ is constant, the elastic analysis of frames considering the $P-\Delta$ effect can be completed with Equation (13.33) without iteration.

It should be noted that the above simplified model can be used not only in elastic analysis, but also in second-order elasto-plastic analysis of steel frames.

13.3.2 Example

To verify the validity of the above simplified model considering the $P-\Delta$ effect, a 20-storey frame as shown in Figure 13.13 is investigated. The vertical and horizontal loads on each floor are identical. The elastic horizontal displacements involving the $P-\Delta$ effect obtained with the simplified model proposed are compared with those obtained by the exact method in Table 13.3, from which it can be seen that the simplified model can be used for making satisfactory prediction on the behaviour of frames with the $P-\Delta$ effect.

Table 13.3 Effect of $P-\Delta$ effect on the 20-storey frame

| | Lateral displacement of floors (cm) | | | Relative errors | |
| | | With $P-\Delta$ effect | | | |
Number of floor	Without $P-\Delta$ effect (1)	Exact analysis (2)	Simplified analysis (3)	$\dfrac{(2)-(3)}{(2)}$	$\dfrac{(2)-(1)}{(2)}$
20th	22.024	24.564	24.270	1.197	10.340
15th	20.256	22.733	22.440	1.289	10.896
10th	15.552	17.650	17.368	1.598	11.887
5th	9.330	10.762	10.487	2.555	13.306
1st	2.596	3.032	2.882	4.947	14.380

14 Seismic Response Analysis of Planar Steel Frames

14.1 GENERAL ANALYSIS METHOD

14.1.1 Kinetic Differential Equation

14.1.1.1 Kinetic differential equation in elastic state

A planar steel frame is shown in Figure 14.1. Let m_i be the mass at the ith floor, u_i be the lateral displacement of the ith floor relative to ground ($i = 1, 2, \ldots, n$, where n is the total storey number of the frame) and u_g the ground movement excited by earthquakes. The ground movement will arouse the dynamic movement of the frame, due to which the inertial forces take place to the mass at each floor of the frame. In elastic state, by regarding the inertial forces as static loads, the stiffness equation of the frame is obtained as

$$[K_e]\{u\} = \{f_I\}, \tag{14.1}$$

in which

$$\{u\} = \{u_1, \quad u_2, \quad \ldots, \quad u_n\}, \tag{14.2}$$

$$\{f_I\} = [-m_1(\ddot{u}_1 + \ddot{u}_g), \quad -m_2(\ddot{u}_1 + \ddot{u}_g), \quad \ldots, \quad -m_n(\ddot{u}_1 + \ddot{u}_g)]^{\mathrm{T}} = -[M](\{\ddot{u}\} + \{1\}\ddot{u}_g), \tag{14.3}$$

$$[M] = \mathrm{diag}[m_1, \quad m_2, \quad \ldots, \quad m_n], \tag{14.4}$$

$$\{1\} = [1, \quad 1, \quad \ldots, \quad 1]^{\mathrm{T}}, \tag{14.5}$$

where $[K_e]$ is the elastic stiffness matrix of the frame, corresponding to $\{u\}$, $[M]$ is the structural mass matrix, and $\ddot{u}$ and $\ddot{u}_g$ are, respectively, the second derivatives of u and u_g with respect to time t.

With substitution of Equation (14.3) into Equation (14.1), the kinetic differential equation for the undamped frame can be obtained as

$$[M]\{\ddot{u}\} + [K_e]\{u\} = -[M]\{1\}\ddot{u}_g. \tag{14.6}$$

If damping is considered, Equation (14.6) becomes

$$[K_e]\{u\} = \{f_I\} + \{f_c\}, \tag{14.7}$$

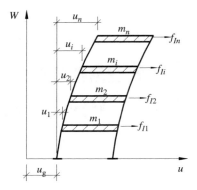

Figure 14.1   A planar frame subjected to action of ground movement

where $\{f_c\}$ is the vector of damping forces at each floor of the frame. If the viscous damping assumption that the damping force is proportional to the velocity of the structural motion is adopted, $\{f_c\}$ can be expressed as

$$\{f_c\} = -[C]\{\dot{u}\}, \tag{14.8}$$

where $[C]$ is the structural damping matrix. With substitution of Equations (14.8) and (14.3) into Equation (14.7), the kinetic differential equation for the damped frame can be obtained as

$$[M]\{\ddot{u}\} + [C]\{\dot{u}\} + [K_e]\{u\} = -[M]\{1\}\ddot{u}_g. \tag{14.9}$$

14.1.1.2 Kinetic differential equation in elasto-plastic state

In Equation (14.9), $[K_e]\{u\}$ is actually the elastic recovery force vector with the structural deformation $\{u\}$. When the frame enters the elasto-plastic state, however, the recovery forces of the frame are not equal to $[K_e]\{u\}$, but depend on the structural motion history $\{u(t)\}$. The kinetic differential equation of the frame in elasto-plastic state is therefore

$$[M]\{\ddot{u}(t)\} + [C]\{\dot{u}(t)\} + \{f(u(t))\} = -[M]\{1\}\ddot{u}_g(t), \tag{14.10}$$

where $\{\ddot{u}(t)\}$, $\{\dot{u}(t)\}$, $\{u(t)\}$ and $\{\ddot{u}_g(t)\}$ are, respectively, the values of $\{\ddot{u}\}$, $\{\dot{u}\}$, $\{u\}$ and $\{\ddot{u}_g\}$ at time t. The same kinetic equation at time $t + \Delta t$ is

$$[M]\{\ddot{u}(t + \Delta t)\} + [C]\{\dot{u}(t + \Delta t)\} + \{f(u(t + \Delta t))\} = -[M]\{1\}\ddot{u}_g(t + \Delta t), \tag{14.11}$$

where Δt is the time incremental.

Subtracting Equation (14.10) from Equation (14.11), one has

$$[M]\{\Delta\ddot{u}\} + [C]\{\Delta\dot{u}\} + \{\Delta f\} = -[M]\{1\}\Delta\ddot{u}_g, \tag{14.12}$$

where

$$\{\Delta f\} = \{f(u(t + \Delta t))\} - \{f(u(t))\}, \tag{14.13}$$

$$\Delta\ddot{u}_g = \ddot{u}_g(t + \Delta t) - \ddot{u}_g(t), \tag{14.14}$$

$$\{\Delta\ddot{u}\} = \{\ddot{u}(t + \Delta t)\} - \{\ddot{u}(t)\}, \tag{14.15a}$$

$$\{\Delta\dot{u}\} = \{\dot{u}(t + \Delta t)\} - \{\dot{u}(t)\}. \tag{14.15b}$$

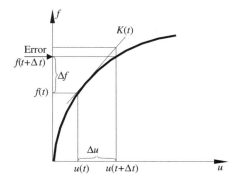

Figure 14.2 Relationship between structural force and displacement incremental

If Δt is small, the change of structural displacement,

$$\{\Delta u\} = \{u(t + \Delta t)\} - \{u(t)\}, \tag{14.15c}$$

is also small and then $\{\Delta f\}$ can be calculated approximately based on the tangent stiffness matrix of the frame at time t, $[K(t)]$, namely (see Figure 14.2)

$$\{\Delta f\} = [K(t)]\{\Delta u\}. \tag{14.16}$$

Substituting Equation (14.16) into Equation (14.12) leads to the incremental elasto-plastic kinetic differential equation for the damped frame as

$$[M]\{\Delta \ddot{u}\} + [C]\{\Delta \dot{u}\} + [K(t)]\{\Delta u\} = -[M]\{1\}\Delta \ddot{u}_\mathrm{g}. \tag{14.17}$$

14.1.2 Solution of Kinetic Differential Equation

14.1.2.1 Seismic response in elastic state

The seismic displacement response of the frame in elastic state can be obtained by solving Equation (14.9). Substituting displacement response into elemental stiffness equations can yield the seismic resultant response.

Equation (14.9) represents a second-order linear simultaneous differential equation in mathematics. For the sake of convenience, the following orthotropic relationships can be used:

$$\{\phi_i\}^\mathrm{T}[M]\{\phi_j\} = 0, \tag{14.18a}$$

$$\{\phi_i\}^\mathrm{T}[K_\mathrm{e}]\{\phi_j\} = 0, \tag{14.18b}$$

$$\{\phi_i\}^\mathrm{T}[C]\{\phi_j\} = 0, \tag{14.18c}$$

where $\{\phi_i\}$ and $\{\phi_j\}$ are, respectively, the ith and jth modes of vibration of the frame.

It is known from the knowledge of structural dynamic mechanism that Equations (14.18a) and (14.18b) are unconditionally correct, whereas Equation (14.18c) is conditionally correct. To satisfy Equation (14.18c), the Rayleigh damping matrix is generally used which will be introduced in the third part of this section.

The vectors of structural vibration modes are independent of each other due to the orthotropic relationships as given in Equation (14.18). The displacement vector can therefore be represented with the linear combination of the vectors of vibration modes, according to theory of linear algebra as

$$\{u\} = \sum_{j=1}^{n} a_j\{\phi_j\}, \tag{14.19}$$

where $a_j (j = 1, 2, \ldots, n)$ are the normalized coordinates of structural displacement vector in the coordinate of vibration modes.

Substituting Equation (14.19) into Equation (14.9) yields

$$\sum_{j=1}^{n} [M]\{\phi_j\}\ddot{a}_j + \sum_{j=1}^{n} [C]\{\phi_j\}\dot{a}_j + \sum_{j=1}^{n} [K_e]\{\phi_j\}a_j = -[M]\{1\}\ddot{u}_g. \tag{14.20}$$

Left-multiplying the above equation with $\{\phi_i\}^{\mathrm{T}}$ and noting the relationships in Equation (14.18), one has

$$M_i\ddot{a}_i + C_i\dot{a}_i + K_i a_i = -\{\phi_i\}^{\mathrm{T}}[M]\{1\}\ddot{u}_g, \tag{14.21}$$

in which

$$M_i = \{\phi_i\}^{\mathrm{T}}[M]\{\phi_i\}, \tag{14.22a}$$

$$K_i = \{\phi_i\}^{\mathrm{T}}[K_e]\{\phi_i\}, \tag{14.22b}$$

$$C_i = \{\phi_i\}^{\mathrm{T}}[C]\{\phi_i\}, \tag{14.22c}$$

where M_i, K_i and C_i are, respectively, the ith generalized mass, stiffness and damping of the frame, mutually related with

$$K_i = \omega_i^2 M_i, \tag{14.23a}$$

$$C_i = 2\xi_i\omega_i M_i, \tag{14.23b}$$

where ω_i and ξ_i are the circular frequency and damping ratio for the ith vibration mode of the frame, respectively.

Substituting Equation (14.23) into Equation (14.21) yields

$$\ddot{a}_i + 2\xi_i\omega_i\dot{a}_i + \omega_i^2 a_i = -\gamma_i\ddot{u}_g, \tag{14.24}$$

where γ_i is the vibration participating factor for the ith vibration mode, given by

$$\gamma_i = \frac{\{\phi_i\}^{\mathrm{T}}[M]\{1\}}{M_i}. \tag{14.25}$$

Equation (14.24) is actually the kinetic differential equation for a single-degree-of-freedom (SDOF) system, the solution of which is the Duhanel integration expressed as

$$a_i(t) = -\frac{\gamma_i}{\omega_{i\mathrm{D}}} \int_0^t \ddot{u}_g(\tau)\, \mathrm{e}^{-\xi_i\omega_i(t-\tau)}\, \sin\omega_{i\mathrm{D}}(t-\tau)\mathrm{d}\tau, \tag{14.26}$$

where $\omega_{i\mathrm{D}}$ is the damped circular frequency for the ith vibration mode of the frame obtained with

$$\omega_{i\mathrm{D}} = \omega_i\sqrt{1 - \xi_i^2}. \tag{14.27}$$

The seismic displacement response of the frame in elastic state can be obtained by substituting Equation (14.26) into Equation (14.19).

Generally, the elastic seismic response of structures is significantly dependent on the first several vibration modes so that the structural seismic displacements can be represented with the linear combination of the first several vibration modes.

14.1.2.2 Seismic response in elasto-plastic state

The kinetic differential equation of the frame in elasto-plastic state, Equation (14.10), is a nonlinear differential equation, the analytical solution of which in theory does not exist. Generally, the incremental kinetic differential equation, Equation (14.17), is used and the integration is performed with respect to time to obtain the numerical results of the structural elasto-plastic time- dependent response.

A Taylor series expansion technique can be used to solve Equation (14.17) by representing the displacement and velocity at time $t + \Delta t$ with displacement t, velocity $\{\dot{u}(t)\}$ and acceleration $\{\ddot{u}(t)\}$ at time t as

$$\{u(t + \Delta t)\} = \{u(t)\} + \{\dot{u}(t)\}\Delta t + \{\ddot{u}(t)\}\frac{\Delta t^2}{2} + \{\dddot{u}(t)\}\frac{\Delta t^3}{6} + \cdots, \tag{14.28a}$$

$$\{\dot{u}(t + \Delta t)\} = \{\dot{u}(t)\} + \{\ddot{u}(t)\}\Delta t + \{\dddot{u}(t)\}\frac{\Delta t^2}{2} + \cdots. \tag{14.28b}$$

It can be assumed that the change of structural acceleration is linear during the incremental time Δt (termed as *linear acceleration assumption*), i.e.

$$\{\dddot{u}(t)\} = \frac{1}{\Delta t}(\{\ddot{u}(t + \Delta t)\} - \{\ddot{u}(t)\}) = \frac{1}{\Delta t}\{\Delta\ddot{u}\}, \tag{14.29a}$$

$$\left\{\frac{\mathrm{d}^r u(t)}{\mathrm{d}t^r}\right\} = 0, \quad \text{for} \quad r = 4, 5, \ldots \tag{14.29b}$$

Substituting Equation (14.29) into Equation (14.28) yields

$$\{\Delta u\} = \{\dot{u}(t)\}\Delta t + \{\ddot{u}(t)\}\frac{\Delta t^2}{2} + \{\dddot{u}(t)\}\frac{\Delta t^3}{6}, \tag{14.30a}$$

$$\{\Delta\dot{u}\} = \{\ddot{u}(t)\}\Delta t + \{\Delta\ddot{u}\}\frac{\Delta t}{2}. \tag{14.30b}$$

The solutions of $\{\Delta\dot{u}\}$ and $\{\Delta\ddot{u}\}$ from the above equations are

$$\{\Delta\ddot{u}\} = \frac{6}{\Delta t^2}\{\Delta u\} - \frac{6}{\Delta t}\{\dot{u}(t)\} - 3\{\ddot{u}(t)\}, \tag{14.31a}$$

$$\{\Delta\dot{u}\} = \frac{3}{\Delta t}\{\Delta u\} - 3\{\dot{u}(t)\} - \frac{\Delta t}{2}\{\ddot{u}(t)\}. \tag{14.31b}$$

Substituting Equation (14.31) into Equation (14.17) leads to

$$[K^*]\{\Delta u\} = \{F^*\}, \tag{14.32}$$

where

$$[K^*] = [K(t)] + \frac{6}{\Delta t^2}[M] + \frac{3}{\Delta t}[C], \tag{14.33}$$

$$\{F^*\} = -[M]\{1\}\Delta\ddot{u}_{\mathrm{g}} + [M]\left(\frac{6}{\Delta t}\{\dot{u}(t)\} + 3\{\ddot{u}(t)\}\right) + [C]\left(3\{\dot{u}(t)\} + \frac{\Delta t}{2}\{\ddot{u}(t)\}\right). \tag{14.34}$$

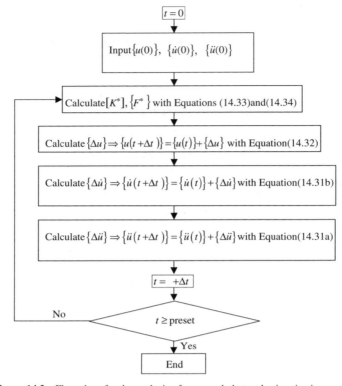

Figure 14.3 Flow chart for the analysis of structural elasto-plastic seismic response

Based on the equations derived above, the flow chart of calculating seismic displacement response of the frame in elasto-plastic state is given in Figure 14.3.

14.1.3 Determination of Mass, Stiffness and Damping Matrices

14.1.3.1 Mass matrix

The mass matrix of frame structures can be determined according to Equation (14.4), where the mass m_i is lumped at the level of the ith floor and is a sum of all the masses within the scope between half-storeys upwards and downwards of the floor.

14.1.3.2 Stiffness matrix

For planar steel frames, the basic variables are the horizontal displacements of floors and the vertical displacements, rotations and shear deformations of joint panels. Based on the incremental stiffness equations at time t for the elements of all the beams, columns, braces and joint panels in a frame, the global incremental stiffness equation of the frame can be assembled as

$$\begin{bmatrix} [K_{uu}(t)] & [K_{ur}(t)] \\ [K_{ru}(t)] & [K_{rr}(t)] \end{bmatrix} \begin{Bmatrix} \Delta u \\ \Delta r \end{Bmatrix} = \begin{Bmatrix} \Delta F_u \\ \Delta F_r \end{Bmatrix}, \tag{14.35}$$

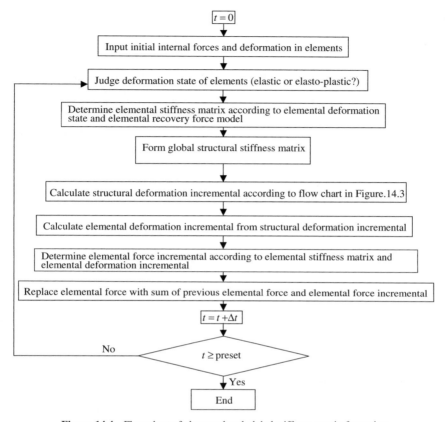

Figure 14.4   Flow chart of elemental and global stiffness matrix formation

where

$\{\Delta u\}$ is the vector of horizontal displacements of all the floors, $\{\Delta r\}$ is the vector of other basic variables, and $\{\Delta F_u\}$ and $\{\Delta F_r\}$ are the force vectors corresponding to the displacement vectors $\{\Delta u\}$ and $\{\Delta r\}$, respectively.

Under the action of horizontal earthquakes, $\{\Delta F_r\} = 0$. And the lateral stiffness matrix of the frame at time t can be derived with static condensation as

$$[K(t)] = [K_{uu}(t)] - [K_{ur}(t)][K_{rr}(t)]^{-1}[K_{ru}(t)]. \tag{14.36}$$

If the frame is in elastic state, i.e. all of the structural members are elastic, at time t, one has

$$[K(t)] = [K_e]. \tag{14.37}$$

In assembling the global stiffness matrix, the elemental stiffness matrix can be determined according to the flow chart given in Figure 14.4.

14.1.3.3 Damping matrix

To satisfy the orthotropic condition in Equation (14.18c), Rayleigh damping matrix can be adopted as

$$[C] = a[M] + b[K_e], \tag{14.38}$$

where a and b are undetermined parameters.

As the mass matrix $[M]$ and stiffness matrix $[K_e]$ satisfy the orthotropic condition, the Rayleigh damping condition defined above must satisfy it as well.

To calculate the parameters a and b taking any two vibration modes to Equation (14.38) results in

$$\{\phi_i\}^T[C]\{\phi_i\} = a\{\phi_i\}^T[M]\{\phi_i\} + b\{\phi_i\}^T[K_e]\{\phi_i\}, \tag{14.39a}$$

$$\{\phi_j\}^T[C]\{\phi_j\} = a\{\phi_j\}^T[M]\{\phi_j\} + b\{\phi_j\}^T[K_e]\{\phi_j\}. \tag{14.39b}$$

Dividing the above two equations with $\{\phi_i\}^T[M]\{\phi_i\}$ and $\{\phi_j\}^T[M]\{\phi_j\}$, respectively, and noting Equations (14.22) and (14.23), one has

$$2\omega_i\xi_i = a + b\omega_i^2, \tag{14.40a}$$

$$2\omega_j\xi_j = a + b\omega_j^2. \tag{14.40b}$$

And the solutions of a and b are

$$a = \frac{2\omega_i\omega_j(\xi_i\omega_j - \xi_j\omega_i)}{\omega_j^2 - \omega_i^2}, \tag{14.41a}$$

$$b = \frac{2(\omega_j\xi_j - \omega_i\xi_i)}{\omega_j^2 - \omega_i^2}. \tag{14.41b}$$

In practice, usually let $i = 1$, $j = 2$ and $\xi_i = \xi_j$.

14.1.4   Numerical Example

Seismic tests on a steel frame were conducted on a 4 m × 4 m shaking table in Tongji University. The frame tested is illustrated in Figure 14.5. The record of EI Centro North–South earthquake waves in 1940 was input along the lateral direction of the frame, and the maximum acceleration of the excitation was modified to 9.6 m/s². To examine the effect of joint-panel shear deformation on the seismic response of steel frames, the following three models are analysed for the frame tested:

- Model I: Treat the joint panel as an isolated element with shear deformation.

- Model II: Treat the joint panel as a rigid body without shear deformation.

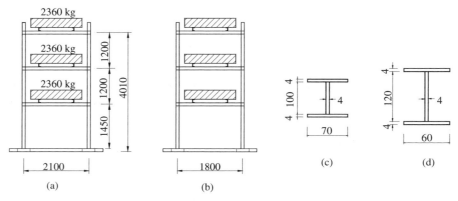

Figure 14.5   Frame tested on the shaking table: (a) front elevation; (b) lateral elevation; (c) section of column; (d) section of beam

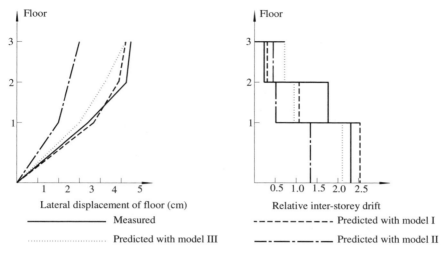

Figure 14.6 Maximum seismic response of the frame tested

- Model III: Neglect the joint panels and extend the lengths of the frame beam and column to the distance between the central lines of the frame components.

The seismic response of the frame, horizontal displacements of the floors and relative inter-storey drifts, obtained with measurements in tests and analyses are compared in Figure 14.6, from which it can be seen that model I (directly considering the joint-panel shear deformation) produces the closest results to test data and model III (the conventional model neglecting joint panels) also predicts the seismic response of H-section frames with good accuracy.

14.2 HALF-FRAME MODEL

The seismic response of structures is dependent on loading history, and the exact analysis should be conducted step by step with respect to time incremental, in each of which the global stiffness matrix of structures must be re-formed if it is in elasto-plastic state. Total computation costs will therefore be huge in such an exact analysis, which appeals that a simplified model with reduced degree of freedoms and economic computation is necessary for the purpose of engineering design and analysis in practice.

A simplified half-frame model is to be introduced in this section for elastic and elasto-plastic response analyses of planar steel frames subjected to earthquakes.

14.2.1 Assumption and Principle of Half-Frame

For the pure steel frames with rigid beam-to-column connections, the following assumptions are made:

(1) neglect axial deformation of frame columns;

(2) the size, rotation and shear deformation of each joint panel at the level of the same floor of the frame applied with horizontal forces are same.

With the above assumptions, any frame can be transformed to a half-frame, as shown in Figure 14.7, to perform simplified seismic analysis.

For a steel frame without any stagger storey, the elastic and elasto-plastic parameters of the corresponding half-frame (see Figure 14.7) at an arbitrary storey with n bays can be determined as follows.

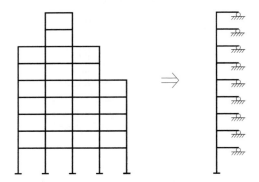

Figure 14.7 An equivalent half-frame of a complete frame

14.2.1.1 *Elastic parameters*

The bending stiffness of the column is

$$(EI)_c = \sum_{i=1}^{n+1} (EI)_{ci}. \tag{14.42}$$

The length of the column is

$$l_c = l_{ci}. \tag{14.43}$$

The shear impact factor of the column section is

$$r_c = \frac{\displaystyle\sum_{i=1}^{n+1} (EI)_{ci} r_{ci}}{\displaystyle\sum_{i=1}^{n+1} (EI)_{ci}}. \tag{14.44}$$

The bending stiffness of the beam is

$$(EI)_g = 2\sum_{i=1}^{n} (EI)_{gi}. \tag{14.45}$$

The length of the beam is

$$l_g = \frac{\displaystyle\sum_{i=1}^{n} (EI)_{gi}}{2\displaystyle\sum_{i=1}^{n} \frac{(EI)_{gi}}{l_{gi}}}. \tag{14.46}$$

The shear impact factor of the beam section is

$$r_g = \frac{\displaystyle\sum_{i=1}^{n} \frac{(EI)_{gi}}{l_{gi}} r_{gi}}{\displaystyle\sum_{i=1}^{n} \frac{(EI)_{gi}}{l_{gi}}}. \tag{14.47}$$

The elastic stiffness of the joint panel is

$$k_{re} = \sum_{i=1}^{n+1} (k_{re})_i. \tag{14.48}$$

All the parameters on the right-hand side of Equations (14.42) –(14.48) are those in the original frame, and the parameters on the left-hand side of the equations are those for the simplified half-frame model. The shear impact factor of beam or column section is defined as

$$r = \frac{12EI\mu}{GAl^2}, \tag{14.49}$$

where E and G are the stretch and shear elastic modulus, I and A are the inertial moment and area of the beam or column section, respectively, l is the length of the beam or column and μ is the shear shape factor of the beam or column section.

14.2.1.2 Elasto-plastic parameters

The initial yielding moment of the column is

$$(M_s)_c = \beta \sum_{i=1}^{n+1} (M_s)_{ci}. \tag{14.50}$$

The ultimate yielding moment of the column is

$$(M_p)_c = \sum_{i=1}^{n+1} (M_p)_{ci}. \tag{14.51}$$

The initial yielding moment of the beam is

$$(M_s)_g = 2\beta \sum_{i=1}^{n} (M_s)_{gi}. \tag{14.52}$$

The ultimate yielding moment of the beam is

$$(M_p)_g = 2 \sum_{i=1}^{n} (M_p)_{gi}. \tag{14.53}$$

The shear yielding moment of the joint panel is

$$M_{\gamma p} = \sum_{i=1}^{n+1} (M_{\gamma p})_i. \tag{14.54}$$

All the parameters on the right-hand side of Equations (14.50) – (14.54) are those in the original frame, and the parameters on the left-hand side of the equations are those for the simplified half-frame model. The factor β is the reduction factor of the initial yielding of beams and columns to include the effects due to nonuniform internal forces in the original frame and asynchronous initial yielding of beams and columns. Generally, $\beta = 0.7 - 0.9$, where if the ratios of the internal forces of frame beams and columns on the storey considered to their yielding capacities are uniform, the upper values can be selected and otherwise lower values are used.

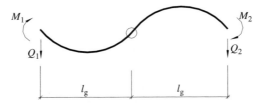

Figure 14.8 Analytical model of the half-frame beam element

14.2.2 Stiffness Equation of Beam Element in Half-Frame

One end of the beam element in a half-frame is pinned and the other is rigid. The stiffness equation of the beam element in the half-frame can be developed as follows.

The corresponding complete-frame beam element to that in the half-frame model can be obtained by elongating the half-frame beam antisymmetrically, as shown in Figure 14.8. From the symmetry, it can be known that $M_1 \equiv M_2$. With analysis of the common beam element, the incremental stiffness equation for the complete-frame beam element is

$$
\begin{Bmatrix} dQ_1 \\ dM_1 \\ dQ_2 \\ dM_2 \end{Bmatrix} = [k] \begin{Bmatrix} d\delta_1 \\ d\theta_1 \\ d\delta_2 \\ d\theta_2 \end{Bmatrix},
\tag{14.55}
$$

where the elemental stiffness matrix $[k]$ is determined according to the approaches presented in Chapter 2 or 4. Assume that

$$
[k] = \begin{bmatrix} k_{11} & k_{12} & k_{13} & k_{14} \\ k_{21} & k_{22} & k_{23} & k_{24} \\ k_{31} & k_{32} & k_{33} & k_{34} \\ k_{41} & k_{42} & k_{43} & k_{44} \end{bmatrix}.
\tag{14.56}
$$

As $d\theta_1 = d\theta_2$ and $d\delta_2 = -d\delta_1$, it can be derived from Equation (14.55) that

$$
\begin{Bmatrix} dQ_1 \\ dM_1 \end{Bmatrix} = [k_n] \begin{Bmatrix} d\delta_1 \\ d\delta_2 \end{Bmatrix},
\tag{14.57}
$$

where $[k_n]$ is the elemental stiffness matrix of the half-frame beam element, given by

$$
[k_n] = \begin{bmatrix} k_{11} - k_{13} & k_{12} + k_{14} \\ k_{21} - k_{23} & k_{22} + k_{24} \end{bmatrix}.
\tag{14.58}
$$

14.2.3 Numerical Examples

To verify the reliability and validity of the simplified half-frame method, a comparative study is conducted between non-simplified and simplified seismic analyses of the steel frames subjected to horizontal earthquakes. Three examples are illustrated where EI Centro North–South wave in 1940 is adopted for seismic ground movement.

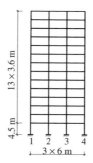

Figure 14.9 A 14-storey regular frame for Example 1

Example 1

This is a 14-storey regular frame as shown in Figure 14.9. The mass on each floor is 6.0×10^4 kg, and the other parameters are given in Table 14.1.

The maximum elasto-plastic lateral floor deflections, inter-storey drifts and inter-storey shears of the frame in this example obtained with both non-simplified and simplified analytical models are compared in Figure 14.10.

Table 14.1 Frame parameters in Example 1

Frame components	Elastic parameter	Elasto-plastic parameter
Perimeter columns (1st–7th storey)	2.92×10^8	1.52×10^6
Perimeter columns (8th–14th storey)	2.40×10^8	1.32×10^6
Inside columns (1st–7th storey)	3.51×10^8	1.75×10^6
Inside columns (8th–14th storey)	2.92×10^8	1.52×10^6
Outside beams (1st–7th storey)	2.92×10^8	1.39×10^6
Outside beams (8th–14th storey)	2.40×10^8	1.20×10^6
Inside beams (1st–7th storey)	3.51×10^8	1.60×10^6
Inside beams (8th–14th storey)	2.92×10^8	1.39×10^6

Notes: (1) The elastic parameter in the table represents the bending stiffness EI in unit of N m^2. (2) The elasto-plastic parameter in the table represents the ultimate yielding moment M_p in unit of N m. (3) The meanings of the parameters in the following tables are same as above.

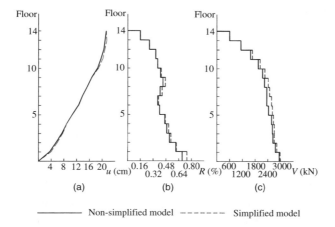

Figure 14.10 Maximum elasto-plastic seismic response in Example 1: (a) lateral deflection; (b) relative inter-storey drift; (c) inter-storey shear

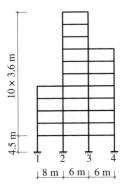

Figure 14.11 An 11-storey irregular frame for Example 2

Example 2

As shown in Figure 14.11, an 11-storey irregular frame is studied in this example. The mass on the floors from the 1st to the 5th is 9.0×10^4 kg, that from the 6th to the 8th is 5.5×10^4 kg and from the 9th to the 11th is 2.5×10^4 kg. The other information of the frame is listed in Table 14.2.

Similarly, the maximum lateral floor deflections, inter-storey drifts and inter-storey shears of the frame in this example obtained with both non-simplified and simplified analytical models are compared in Figure 14.12.

Table 14.2 Frame parameters in Example 2

Frame components	Elastic parameter	Elasto-plastic parameter
Columns at axis 1	2.40×10^8	1.25×10^4
Columns at axes 2–4	4.98×10^8	2.16×10^4
Beams at left span	4.98×10^8	1.73×10^4
Beams at middle and right spans	2.40×10^8	1.00×10^4

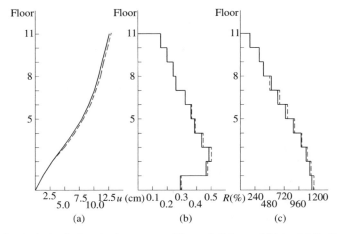

Figure 14.12 Maximum elasto-plastic seismic response of Example 2 frame: (a) lateral deflection; (b) relative inter-storey drift; (c) inter-storey shear

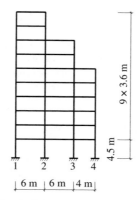

Figure 14.13 A 10-storey irregular frame for Example 3

Example 3

A 10-storey irregular frame in Figure 14.13 is studied as the third example. The mass on the floors from the 1st to the 6th is 3.5×10^4 kg, that from the 7th to the 8th is 2.6×10^4 kg and from the 9th to the 10th is 1.3×10^4 kg. The other information of the frame is listed in Table 14.3.

The maximum lateral floor deflections, inter-storey drifts and inter-storey shears of the frame in this example obtained with both non-simplified and simplified analytical models are compared in Figure 14.14.

Table 14.3 Frame parameters in Example 3

Frame components	Elastic parameter	Elasto-plastic parameter
Columns at axes 1–3	2.40×10^8	8.75×10^5
Columns at axis 4	1.57×10^8	6.37×10^5
Beams	2.40×10^8	7.00×10^5

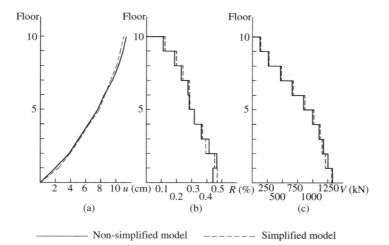

——— Non-simplified model   – – – – – – Simplified model

Figure 14.14 Maximum elasto-plastic seismic response of Example 3 frame: (a) lateral deflection; (b) relative inter-storey drift; (c) inter-storey shear

The above examples indicate that the half-frame model has good adaptability because it retains the characteristics of the component-representing frame system. This simplified model is applicable to strong-beam–weak-column frames or weak-beam–strong-column frames and to regular or irregular frames. Satisfactory results can be obtained by the simplified model in elastic and elasto-plastic analyses of the steel frames subjected to earthquakes. Generally, the relative error of the maximum lateral floor deflections and inter-storey shears is under 10 % whereas that of inter-storey drifts is under 15 %.

14.3 SHEAR-BENDING STOREY MODEL

As axial deformation of frame columns is ignored in the half-frame model, it may be suspicious when the half-frame model is used for seismic analysis of high-rise frame buildings. Another simplified model, the shear-bending storey model (see Figure 14.15), is proposed especially for considering the characteristics of high-rise frames. Equivalent storey-type structure is used in the shear-bending storey model, where lateral displacement of the original frame is decoupled into the bending deflection and shear deformation of the storey-type structure. The bending stiffness of the equivalent structure is determined by the axial stiffness of the original frame columns, whereas the shear stiffness of the storey-type structure by the characteristics of inter-storey recovery shear force of the original frame. For the sake of convenient application, the following assumptions are made:

(1) Yielding of joint panels is not earlier than that of beams or columns;

(2) Effects of joint-panel shear deformations are approximately considered by neglecting joint-panel sizes and adopting the length of beams and columns as the distance between their central lines.

14.3.1 Equivalent Stiffness

14.3.1.1 Equivalent bending stiffness

The global bending stiffness of the high-rise frame is due to the axial deformation of the frame columns, which can be assumed to be always in elastic state.

The axial deformation lags of the frame columns will happen due to the finite stiffness of the frame beams, which leads to a nonlinear distribution of axial deformations in the frame columns at the same storey as shown in Figure 14.16. Based on this, the equivalent inter-storey bending stiffness can be expressed approximately as

$$(EI)_i = \sum_j EI_{ij} + \alpha_i \sum_j EA_{ij} l_{ij}^2, \tag{14.59}$$

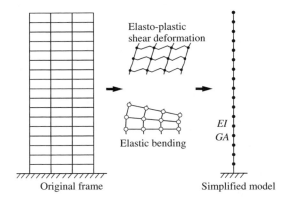

Figure 14.15 Diagram for the bending-shear storey model

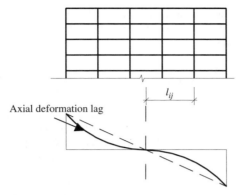

Figure 14.16 Distribution of axial deformation of frame columns

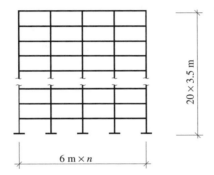

Figure 14.17 Standard frame

where I_{ij}, A_{ij} and l_{ij} are, respectively, the inertial moment, section area and the distance to the neutral axis (see Figure 14.16) of the jth column on the ith storey of the frame and α_i is the convert parameter considering the axial deformation lag in the frame columns.

The values of α_i can be calculated with trial. The horizontal displacement of the original frame, x, due to axial deformation under different ratios of beam to column linear stiffness is calculated at first, and then a single cantilever model with equivalent bending stiffness is used to simulate the horizontal displacement of the frame under the same loads. When the horizontal displacement of the equivalent cantilever, x', is equal to x, α_i is obtained. A standard regular frame, the size of which is shown in Figure 14.17, is used in such trial calculations. A certain ratio of beam to column linear stiffness is adopted in the trial calculations for all the beams and columns in the frame. The values of α_i in different conditions are listed in Table 14.4.

In the calculation of Equation (14.59), if the linear stiffness ratios of the beams to the columns on one storey of the frame are not the same, the average of the linear bending stiffnesses, i_g and i_c, of the beams and

Table 14.4 Values of α_i in Equation (14.59)

Number of spans	i_g/i_c										
	0.1	0.3	0.5	0.7	0.9	1.0	2.0	3.0	5.0	8.0	10.0
3	0.956	0.960	0.963	0.963	0.965	0.965	0.968	0.970	0.975	0.980	0.983
4	0.900	0.903	0.903	0.905	0.910	0.910	0.915	0.915	0.920	0.925	0.925
5	0.800	0.803	0.805	0.810	0.815	0.820	0.825	0.825	0.830	0.835	0.840
6	0.700	0.725	0.735	0.735	0.740	0.745	0.750	0.750	0.760	0.765	0.785

the columns on the storey can be used to check out α_i in Table 14.4. For one-bay or two-bay symmetric frames, it simply takes $\alpha_i = 1$, whereas for two-bay asymmetric frames, α_i can be calculated according to the condition of the three-bay frame.

14.3.1.2 Equivalent shear stiffness

The shear recovery force model for an arbitrary storey of a steel frame with the inter-storey shear deformation under cyclic loading is given in Figure 14.18, where V_s and V_p are the shear forces corresponding to initial and ultimate yielding states of the frame storey, respectively, δ_s and δ_p are the corresponding inter-storey drifts, and V_u and δ_u are the inter-storey shear and inter-storey drift at the time of unloading, respectively.

When the storey is in loading state, if storey shear $|V| < V_s$, the storey is in elastic, and its shear stiffness can be calculated with the 'D-value method' (Long and Bao, 1981; Yang, 1979); if $|V| > V_p$, the storey is in hardening state; and if $V_s \leq |V| \leq V_p$, the storey is in elasto-plastic state, and its shear stiffness can be calculated with linear interpolation between those in elastic and hardening states. The equivalent shear stiffness of the storey in loading state can be expressed as

$$(GA)_i = \frac{\mathrm{d}V_i}{\mathrm{d}\delta_i}H_i = RH_i \sum D_i, \tag{14.60}$$

in which

$$\begin{aligned}
R &= 1, \quad \text{for} \quad |V| < V_s, \\
R &= q, \quad \text{for} \quad |V| > V_p, \\
R &= 1 - (1-q)\frac{|V| - V_s}{V_p - V_s}, \quad \text{for} \quad V_s \leq |V| \leq V_p,
\end{aligned} \tag{14.61}$$

where $\sum D_i$ is the total inter-storey sway stiffness of the storey, determined as the sum of the D-values of every column at the storey of the frame (Long and Bao, 1981; Yang, 1979), H_i is height of the storey and q is the hardening factor and generally $q = 0.025$.

When the storey is in unloading state, it is always elastic, and its shear stiffness can be obtained by

$$(GA)_i = H_i \sum D_i. \tag{14.62}$$

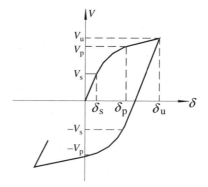

Figure 14.18   Inter-storey shear recovery force model of frame

14.3.2 Inter-Storey Shear Yielding Parameters

To determine the equivalent shear stiffness in the simplified model, the inter-storey shear forces V_s and V_p corresponding to the initial and ultimate yielding states of the frame storey are needed.

14.3.2.1 Calculation of V_s

Assume that the horizontal load under horizontal earthquakes applied on the frame is converse-triangular distributed (see Figure 14.19). The end moments of the frame beams and columns, M_g and M_c, can be calculated with the 'D-value method' under this load. Denote the initial yielding moments of the beams and columns with M_{gs} and M_{cs}, then

$$V_s = \min\left(\min\frac{M_{gs}}{M_g}, \min\frac{M_{cs}}{M_c}\right)V_i, \tag{14.63}$$

where V_i is the inter-storey shear force induced by the converse-triangular distributed load, $\min(M_{gs}/M_g)$ is the minimum value of M_{gs}/M_g in the beams on the ith storey and $\min(M_{cs}/M_c)$ is the minimum value of M_{cs}/M_c in the columns on the ith storey.

14.3.2.2 Calculation of V_p

For the frame where columns yield in advance of beams, V_p is the summation of the shear forces resisted by all the columns on the storey when the end moments of the columns approach ultimate yielding moment M_{cp}, namely

$$V_p = \frac{2\sum M_{cp}}{H_i}. \tag{14.64}$$

For the frame where beams yield in advance of columns and if the yielding is as shown in Figure 14.19(a), by the virtual work principle one has

$$(m+1)M_{cp}\theta + 2nmM_{gp} = P_0H_1\theta + 2P_0(H_1+H_2)\theta + \cdots + iP_0(H_1+H_2+\cdots+H_i)\theta + \cdots$$

$$+ nP_0(H_1+H_2+\cdots+H_n)\theta, +\cdots+ iP_0(H_1+H_2+\cdots+H_i)\theta + \cdots$$

$$+ nP_0(H_1+H_2+\cdots+H_n)\theta \tag{14.64}$$

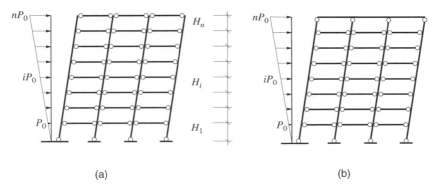

(a)   (b)

Figure 14.19 Frame where beams yield in advance of columns

where m is the number of frame bays, n is the number of frame storeys, θ is the rotation of plastic hinges, and M_{gp} and M_{cp} are the ultimate yielding moments of frame beams and columns, respectively.

Equation (14.64) can be solved as

$$P_0 = \frac{(m+1)M_{cp} + 2nmM_{gp}}{H_A},\tag{14.65}$$

where

$$H_A = H_1 + 2(H_1 + H_2) + \cdots + iP_0(H_1 + H_2 + \cdots + H_i) + \cdots + n(H_1 + H_2 + \cdots + H_n).\tag{14.66}$$

If the yielding of the frame is as shown in Figure 14.19(b), the expression for P_0 by similar derivation is

$$P_0 = \frac{2(m+1)M_{cp} + 2(n-1)M_{gp}}{H_A}.\tag{14.67}$$

For the frame of arbitrary yielding type, i.e. the frame can be irregular and the ultimate yielding moments of the frame beams and columns can be arbitrary, if the distribution of the plastic hinges of the frame is determined, one has

$$P_0 = \frac{\sum_j M_{cp} + \sum_j M_{gp}}{H_A},\tag{14.68}$$

where $\sum_j M_{cp}$ and $\sum_j M_{gp}$ are the summation of the moment resistances by all the beam and column plastic hinges, respectively.

The ultimate shear capacity of the ith storey of the frame can be expressed with P_0 as

$$V_p = \sum_{j=1}^{n} jP_0.\tag{14.69}$$

14.3.3 Examples

A two-bay 20-storey weak-beam–strong-column regular frame, as shown in Figure 14.20, is selected as an illustrative example to the application of the shear-bending storey model for elastic and elasto-plastic

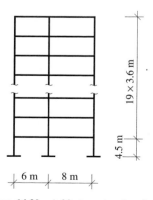

Figure 14.20   A 20-storey two-bay frame

Table 14.5 Major parameters in the example structure

Parameters	Components				
	Columns at axis 1	Columns at axis 2	Columns at axis 3	Beams at left span	Beams at right span
Compression stiffness EA (N)	2.975×10^9	4.45×10^9	3.42×10^9	—	—
Bending stiffness EI (N m²)	1.57×10^8	4.83×10^8	2.40×10^8	1.57×10^8	2.40×10^8
Initial yielding moment M_s (N m)	0.77×10^6	1.84×10^6	1.06×10^6	0.62×10^6	0.85×10^6
Ultimate bending moment M_p (N m)	0.91×10^6	2.16×10^6	1.25×10^6	0.73×10^6	1.00×10^6

seismic analyses. The mass on each floor of the frame is constantly 2.7×10^4 kg, and other parameters are listed in Table 14.5. The input earthquake movement is the EI Centro North–South wave of 1940.

14.3.3.1 Calculation of $(EI)_i$

The average linear stiffness of the frame beams is

$$i_g = \frac{i_{g1} + i_{g2}}{2} = \frac{1.57 \times 10^8/6 + 2.40 \times 10^8/8}{2} = 7.8 \times 10^7 \ (\text{N m}).$$

The average linear stiffness of the columns is

• for the first storey:

$$i_{c0} = \frac{i_{c10} + i_{c20} + i_{c30}}{3} = 6.5 \times 10^7 \ (\text{N m});$$

• for the other storeys:

$$i_c = \frac{i_{c1} + i_{c2} + i_{c3}}{3} = 8.1 \times 10^7 \ (\text{N m}).$$

The ratio of beam to column linear stiffness is

$$i_g/i_{c0} = \frac{2.8 \times 10^7}{6.5 \times 10^7} = 0.43, \quad \text{by Table 14.4,} \quad \alpha = 0.962;$$

$$i_g/i_c = \frac{2.8 \times 10^7}{8.1 \times 10^7} = 0.35, \quad \text{by Table 14.4,} \quad \alpha = 0.961.$$

Assume the neutral axis of the global bending of the frame is at the middle of overall span, and then the equivalent bending stiffness by Equation (14.59) is

• for the first storey:

$$(EI)_0 = 8.8 \times 10^8 + 0.962 \times 3.18 \times 10^{11} = 3.07 \times 10^{11} \ (\text{N m});$$

• for the other storeys:

$$(EI)_i = \sum EI_{ij} + \alpha \sum EA_{ij} l_{ij}^2 = (1.57 \times 10^8 + 4.83 \times 10^8 + 2.40 \times 10^8)$$
$$+ 0.961(2.975 \times 10^8 \times 7^2 + 4.45 \times 10^9 \times 1^2 + 3.42 \times 10^9 \times 7^2)$$
$$= 8.8 \times 10^8 + 0.961 \times 3.18 \times 10^{11}$$
$$= 3.06 \times 10^{11} \ (\text{N m}).$$

14.3.3.2 Calculation of (GA)ᵢ

The 'D-value method' is first used to calculate the D-value, i.e. the lateral stiffness of the frame columns, which is not repeated here. And then

- for the first storey: $\sum D_{c0} = 4.46 \times 10^7 (\text{N/m})$;
- for the other storeys: $\sum D_c = 4.02 \times 10^7 (\text{N/m})$.

By Equation (14.60), the equivalent inter-storey shear stiffness of the frame is obtained as

- for the first storey: $(GA)_0 = R \times 4.5 \times 4.46 \times 10^7 = 2.01 \times 10^8 R(\text{N})$;
- for the other storeys: $(GA)_i = R \times 3.6 \times 4.02 \times 10^7 = 1.45 \times 10^8 R(\text{N})$,

where R is determined according to Equation (14.61) for loading state on the basis of the inter-storey shear force at different time under earthquake action, whereas $R = 1$ for unloading state.

In the calculation of R with Equation (14.61), the initial inter-storey yielding shear V_s is determined by Equation (14.63) and the ultimate yielding shear V_p by Equation (14.69). (Note that this example is the case that frame beams yield in advance of frame columns.) Then, P_0 is obtained as

$$P_0 = [2 \times 20 \times (0.73 \times 10^6 + 1.00 \times 10^6) + (0.91 \times 10^6 + 2.16 \times 10^6 + 1.25 \times 10^6)]$$
$$\div [4.5 \times 20 + 3.6 \times (2 \times 3 \times 2 + 4 \times 3 + \cdots + 20 \times 19)]$$
$$= 7606 \, \text{N}.$$

14.3.3.3 Analysis and results

After the equivalent inter-storey bending stiffness and equivalent shear stiffness are obtained, the frame can be simplified to a bending-shear cantilever where only two degrees of freedom, horizontal displacement and rotation, exist at the location of each floor. Computation effort decreases dramatically due to the largely reduced structural degrees of freedom. The maximum horizontal displacement, inter-storey drift and inter-storey shear results of the frame under seismic excitation determined by non-simplified and simplified models are compared in elastic scope (Figure 14.21) and elasto-plastic scope (Figure 14.22).

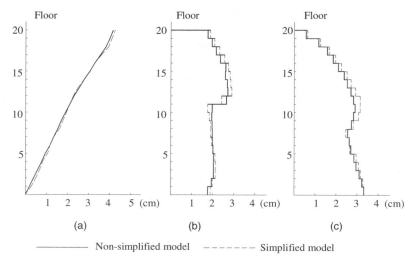

Figure 14.21 Maximum elastic seismic response of the frame: (a) lateral deflection; (b) relative inter-storey drift; (c) inter-storey drift

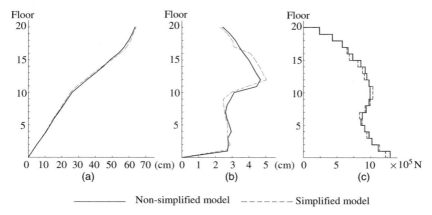

Figure 14.22 Maximum elasto-plastic seismic response of the frame: (a) lateral deflection; (b) relative inter-storey drift; (c) inter-storey drift

The above comparisons indicate good agreement between the elastic results of the simplified bending-shear storey model and the non-simplified complete component-representing frame model, and the maximum relative error is less than 10 %. The relative error of the elasto-plastic results is larger than that of elastic ones, but within 20 %. However, usage of the simplified model reduces hugely the computation effort and facilitates data input. Moreover, the maximum relative error happens generally at the storey where the inter-storey drift is small, which will not play a controllable role in the practical design of frames against earthquakes. As an approximate approach, therefore, the bending-shear storey model is applicable to the seismic analysis of high-rise frames in practice.

14.4 SIMPLIFIED MODEL FOR BRACED FRAME

Large error is possibly produced by the half-frame model to high-rise frame seismic analysis because of neglecting axial deformation of frame columns. Meanwhile, the bending-shear storey model may not be satisfactory in the elasto-plastic seismic analysis of steel frames because it departs far from the component-representing frame system. Combining the advantages of the half-frame model and the bending-shear storey model may produce a more general and efficient simplified model for seismic analysis of arbitrary steel frames, including braced frames.

14.4.1 Decomposition and Simplification of Braced Frame

Decompose a braced frame into two parts working together, a pure frame and a pure bracing system, as shown in Figure 14.23. The pure frame part can be simplified with the half-frame model to reduce structural degrees of freedom, whereas the pure bracing system can be transferred to a truss system for the sake of analysis convenience. The geometric nonlinearity and $P - \Delta$ effect can be simulated with the model of a series of hinged rigid masts (see Section 13.3), where the vertical loads P_i $(i = 1, 2, ..., n)$ applied at hinged nodes are the total gravity loads (dead load plus live load) on the ith floor of the frame. For the simplified model in Figure 14.23, the stiffness matrix of the frame corresponding to the lateral floor deflections can be

$$[K] = [K_f] + [K_b] - [K_G], \tag{14.70}$$

where $[K_f]$ is the stiffness matrix of the pure frame, $[K_b]$ the stiffness matrix of the pure bracing system and $[K_G]$ the geometric matrix of the frame, determined by Equation (13.32).

The calculation of $[K_f]$ and $[K_b]$ is described in the following.

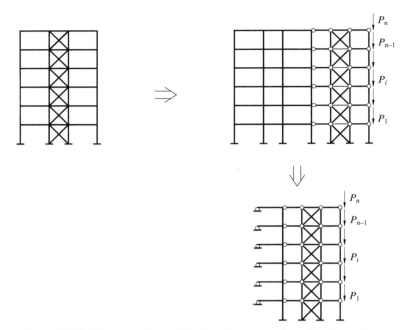

Figure 14.23   Diagram of the simplified model for the analysis of the braced frame

14.4.2   Stiffness Matrix of Pure Frame

The elastic and elasto-plastic stiffness matrices of the pure frame can be obtained by the half-frame model. But it should be noted that the half-frame model is based on the assumption that the axial deformations of frame columns are neglected. To involve the effect of axial deformation of columns, an elastic stepped bending bar (see Figure 14.24) can be used to approximate the global flexure due to axial deformation of frame columns. This equivalent bending bar works together with the half-frame in a serial manner, and the stiffness matrix of the pure frame can be expressed as

$$[K_f] = ([K_{hf}]^{-1} + [K_{af}]^{-1})^{-1}, \tag{14.71}$$

where $[K_{hf}]$ is the elastic or elasto-plastic stiffness matrix of the half-frame and is determined by the method in Section 14.2, and $[K_{af}]$ is the elastic stiffness matrix of the equivalent bending bar, considering that the axial deformation of frame columns is always in elastic state.

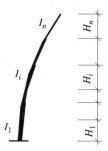

Figure 14.24   An equivalent bending bar

By the elemental stiffness equations of the equivalent bending bar for each storey of the frame, the incremental global stiffness equation of the equivalent bending bar can be assembled as

$$
\left\{ \begin{array}{c} \{dF_u\} \\ \{dF_\theta\} \end{array} \right\} = \left[\begin{array}{cc} [K_{auu}] & [K_{au\theta}] \\ [K_{a\theta u}] & [K_{a\theta\theta}] \end{array} \right] \left\{ \begin{array}{c} \{du\} \\ \{d\theta\} \end{array} \right\},
\tag{14.72}
$$

where $\{du\}$ and $\{d\theta\}$ are the incremental vectors of the horizontal displacements and rotations of the nodes at the level of all the floors, respectively, and $\{dF_u\}$ and $\{dF_\theta\}$ are the incremental vectors of the nodal horizontal forces and moments, respectively.

Under the action of horizontal earthquakes, $\{dF_\theta\} = 0$, the stiffness matrix of the equivalent bending bar can be obtained by static condensation as

$$
\{dF_u\} = [K_{af}]\{du\},
\tag{14.73}
$$

where

$$
[K_{af}] = [K_{auu}] - [K_{au\theta}][K_{a\theta\theta}]^{-1}[K_{a\theta u}].
\tag{14.74}
$$

14.4.3 Stiffness Matrix of Pure Bracing System

The elastic and elasto-plastic stiffness matrices of the pure bracing system can be determined according to the model of the hinged truss system, where the axial deformation of truss columns is considered but that of truss beams is neglected. The axial stiffness of truss columns remains elastic, whereas the analysis of the bracing element can be referred to in Chapter 10. By the elemental axial stiffness equations of truss columns and braces, the incremental global stiffness equation of the pure bracing system can be assembled as

$$
\left\{ \begin{array}{c} \{dF_u\} \\ \{dF_v\} \end{array} \right\} = \left[\begin{array}{cc} [K_{buu}] & [K_{buv}] \\ [K_{bvu}] & [K_{bvv}] \end{array} \right] \left\{ \begin{array}{c} \{du\} \\ \{dv\} \end{array} \right\},
\tag{14.75}
$$

where $\{du\}$ is the incremental vector of horizontal displacements of all the floors, $\{dv\}$ is the incremental vector of vertical displacements of the nodes in the bracing system, and $\{dF_u\}$ and $\{dF_v\}$ are the incremental vectors corresponding to $\{du\}$ and $\{dv\}$, respectively.

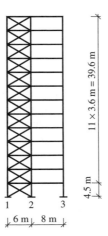

Figure 14.25   A 12-storey two-bay braced frame

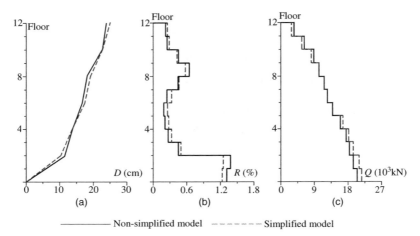

Figure 14.26 Maximum elasto-plastic seismic response of the frame: (a) lateral deflection; (b) relative inter-storey drift; (c) inter-storey drift

Table 14.6 Beam and column parameters in the example structure

Components	Columns at axis 1	Columns at axis 2	Columns at axis 3	Beams at left span	Beams at right span
Bending stiffness EI (N m^2)	1.39×10^8	3.33×10^8	1.93×10^8	2.03×10^8	2.83×10^8
Plastic moment (N m)	0.73×10^6	1.30×10^6	0.90×10^6	0.87×10^6	1.08×10^6

Under the action of horizontal earthquakes, $\{dF_v\} = 0$, the stiffness matrix of the pure bracing system can be obtained by static condensation as

$$\{dF_u\} = [K_b]\{du\}, \tag{14.76}$$

where

$$[K_b] = [K_{buu}] - [K_{buv}][K_{bvv}]^{-1}[K_{bvu}]. \tag{14.77}$$

14.4.4 Example

To illustrate the validity of the simplified model proposed above, a 12-storey two-bay braced frame is studied. The non-simplified complete component-representing model and the simplified model are comparably used in this example for elasto-plastic seismic analysis.

The frame for this example study is shown in Figure 14.25, and the mass on each floor is constantly 4.0×10^4 kg. Along the whole frame height of the first bay, the X-type braces are placed. The sectional area of the brace on the first storey is 36.24×10^{-4} m^2 and the slenderness ratio is 44.5, whereas the sectional area and slenderness ratio for braces on all the other storeys are 29.29×10^{-4} m^2 and 51.1, respectively. The other necessary parameters of the frame beams and columns are listed in Table 14.6. EI Centro North–South wave in 1940 is also selected as the earthquake input. The maximum seismic response results by both simplified and non-simplified models are given in Figure 14.26, where the comparison indicates a good coincidence.

15 Analysis Model for Space Steel Frames

No matter what kind of structural analysis is performed, such as elastic stability analysis, nonlinear analysis and seismic response analysis of steel frames as described in Chapters 12, 13 and 14, the important task is to develop the global structural stiffness equation based on the analysis model of frames.

Space steel frames generally have the following characteristics (Figure 15.1):

(1) Frame columns are perpendicular to the horizontal (ground) plane, whereas frame beams are parallel to the horizontal plane.

(2) The in-plane stiffness of each floor in the frames is very large and can be idealized to be infinite, due to the existence of floor slabs and/or horizontal floor bracing.

Several analysis models considering the above characteristics are to be introduced in this chapter.

15.1 SPACE BAR MODEL

In the space bar model, the columns, beams, braces and joint panels in a steel frame are treated directly as basic elements.

15.1.1 Transformation from Local to Global Coordinates

The elemental stiffness equations of the beams, columns, braces and joint panels can be established in their own local coordinates . The first step in the analysis of a space frame is to transform force and displacement vectors in the local coordinates to those in the global coordinates of the frame and establish the stiffness matrix of the frame in the global coordinates.

In the following discussion, $x - y - z$ denote the local coordinates and $u - v - w$ the global coordinates, where x, y and u, v are the axes in the horizontal plane, whereas z and w are the axes perpendicular to the horizontal plane, as shown in Figure 15.2.

15.1.1.1 Coordinate transformation of beam element

An arbitrary beam element in the global coordinates is shown in Figure 15.3, where φ_g is the angle between the length axis of the beam element and the global axis ou. In the local coordinates, the displacement vectors

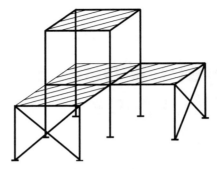

Figure 15.1 A space steel frame

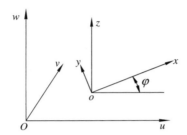

Figure 15.2 The global and local coordinate systems

at ends i and j of the beam element with the joint panels are

$$\{\delta_{gyi}\} = \begin{bmatrix} \delta_{zi}, & \theta_{yi}, & \gamma_{yi} \end{bmatrix}^{\mathrm{T}}, \tag{15.1a}$$

$$\{\delta_{gyj}\} = \begin{bmatrix} \delta_{zj}, & \theta_{yj}, & \gamma_{yj} \end{bmatrix}^{\mathrm{T}}. \tag{15.1b}$$

The displacement vectors at ends i and j of the element in the global coordinates are

$$\{D_{gyi}\} = \begin{bmatrix} \delta_{wi}, & \theta_{vi}, & \theta_{ui}, & \gamma_{vi}, & \gamma_{ui} \end{bmatrix}^{\mathrm{T}}, \tag{15.2a}$$

$$\{D_{gyj}\} = \begin{bmatrix} \delta_{wj}, & \theta_{vj}, & \theta_{uj}, & \gamma_{vj}, & \gamma_{uj} \end{bmatrix}^{\mathrm{T}}. \tag{15.2b}$$

By geometric derivation, the relationship between the displacement vectors of the beam element in the local and global coordinates is obtained as

$$\{\delta_{gys}\} = [B_g]\{D_{gys}\}, \quad s = i, j, \tag{15.3}$$

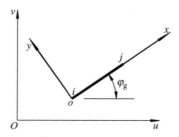

Figure 15.3 Beam element and global coordinates

where

$$[B_g] = \begin{bmatrix} 1 & 0 & 0 & 0 & 0 \\ 0 & \cos\varphi_g & \sin\varphi_g & 0 & 0 \\ 0 & 0 & 0 & \cos\varphi_g & \sin\varphi_g \end{bmatrix}. \tag{15.4}$$

Combining the displacement vectors at ends i and j of the element leads to the transformation equation as

$$\{\delta_{g\gamma}\} = \left\{ \begin{matrix} \{\delta_{g\gamma i}\} \\ \{\delta_{g\gamma j}\} \end{matrix} \right\} = \begin{bmatrix} [B_g] & 0 \\ 0 & [B_g] \end{bmatrix} \left\{ \begin{matrix} \{D_{g\gamma i}\} \\ \{D_{g\gamma j}\} \end{matrix} \right\} = [T_g]\{D_{g\gamma}\}. \tag{15.5}$$

By a similar derivation, the transformation equation for the force vectors from the local to the global coordinates is

$$\{F_{g\gamma}\} = [T_g]^{\mathrm{T}}\{f_{g\gamma}\}, \tag{15.6}$$

where $\{f_{g\gamma}\}$ and $\{F_{g\gamma}\}$ are the force vectors corresponding to $\{\delta_{g\gamma}\}$ and $\{D_{g\gamma}\}$, respectively.

Assume that the incremental stiffness equation of the beam element with joint panels in the local coordinates is

$$\{\mathrm{d}f_{g\gamma}\} = [k_{g\gamma}]\{\mathrm{d}\delta_{g\gamma}\}. \tag{15.7}$$

Substituting Equations (15.5) and (15.6) into Equation (15.7) yields the incremental stiffness equation of the beam element with joint panels in the global coordinates as

$$\{\mathrm{d}F_{g\gamma}\} = [K_{g\gamma}]\{\mathrm{d}D_{g\gamma}\}, \tag{15.8}$$

where $[K_{g\gamma}]$ is the stiffness matrix of the beam element with joint panels in the global coordinates, given by

$$[K_{g\gamma}] = [T_g]^{\mathrm{T}}[k_{g\gamma}][T_g]. \tag{15.9}$$

15.1.1.2 *Coordinate transformation of column element*

An arbitrary column element in the global coordinates is shown in Figure 15.4, where φ_c is the angle between the first principal axis of the column section, ox, and the global axis ou.

In the local coordinates, the displacement vectors at ends i and j of the column element with joint panels are

$$\{\delta_{cys}\} = [\delta_{zs}, \ \delta_{xs}, \ \theta_{ys}, \ \gamma_{ys}, \ \delta_{ys}, \ \theta_{xs}, \ \gamma_{xs}, \ \theta_{zs}]^{\mathrm{T}}, \quad s = i,j. \tag{15.10}$$

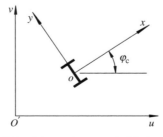

Figure 15.4   Column element and global coordinates

The displacement vectors at ends i and j of the element in the global coordinates are

$$\{D_{c\gamma s}\} = [\delta_{ws}, \quad \delta_{us}, \quad \theta_{vs}, \quad \gamma_{vs}, \quad \delta_{vs}, \quad \theta_{us}, \quad \gamma_{us}, \quad \theta_{ws}]^{\mathrm{T}}, \quad s = i, j. \tag{15.11}$$

The transformation equation of the displacement vectors from the local to the global coordinates is

$$\{\delta_{c\gamma}\} = \left\{ \begin{array}{c} \{\delta_{c\gamma i}\} \\ \{\delta_{c\gamma j}\} \end{array} \right\} = \left[\begin{array}{cc} [B_c] & 0 \\ 0 & [B_c] \end{array} \right] \left\{ \begin{array}{c} \{D_{c\gamma i}\} \\ \{D_{c\gamma j}\} \end{array} \right\} = [T_c]\{D_{c\gamma}\}, \tag{15.12}$$

where

$$[B_c] = \begin{bmatrix} 1 & 0 & 0 & 0 & 0 & 0 & 0 & 0 \\ 0 & \cos\varphi_c & 0 & 0 & \sin\varphi_c & 0 & 0 & 0 \\ 0 & 0 & \cos\varphi_c & 0 & 0 & \sin\varphi_c & 0 & 0 \\ 0 & 0 & 0 & \cos\varphi_c & 0 & 0 & \sin\varphi_c & 0 \\ 0 & -\sin\varphi_c & 0 & 0 & \cos\varphi_c & 0 & 0 & 0 \\ 0 & 0 & -\sin\varphi_c & 0 & 0 & \cos\varphi_c & 0 & 0 \\ 0 & 0 & 0 & -\sin\varphi_c & 0 & 0 & \cos\varphi_c & 0 \\ 0 & 0 & 0 & 0 & 0 & 0 & 0 & 1 \end{bmatrix}. \tag{15.13}$$

The transformation equation for the force vectors from the local to the global coordinates is

$$\{F_{c\gamma}\} = [T_c]^{\mathrm{T}}\{f_{c\gamma}\}, \tag{15.14}$$

where $\{f_{c\gamma}\}$ and $\{F_{c\gamma}\}$ are the force vectors corresponding to $\{\delta_{c\gamma}\}$ and $\{D_{c\gamma}\}$, respectively.

Assume that the incremental stiffness equation of the column element with joint panels in the local coordinates is

$$\{df_{c\gamma}\} = [k_{c\gamma}]\{d\delta_{c\gamma}\}. \tag{15.15}$$

Substituting Equations (15.12) and (15.14) into Equation (15.15) yields the incremental stiffness equation of the column element with joint panels in the global coordinates as

$$\{dF_{c\gamma}\} = [K_{c\gamma}]\{dD_{c\gamma}\}, \tag{15.16}$$

where $[K_{c\gamma}]$ is the stiffness matrix of the column element with joint panels in the global coordinates, given by

$$[K_{c\gamma}] = [T_c]^{\mathrm{T}}[k_{c\gamma}][T_c]. \tag{15.17}$$

15.1.1.3 Coordinate transformation of brace element

An arbitrary brace element in the global coordinates is shown in Figure 15.5, where φ_b is the angle between the projection of the brace element into the horizontal plane and the global axis ou.

In the local coordinates, the displacement vectors at ends i and j of the brace element with joint panels are

$$\{\delta_{b\gamma s}\} = [\delta_{xs}, \quad \delta_{zs}, \quad \theta_{ys}, \quad \gamma_{ys}]^{\mathrm{T}}, \quad s = i, j, \tag{15.18}$$

The displacement vectors at ends i and j of the element in the global coordinates are

$$\{D_{b\gamma s}\} = [\delta_{us}, \quad \delta_{vs}, \quad \delta_{ws}, \quad \theta_{vs}, \quad \theta_{us}, \quad \gamma_{vs}, \quad \gamma_{us}]^{\mathrm{T}}, \quad s = i, j, \tag{15.19}$$

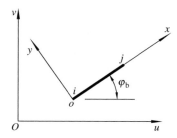

Figure 15.5   Brace element and global coordinates

By the geometric derivation, the relationship between the displacement vectors of the beam element in the local and global coordinates is obtained as

$$\{\delta_{b\gamma s}\} = [B_b]\{D_{b\gamma s}\}, \quad s = i, j, \tag{15.20}$$

where

$$[B_b] = \begin{bmatrix} \cos\varphi_b & \sin\varphi_b & 0 & 0 & 0 & 0 & 0 \\ 0 & 0 & 1 & 0 & 0 & 0 & 0 \\ 0 & 0 & 0 & \cos\varphi_b & \sin\varphi_b & 0 & 0 \\ 0 & 0 & 0 & 0 & 0 & \cos\varphi_b & \sin\varphi_b \end{bmatrix}. \tag{15.21}$$

Combining the displacement vectors at ends i and j of the element leads to the transformation equation as

$$\{\delta_{b\gamma}\} = \left\{ \begin{array}{c} \{\delta_{b\gamma i}\} \\ \{\delta_{b\gamma j}\} \end{array} \right\} = \begin{bmatrix} [B_b] & 0 \\ 0 & [B_b] \end{bmatrix} \left\{ \begin{array}{c} \{D_{b\gamma i}\} \\ \{D_{b\gamma j}\} \end{array} \right\} = [T_b]\{D_{b\gamma}\}. \tag{15.22}$$

In a similar manner, the transformation equation for the force vectors from the local to the global coordinates is obtained as

$$\{F_{b\gamma}\} = [T_b]^{\mathrm{T}}\{f_{b\gamma}\}, \tag{15.23}$$

where $\{f_{b\gamma}\}$ and $\{F_{b\gamma}\}$ are the force vectors corresponding to $\{\delta_{b\gamma}\}$ and $\{D_{b\gamma}\}$, respectively.

Assume that the incremental stiffness equation of the brace element with joint panels in the local coordinates is

$$\{df_{b\gamma}\} = [k_{b\gamma}]\{d\delta_{b\gamma}\}. \tag{15.24}$$

Substituting Equations (15.22) and (15.23) into Equation (15.24) yields the incremental stiffness equation of the brace element with joint panels in the global coordinates as

$$\{dF_{b\gamma}\} = [K_{b\gamma}]\{dD_{b\gamma}\}, \tag{15.25}$$

where $[K_{b\gamma}]$ is the stiffness matrix of the brace element with joint panels in the global coordinates, given by

$$[K_{b\gamma}] = [T_b]^{\mathrm{T}}[k_{b\gamma}][T_b]. \tag{15.26}$$

15.1.1.4   *Coordinate transformation of joint-panel element*

An arbitrary joint-panel element in the global coordinates is shown in Figure 15.6, where φ_γ is the angle between the joint-panel plane and the global axis *ou*.

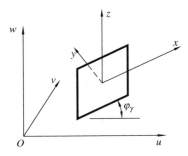

Figure 15.6 Joint-panel element and global coordinates

Assume that the incremental stiffness equation of the joint-panel element in the local coordinates is

$$dM_{\gamma y} = k_\gamma \, d\gamma_y. \tag{15.27}$$

And by the relationship between the force and the deformation vector of the joint-panel element in the local and global coordinates,

$$\left\{ \begin{matrix} dM_{\gamma v} \\ dM_{\gamma u} \end{matrix} \right\} = \begin{bmatrix} \cos \varphi_\gamma \\ \sin \varphi_\gamma \end{bmatrix} dM_{\gamma y}, \tag{15.28a}$$

$$d\gamma_y = [\cos \varphi_\gamma \ \sin \varphi_\gamma] \left\{ \begin{matrix} d\gamma_v \\ d\gamma_u \end{matrix} \right\}, \tag{15.28b}$$

one can obtain the incremental stiffness equation of the joint-panel element in the global coordinates as

$$\{dF_\gamma\} = [K_\gamma]\{dD_\gamma\}, \tag{15.29}$$

where

$$\{dF_\gamma\} = [dM_{\gamma v} \ dM_{\gamma u}]^{\mathrm{T}}, \tag{15.30a}$$

$$\{dD_\gamma\} = [d\gamma_v \ d\gamma_u]^{\mathrm{T}}, \tag{15.30b}$$

and $[K_\gamma]$ is the stiffness matrix of the joint-panel element in the global coordinates, given by

$$[K_\gamma] = \begin{bmatrix} \cos \varphi_\gamma \\ \sin \varphi_\gamma \end{bmatrix} k_\gamma [\cos \varphi_\gamma \ \sin \varphi_\gamma] = k_\gamma \begin{bmatrix} \cos^2 \varphi_\gamma & \sin \varphi_\gamma \cos \varphi_\gamma \\ \sin \varphi_\gamma \cos \varphi_\gamma & \sin^2 \varphi_\gamma \end{bmatrix}. \tag{15.31}$$

15.1.2 Requirement of Rigid Floor

Considering the infinite stiffness of the floors in a steel frame building, which are denoted as rigid floors, the horizontal displacements and rotations about the vertical axis of the nodes for structural analysis on an arbitrary floor relate to each other and can be represented with the horizontal displacements δ_{uok}, δ_{vok} and rotation θ_{wok} of a reference node on the floor (see Figure 15.7), namely

$$\delta_{ui} = \delta_{uok} - b_{ik}\theta_{wok},$$
$$\delta_{vi} = \delta_{vok} + a_{ik}\theta_{wok}, \tag{15.32}$$
$$\theta_{wi} = \theta_{wok},$$

where a_{ik} and b_{ik} are the global coordinates of node i relative to the reference node O_k of the kth floor.

Horizontal displacements and rotations about the vertical axis of all the columns and braces on a floor are therefore no longer independent . And such horizontal displacements and rotations can be incorporated and

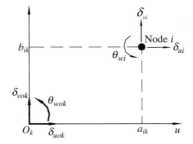

Figure 15.7   Horizontal displacements of a floor and node on the floor

represented with displacements δ_{uok}, δ_{vok} and rotation θ_{wok}, by which the total structural degrees of freedom are greatly reduced.

15.1.2.1 Stiffness equation of column element considering rigid floor

Assume that the ends i and j of an arbitrary column element are located on the $(k-1)$th and kth floors, respectively, and the global coordinates of these two ends relative to the reference nodes of the two floors are a_{ik-1}, b_{ik-1} and a_{ik}, b_{ik}, respectively. Denote the displacements of the column element in the global coordinates with the floor displacements as

$$\{D_{c\gamma ik-1}\} = [\delta_{wi}, \quad \delta_{uok-1}, \quad \theta_{vi}, \quad \gamma_{vi}, \quad \delta_{vok-1}, \quad \theta_{ui}, \quad \gamma_{ui}, \quad \theta_{uok-1}]^{\mathrm{T}}, \tag{15.33a}$$

$$\{D_{c\gamma jk}\} = [\delta_{wj}, \quad \delta_{uok}, \quad \theta_{vj}, \quad \gamma_{vj}, \quad \delta_{vok}, \quad \theta_{uj}, \quad \gamma_{uj}, \quad \theta_{uok}]^{\mathrm{T}}. \tag{15.33b}$$

By Equation (15.32), the relationship between the displacement vector of the column element itself in the global coordinates and that considering the rigid floor can be expressed as

$$\{D_{c\gamma}\} = \left\{ \begin{matrix} \{D_{c\gamma i}\} \\ \{D_{c\gamma j}\} \end{matrix} \right\} = \begin{bmatrix} [B_{cik-1}] & 0 \\ 0 & [B_{cjk}] \end{bmatrix} \left\{ \begin{matrix} \{D_{c\gamma k-1i}\} \\ \{D_{c\gamma jk}\} \end{matrix} \right\} = [T_{ck}]\{D_{c\gamma k}\}, \tag{15.34}$$

where

$$[B_{cik-1}] = \begin{bmatrix} 1 & 0 & 0 & 0 & 0 & 0 & 0 & 0 \\ 0 & 1 & 0 & 0 & 0 & 0 & 0 & -b_{ik-1} \\ 0 & 0 & 1 & 0 & 0 & 0 & 0 & 0 \\ 0 & 0 & 0 & 1 & 0 & 0 & 0 & 0 \\ 0 & 0 & 0 & 0 & 1 & 0 & 0 & a_{ik-1} \\ 0 & 0 & 0 & 0 & 0 & 1 & 0 & 0 \\ 0 & 0 & 0 & 0 & 0 & 0 & 1 & 0 \\ 0 & 0 & 0 & 0 & 0 & 0 & 0 & 1 \end{bmatrix}, \tag{15.35a}$$

$$[B_{cjk}] = \begin{bmatrix} 1 & 0 & 0 & 0 & 0 & 0 & 0 & 0 \\ 0 & 1 & 0 & 0 & 0 & 0 & 0 & -b_{jk} \\ 0 & 0 & 1 & 0 & 0 & 0 & 0 & 0 \\ 0 & 0 & 0 & 1 & 0 & 0 & 0 & 0 \\ 0 & 0 & 0 & 0 & 1 & 0 & 0 & a_{jk} \\ 0 & 0 & 0 & 0 & 0 & 1 & 0 & 0 \\ 0 & 0 & 0 & 0 & 0 & 0 & 1 & 0 \\ 0 & 0 & 0 & 0 & 0 & 0 & 0 & 1 \end{bmatrix}. \tag{15.35b}$$

In a similar manner, the horizontal forces and moments about the vertical axis can be represented with the force vector of the column element in the global coordinates as

$$\{F_{c\gamma k}\} = [T_{ck}]^{\mathrm{T}}\{F_{c\gamma}\} \tag{15.36}$$

where $\{F_{c\gamma k}\}$ is the force vector corresponding to $\{D_{c\gamma k}\}$.

Substituting Equations (15.34) and (15.36) into Equation (15.16) leads to the incremental stiffness equation of the column element in the global coordinates with consideration of the rigid floor as

$$\{\mathrm{d}F_{c\gamma k}\} = [T_{ck}]^{\mathrm{T}}[K_{c\gamma}][T_{ck}]\{\mathrm{d}D_{c\gamma k}\}. \tag{15.37}$$

15.1.2.2 Stiffness equation of brace element considering rigid floor

By similar derivation of the column element above, the incremental stiffness equation of a brace element in the global coordinates with consideration of the rigid floor can be obtained as

$$\{\mathrm{d}F_{b\gamma k}\} = [T_{bk}]^{\mathrm{T}}[K_{b\gamma}][T_{bk}]\{\mathrm{d}D_{b\gamma k}\}, \tag{15.38}$$

where

$$\{D_{b\gamma k}\} = \left\{ \begin{array}{c} \{D_{b\gamma k-1i}\} \\ \{D_{b\gamma jk}\} \end{array} \right\}, \tag{15.39}$$

$$\{D_{b\gamma ik-1}\} = [\delta_{uok-1}, \quad \delta_{vok-1}, \quad \theta_{wok-1}, \quad \delta_{wi}, \quad \theta_{vi}, \quad \theta_{ui}, \quad \gamma_{vi}, \quad \gamma_{ui}]^{\mathrm{T}}, \tag{15.40a}$$

$$\{D_{b\gamma jk}\} = [\delta_{uok}, \quad \delta_{vok}, \quad \theta_{wok}, \quad \delta_{wj}, \quad \theta_{vj}, \quad \theta_{uj}, \quad \gamma_{vj}, \quad \gamma_{uj}]^{\mathrm{T}}, \tag{15.40b}$$

$$[T_{bk}] = \begin{bmatrix} [B_{bik-1}] & 0 \\ 0 & [B_{bjk}] \end{bmatrix}, \tag{15.41}$$

$$[B_{bik-1}] = \begin{bmatrix} 1 & 0 & -b_{ik-1} & 0 & 0 & 0 & 0 & 0 \\ 0 & 1 & a_{ik-1} & 0 & 0 & 0 & 0 & 0 \\ 0 & 0 & 0 & 1 & 0 & 0 & 0 & 0 \\ 0 & 0 & 0 & 0 & 1 & 0 & 0 & 0 \\ 0 & 0 & 0 & 0 & 0 & 1 & 0 & 0 \\ 0 & 0 & 0 & 0 & 0 & 0 & 1 & 0 \\ 0 & 0 & 0 & 0 & 0 & 0 & 0 & 1 \end{bmatrix}, \tag{15.42a}$$

$$[B_{bjk}] = \begin{bmatrix} 1 & 0 & -b_{jk} & 0 & 0 & 0 & 0 & 0 \\ 0 & 1 & a_{jk} & 0 & 0 & 0 & 0 & 0 \\ 0 & 0 & 0 & 1 & 0 & 0 & 0 & 0 \\ 0 & 0 & 0 & 0 & 1 & 0 & 0 & 0 \\ 0 & 0 & 0 & 0 & 0 & 1 & 0 & 0 \\ 0 & 0 & 0 & 0 & 0 & 0 & 1 & 0 \\ 0 & 0 & 0 & 0 & 0 & 0 & 0 & 1 \end{bmatrix} \tag{15.42b}$$

15.1.3   Global Stiffness Equation of Frame and Static Condensation

The incremental global stiffness equation of a space frame with rigid floors can be obtained by assembling all the incremental stiffness equations of the beam and joint-panel elements in the frame and all the incremental stiffness equations of the column and brace elements considering the rigid floor as

$$\left\{ \begin{array}{c} \{dF_H\} \\ \{dF_R\} \end{array} \right\} = \left[\begin{array}{cc} [K_{HH}] & [K_{HR}] \\ [K_{RH}] & [K_{RR}] \end{array} \right] \left\{ \begin{array}{c} \{dD_H\} \\ \{dD_R\} \end{array} \right\}, \tag{15.43}$$

where $\{dD_H\}$ is the incremental vector of the horizontal displacements and rotations about the vertical axis of all the reference nodes of floors, $\{dD_R\}$ is the incremental vector of the vertical displacements, rotations about the horizontal axis and joint-panel shear deformation of all the nodes in the frame for structural analysis, and $\{dF_H\}$ and $\{dF_R\}$ are the force vectors corresponding to $\{dD_H\}$ and $\{dD_R\}$, respectively.

If only vertical loads are applied on the frame, one has $\{dF_H\} = 0$. Then, Equation (15.43) can be rewritten as

$$\{dD_H\} = -[K_{HH}]^{-1}[K_{HR}]\{dD_R\} \tag{15.44}$$

and

$$\{dF_R\} = [K_R]\{dD_R\}, \tag{15.45}$$

where

$$[K_R] = [K_{RR}] - [K_{RH}][K_{HH}]^{-1}[K_{HR}]. \tag{15.46}$$

If only horizontal loads due to wind or horizontal earthquakes loads are applied on the frame, $\{dF_R\} = 0$. Then, Equation (15.43) can be condensed as

$$\{dD_R\} = -[K_{RR}]^{-1}[K_{RH}]\{dD_H\} \tag{15.47}$$

and

$$\{dF_H\} = [K_H]\{dD_H\}, \tag{15.48}$$

where

$$[K_H] = [K_{HH}] - [K_{HR}][K_{RR}]^{-1}[K_{RH}]. \tag{15.49}$$

It should be noted that as the stiffness equations of column elements and brace elements considering the rigid floor relate to the position of the reference node on each floor, the global stiffness of a space frame, Equation (15.43), depends on the selection of the reference node on each floor. Different positions of reference nodes on floors produce different global stiffness equations of space frames. The horizontal force vector $\{F_H\}$ in the global stiffness equation consists of the horizontal forces applied on the reference nodes and the moments about the vertical axis through the reference nodes, whereas the corresponding displacement vector $\{D_H\}$ consists of the horizontal displacements of the reference nodes and the torsion angles about the vertical axis through the reference nodes. For the sake of convenience in dynamic analysis, it is best to select the mass centre of each floor as the reference node, i.e. the origin of the floor global coordinates. Despite that all the reference nodes on the floors of the frame may not be on the same vertical line by such a selection (see Figure 15.8), the mass matrix of the frame corresponding to the floor horizontal movements, two orthotropic horizontal displacements and torsion about the vertical axis is diagonal. Otherwise, such a mass matrix will be nondiagonal.

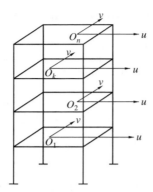

Figure 15.8   The global floor coordinate systems

15.2 PLANAR SUBSTRUCTURE MODEL

When the space bar model is used in the analysis of space steel frames, results have high accuracy, but a huge number of degrees of freedom should be dealt with. Generally, a structural node in the space bar model has eight deformation variables if joint-panel shear deformation is involved, and even if the rigid floor is considered, five independent variables still exist for each node. The computation effort is therefore extremely large especially in nonlinear structural analysis and elasto-plastic seismic response analysis of space steel frames. A compromised model between accuracy and computation effort for structural analysis of space steel frames is necessary, and the planar substructure model is one of them. In the planar substructure model, a space frame is divided into a number of planar subframes (planar substructures) based on the assumption that loads are applied in the plane of the planar subframes. Meanwhile, the rigid floor assumption is also adopted, by which all the planar subframes can resist horizontal loads together. Structural degrees of freedom and computation effort in such a planar substructure model are evidently reduced, and hence the model is applicable to the analysis of space steel frames subjected to horizontal forces applied at the frame floors in practice.

15.2.1 Stiffness Equation of Planar Substructure in Global Coordinates

As shown in Figure 15.9, denote the angle between the ith planar frame in a space frame and the ou axis in the global coordinates with φ_i and denote the coordinates of one point in this planar frame, in the global

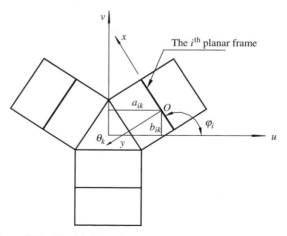

Figure 15.9   The kth floor of the space frame and coordinate systems

coordinate system of the kth floor, with a_{ik} and b_{ik}, where $k = 1, 2, \ldots, n$ and n is the total number of storeys of the frame. The sway stiffness equation of the ith planar frame can be established according to the method described in Chapter 13 and written as

$$[K_x]_i \{dD_x\}_i = \{dF_x\}_i, \qquad (15.50)$$

where $\{dD_x\}_i$ and $\{dF_x\}_i$ are, respectively, the incremental vector of the horizontal displacements of the frame floors along the x-axis and the corresponding incremental force vector and $[K_x]_i i$ is the sway stiffness matrix of the ith planar frame in its own plane.

Expand Equation (15.50) to the spatial sway stiffness equation of the planar frame in its own coordinates as

$$\begin{bmatrix} [K_x]_i & [0] & [0] \\ [0] & [0] & [0] \\ [0] & [0] & [0] \end{bmatrix} \begin{Bmatrix} \{dD_x\}_i \\ \{dD_y\}_i \\ \{d\theta_z\}_i \end{Bmatrix} = \begin{Bmatrix} \{dF_x\}_i \\ \{0\} \\ \{0\} \end{Bmatrix}, \qquad (15.51)$$

where $\{dD_y\}_i$ and $\{d\theta_z\}_i$ are, respectively, the incremental vectors of the horizontal floor displacements along the y-axis and floor torsions.

By the transition relationship between the global coordinates and the local coordinates of the planar frame, one has

$$\begin{Bmatrix} \{dD_x\}_i \\ \{dD_y\}_i \\ \{d\theta_z\}_i \end{Bmatrix} = [T_i] \begin{Bmatrix} \{dD_u\}_i \\ \{dD_v\}_i \\ \{d\theta_w\}_i \end{Bmatrix}, \qquad (15.52)$$

$$\begin{Bmatrix} \{dF_u\}_i \\ \{dF_v\}_i \\ \{0\} \end{Bmatrix} = [T_i]^{\mathrm{T}} \begin{Bmatrix} \{dF_x\}_i \\ \{0\} \\ \{0\} \end{Bmatrix}, \qquad (15.53)$$

where $\{dD_u\}_i$, $\{dD_v\}_i$ and $\{d\theta_w\}_i$ are, respectively, the incremental vectors of the horizontal floor displacements along the u-axis and v-axis, and floor torsions in the global coordinates, $\{dF_u\}_i$ and $\{dF_v\}_i$ are the incremental vectors corresponding to $\{dD_u\}_i$ and $\{dD_v\}_i$, respectively, and $[T_i]$ is the coordinate transition matrix of the ith planar frame, which can be expressed as

$$[T_i] = \begin{bmatrix} [c_i] & [s_i] & [0] \\ [-s_i] & [c_i] & [0] \\ [0] & [0] & [I] \end{bmatrix}, \qquad (15.54)$$

in which

$$[c_i] = \cos \varphi_i[I] = c_i[I], \qquad (15.55a)$$

$$[s_i] = \sin \varphi_i[I] = s_i[I], \qquad (15.55b)$$

where $[I]$ is the unit matrix.

Substituting Equations (15.52) and (15.53) into Equation (15.51) leads to

$$\begin{bmatrix} c_i^2[K_x]_i & c_i s_i[K_x]_i & [0] \\ c_i s_i[K_x]_i & s_i^2[K_x]_i & [0] \\ [0] & [0] & [0] \end{bmatrix} \begin{Bmatrix} \{dD_u\}_i \\ \{dD_v\}_i \\ \{d\theta_w\}_i \end{Bmatrix} = \begin{Bmatrix} \{dF_u\}_i \\ \{dF_v\}_i \\ \{0\} \end{Bmatrix}. \qquad (15.56)$$

Express the displacement vectors at the origin of the subframe coordinates with those at the origin of the global coordinates in the global coordinate system as

$$\{dD_u\}_i = \{dD_u\}_i - [b]_i\{d\theta_{w0}\},$$

$$\{dD_v\}_i = \{dD_v\}_i - [a]_i\{d\theta_{w0}\}, \tag{15.57}$$

$$\{d\theta_w\}_i = \{d\theta_{w0}\},$$

where $\{dD_{u0}\}$, $\{dD_{v0}\}$ and $\{d\theta_{w0}\}$ are, respectively, the incremental vectors of the horizontal floor displacements at the origin along the u-axis and v-axis, and floor torsions in the global coordinates, $[a_i]$ and $[b_i]$ are diagonal matrices given by

$$[a]_i = \text{diag}[a_{i1}, a_{i2}, \ldots, a_{in}], \tag{15.58a}$$

$$[b]_i = \text{diag}[b_{i1}, b_{i2}, \ldots, b_{in}], \tag{15.58b}$$

and a_{ik} and b_{ik} ($k = 1, 2, \ldots, n$) are the coordinates of the origin of the local coordinate system for the ith subframe in the global coordinate system of the kth floor (see Figure 15.9). Note that each floor has its own global coordinate system (see Figure 15.8), and the origin of each floor coordinates may not be in the same vertical line but the directions of the u-axis and v-axis coincide, and the values of a_{ik} and b_{ik} on different floors may be different.

Substituting Equation (15.57) into Equation (15.56) yields

$$\begin{bmatrix} c_i^2[K_x]_i & c_is_i[K_x]_i & [0] \\ c_is_i[K_x]_i & s_i^2[K_x]_i & [0] \\ [0] & [0] & [0] \end{bmatrix} \left\{ \begin{array}{c} \{dD_{u0}\} - [b_i]\{d\theta_{w0}\} \\ \{dD_{v0}\} - [a_i]\{d\theta_{w0}\} \\ \{d\theta_{w0}\} \end{array} \right\} = \left\{ \begin{array}{c} \{dF_u\}_i \\ \{dF_v\}_i \\ \{0\} \end{array} \right\} \tag{15.59}$$

or

$$\begin{bmatrix} c_i^2[K_x]_i & c_is_i[K_x]_i & (-c_i^2[b]_i + c_is_i[a]_i)[K_x]_i \\ c_is_i[K_x]_i & s_i^2[K_x]_i & (-c_is_i[b]_i + c_i^2[a]_i)[K_x]_i \\ [0] & [0] & [0] \end{bmatrix} \left\{ \begin{array}{c} \{dD_{u0}\} \\ \{dD_{v0}\} \\ \{d\theta_{w0}\} \end{array} \right\} = \left\{ \begin{array}{c} \{dF_u\}_i \\ \{dF_v\}_i \\ \{0\} \end{array} \right\}. \tag{15.60}$$

Left-multiplying $[b]_i$ and $[a]_i$ in Equation (15.60), respectively, in the lines corresponding to $\{dD_{u0}\}$ and $\{dD_{v0}\}$, and then adding them to the line corresponding to $\{d\theta_{w0}\}$ can result in

$$[K_H]_i \left\{ \begin{array}{c} \{dD_{u0}\} \\ \{dD_{v0}\} \\ \{d\theta_{w0}\} \end{array} \right\} = \left\{ \begin{array}{c} \{dF_{u0}\}_i \\ \{dF_{v0}\}_i \\ \{dM_{w0}\}_i \end{array} \right\}, \tag{15.61}$$

in which

$$\{dF_{u0}\}_i = \{dF_u\}_i,$$

$$\{dF_{v0}\}_i = \{dF_v\}_i, \tag{15.62}$$

$$\{dM_{w0}\}_i = -[b]_i\{dF_u\}_i + [a]_i\{dF_v\}_i,$$

where $\{dF_{u0}\}_i$ and $\{dF_{v0}\}_i$ are, respectively, the incremental horizontal force vectors along the u-axis and v-axis, which result from the horizontal forces of the ith planar subframe and act in the origin of the floor global coordinates. $\{dM_{w0}\}_i$ is the torque moment about the vertical axis through the origin of the floor global coordinates, which also results from the horizontal forces of the ith planar subframe, and $[K_H]_i$ is the incremental stiffness matrix of the ith planar subframe in the global coordinates, given by

$$[K_H]_i = \begin{bmatrix} [K_{uu}]_i & [K_{uv}]_i & [K_{u\varphi}]_i \\ [K_{vu}]_i & [K_{vv}]_i & [K_{uu}]_i \\ [K_{\varphi u}]_i & [K_{\varphi u}]_i & [K_{\varphi\varphi}]_i \end{bmatrix}, \tag{15.63}$$

where

$$[K_{uu}]_i = c_i^2 [K_x]_i, \tag{15.64a}$$

$$[K_{uv}]_i = [K_{vu}]_i^T = c_i s_i [K_x]_i, \tag{15.64b}$$

$$[K_{vv}]_i = s_i^2 [K_x]_i, \tag{15.64c}$$

$$[K_{u\varphi}]_i = [K_{\varphi u}]_i^T = c_i(-c_i[b]_i + s_i[a]_i)[K_x]_i, \tag{15.64d}$$

$$[K_{v\varphi}]_i = [K_{\varphi v}]_i^T = s_i(-c_i[b]_i + s_i[a]_i)[K_x]_i, \tag{15.64e}$$

$$[K_{\varphi\varphi}]_i = (c_i[b]_i - s_i[a]_i)^2 [K_x]_i. \tag{15.64f}$$

15.2.2   Global Stiffness Equation of Spatial Frame

Assembling the stiffness equations of all the planar subframes can produce the global stiffness equation for the whole space frame as

$$[K_H]\{dD_H\} = \{dF_H\}, \tag{15.65}$$

in which

$$\{dD_H\} = [\{dD_{u0}\}^T, \{dD_{v0}\}^T, \{dD_{w0}\}^T]^T, \tag{15.66a}$$

$$\{dF_H\} = [\{dF_{u0}\}^T, \{dF_{v0}\}^T, \{dM_{w0}\}^T]^T, \tag{15.66b}$$

$$\{dF_{u0}\} = \sum_i \{dF_{u0}\}_i,$$

$$\{dF_{v0}\} = \sum_i \{dF_{v0}\}_i, \tag{15.67}$$

$$\{dM_{w0}\} = \sum_i \{dM_{w0}\}_i,$$

$$[K_H] = \sum_i [K_H]_i. \tag{15.68}$$

where $\{dF_{u0}\}$ and $\{dF_{v0}\}$ are the incremental horizontal force vectors, respectively, along the u-axis and v-axis acting at the origins of each floor global coordinates, $\{dM_{w0}\}$ is the torque moment vector about the vertical axis through the origins of each floor global coordinates, and $[K_H]_i$ and $[K_H]$ are, respectively, the space stiffness matrices of the planar subframes and the global space frame.

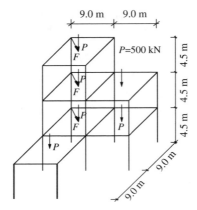

Figure 15.10   Space frame for example

Obviously, the simplification is achieved in the planar substructure model by transforming the analysis of a space frame to the analysis of several planar subframes, and the number of degrees of freedom for the structural analysis is reduced largely. However, it should be noted that the following two approximations exist in the planar substructure model:

(1) Only consistency of the horizontal displacements of the same nodes in different planar subframes is considered, and that of the vertical displacements is ignored.

(2) In the elasto-plastic state of the space frame, the sway stiffness of the frame columns along the two sectional principal axes will become coupled, whereas the planar substructure model neglects this effect.

15.2.3 Numerical Example

Figure 15.10 illustrates a space steel frame, and in the plan view of the frame shown in Figure 15.11, the member sections are identified (US wide-flange standard section used). Each floor of the frame has slab so that the rigid floor assumption is adopted. The loading sequence is that vertical loads, the values of which are given in Figure 15.10, are applied first and then the horizontal loads F (see Figures 15.10 and 15.11) are applied, keeping the vertical loads constant. With the planar substructure model, the curves shown in Figure 15.12 can be obtained, which represent the relationship between the horizontal force F and the horizontal displacement u_3, vertical displacement v_3 and the slab torsion φ_3 of the top floor at the location where the horizontal force acts. The peak point of the curves corresponds to the ultimate horizontal force $F_{\max}$ of the space frame.

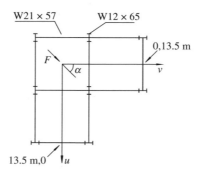

Figure 15.11   The plan of the space frame for example

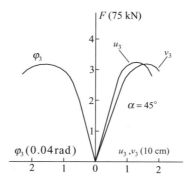

Figure 15.12   Horizontal load–deformation curves

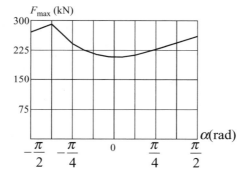

Figure 15.13   Relation between F_{max} and α

The results obtained by the space bar model coincide with those given in Figure 15.12 obtained by the planar subframe model, which indicates that the inconsistency of the vertical deformation of frame columns and the coupling effect of sway stiffness of frame columns in different directions are negligible when the storey number of space frames is large and proportional loading sequence is applied.

Effects of the direction and position of the horizontal loads on the ultimate load-carrying capacity of the frame F_{max} are examined. Let the angle α between the horizontal force direction and the v-axis vary from $-\frac{\pi}{2}$ to $\frac{\pi}{2}$; the corresponding variation of F_{max} is given in Figure 15.13. And let the positions of the horizontal forces move in scope from $-2m$ to $2m$ along the u-axis when $\alpha = 0$; the corresponding variation of F_{max} is shown in Figure 15.14. Obviously, the direction and position of the horizontal forces significantly affect the ultimate load-carrying capacity of the frame .

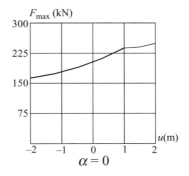

Figure 15.14   Relation between F_{max} and u

15.3 COMPONENT MODE SYNTHESIS METHOD

15.3.1 Principle of Component Mode Synthesis Method

The component mode synthesis method is a combination of the finite element method and the Ritz method. This method effectively reduces the number of structural degrees of freedom and makes it possible to use a microcomputer to analyse large and complex structures. The basic principle of the component mode synthesis method is that a displacement mode is assigned to each structural node, which may be a linear combination of functions satisfying structural boundary constrains. In the component mode synthesis method, the nodal displacement mode in addition to the elemental displacement field for structural analysis is established based on the finite element method. Similar to the Ritz method, variables with the number of structural degrees of freedom much less than that needed in the finite element method can be solved in the component mode synthesis method because the number of independent parameters controlling the displacement field of the global structure is limited. As this method is based on the finite element method, it also has some advantages of the finite element method.

Using the component mode synthesis method in the analysis of space steel frames requires a series of functions with undetermined parameters to approach the displacement field along the height of the frame under analysis. By the viewpoint of function approach theory, the function series $\{f_m(w)\}$ ensuring approaching to the real solution should satisfy three conditions: continuity, independence and completeness. Mathematical investigation indicates that orthotropic function series are advantageous in stability and fast convergence in addition to satisfying the above three conditions . So, the following orthotropic polynomial function is selected as the displacement mode along the height of the frame:

$$\{f_m(w)\} = \sum_{j=1}^{m} (-1)^{j-1} \frac{(m+1)!}{(j-1)!(j+1)!(m-1)!} \left(\frac{w}{H}\right)^j, \tag{15.69}$$

where w is the height of the node considered and H is the total height of the frame.

As for the space steel frame shown in Figure 15.15, the basic variables in the global coordinate system and considering the rigid floor include the horizontal displacements D_{uok} and D_{vok} at the origin of the global coordinates along the u-axis and v-axis and torsion θ_{wok} on the arbitrary kth floor, the vertical displacement D_{wik}, rotations θ_{wik} and θ_{vik}, and joint-panel shear deformations γ_{uik} and γ_{vik} of the ith column on the kth floor. All of the basic variables can be expressed as

$$D_{uok} = \sum_{m=1}^{r} a_{uom}f_m(w_k),$$

$$D_{vok} = \sum_{m=1}^{r} a_{vom}f_m(w_k),$$

$$\theta_{wok} = \sum_{m=1}^{r} b_{wom}f_m(w_k),$$

$$D_{wik} = \sum_{m=1}^{r} a_{wim}f_m(w_k),$$

$$\theta_{uik} = \sum_{m=1}^{r} b_{uim}f_m(w_k), \tag{15.70}$$

$$\theta_{vik} = \sum_{m=1}^{r} b_{vim}f_m(w_k),$$

$$\gamma_{uik} = \sum_{m=1}^{r} c_{uim}f_m(w_k),$$

$$\gamma_{vik} = \sum_{m=1}^{r} c_{vim}f_m(w_k),$$

where w_k is the height of the kth floor.

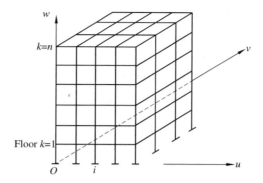

Figure 15.15　Space frame and global coordinate system

The relation between the displacements of the node at the ith column on the kth floor and the origin of the floor global coordinates is (see Figure 15.16)

$$
\begin{aligned}
D_{uik} &= D_{uok} - v_{ik}\theta_{wok}, \\
D_{vik} &= D_{vok} + u_{ik}\theta_{wok}, \\
\theta_{wik} &= \theta_{wok},
\end{aligned}
\tag{15.71}
$$

where u_{ik} are v_{ik} are the coordinate values of the node at the ith column on the kth floor in the floor global coordinate system.

The displacement vector of the node at the ith column on the kth floor can be expressed as

$$
\{D_{ik}\} = \sum_{m=1}^{r} [N_k]_{im}\{e\}_{im} = [N_k]_i\{e\}_i,
\tag{15.72}
$$

in which

$$
\{D_{ik}\} = [D_{uik},\quad D_{vik},\quad \theta_{wik},\quad D_{wik},\quad \theta_{vik},\quad \theta_{uik},\quad \gamma_{vik},\quad \gamma_{uik}]^{\mathrm{T}},
\tag{15.73}
$$

$$
\{e\}_i = [\{e\}_{i1}^{\mathrm{T}},\quad \{e\}_{i2}^{\mathrm{T}},\quad \cdots\{e\}_{ir}^{\mathrm{T}}]^{\mathrm{T}},
\tag{15.74}
$$

$$
\{e\}_{im} = [a_{uom},\quad a_{vom},\quad b_{wom},\quad a_{wim},\quad b_{vim},\quad b_{uim},\quad c_{vim},\quad c_{uim}]^{\mathrm{T}},
\tag{15.75}
$$

$$
[N_k]_i = [[N_k]_{i1},\quad [N_k]_{i2},\quad \cdots, [N_k]_{ir}]^{\mathrm{T}},
\tag{15.76}
$$

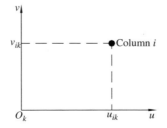

Figure 15.16　The floor global coordinate system

$$[N_k]_{im} = \begin{bmatrix} f_m(w_k) & -v_{ik}f_m(w_k) & & & & & & 0 \\ & f_m(w_k) & u_{ik}f_m(w_k) & & & & & \\ & & f_m(w_k) & & & & & \\ & & & f_m(w_k) & & & & \\ & & & & f_m(w_k) & & & \\ & & & & & f_m(w_k) & & \\ & & & & & & f_m(w_k) & \\ 0 & & & & & & & f_m(w_k) \end{bmatrix}, \quad (15.77)$$

where $\{D_{ik}\}$ is the nodal displacement vector, $\{e\}_i$ is the vector of undetermined parameters (namely the generalized displacement vector) and $[N_k]_i$ is the transition matrix between the nodal displacement vector and the generalized displacement vector.

15.3.2 Analysis of Generalized Elements

15.3.2.1 Generalized column element

A generalized column element consists of all the columns on one vertical line, i.e. the columns from bottom to top storeys, as shown in Figure 15.17. By Equation (15.72), the relation between the end displacements of the column at the kth floor and the generalized displacements is

$$\{D_{ci}\}_k = \left\{ \begin{array}{c} \{D_{ik-1}\} \\ \{D_{ik}\} \end{array} \right\} = \left[\begin{array}{c} [N_{k-1}]_i \\ [N_k]_i \end{array} \right] \{e\}_i = [N_{k-1,k}]_i \{e\}_i. \quad (15.78)$$

It is known that the incremental stiffness equation of the column element with joint panels in the global coordinates is Equation (15.16), and the relation between the elemental displacement vectors $\{D_{c\gamma}\}$ and $\{D_{ci}\}_k$ is

$$\{D_{c\gamma}\} = [R_c]\{D_{ci}\}_k, \quad (15.79)$$

where

$$[R_c] = \begin{bmatrix} [r_c] & [0] \\ [0] & [r_c] \end{bmatrix}, \quad (15.80)$$

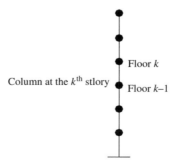

Column at the kth stlory

Floor k

Floor $k-1$

Figure 15.17 The generalized column element

$$[r_c] = \begin{bmatrix} 0 & 0 & 0 & 1 & 0 & 0 & 0 & 0 \\ 1 & 0 & 0 & 0 & 0 & 0 & 0 & 0 \\ 0 & 0 & 0 & 0 & 0 & 0 & 0 & 0 \\ 0 & 0 & 0 & 0 & 0 & 0 & 1 & 0 \\ 0 & 1 & 0 & 0 & 0 & 0 & 0 & 0 \\ 0 & 0 & 0 & 0 & 0 & 1 & 0 & 0 \\ 0 & 0 & 0 & 0 & 0 & 0 & 0 & 1 \\ 0 & 0 & 1 & 0 & 0 & 0 & 0 & 0 \end{bmatrix}. \tag{15.81}$$

By Equations (15.79), (15.78) and (15.16), the incremental relation between the generalized forces and the generalized displacements of the node at the ith column on the kth floor is obtained as

$$\{df_{ci}\}_k = [K_{ci}]_k \{de\}_i, \tag{15.82}$$

where

$$\{df_{ci}\}_k = [N_{k-1,k}]_i^T [R_c]^T \{dF_{cy}\}, \tag{15.83}$$

$$[K_{ci}]_k = [N_{k-1,k}]_i^T [R_c]^T [K_{cy}][R_c][N_{k-1,k}]_i. \tag{15.84}$$

Hence, the incremental stiffness equation of the ith generalized column element can be written as

$$\{df_{ci}\} = [K_{ci}]\{de\}_i, \tag{15.85}$$

in which

$$\{df_{ci}\} = \sum_{k=1}^{n} \{df_{ci}\}_k, \tag{15.86}$$

$$[K_{ci}] = \sum_{k=1}^{n} [K_{ci}]_k, \tag{15.87}$$

where $\{df_{ci}\}$ is the generalized force vector in the generalized column element corresponding to the generalized displacement vector $\{e\}_i$ and $[K_{ci}]$ is the stiffness matrix of the generalized column element.

15.3.2.2 *Generalized beam element*

A generalized beam element consists of all the beams between two adjacent column lines, as shown in Figure 15.18. By Equation (15.72), the relation between the end displacements of the beam at the kth floor and the generalized displacements is

$$\{D_{gij}\}_k = \left\{ \begin{matrix} \{D_{ik}\} \\ \{D_{jk}\} \end{matrix} \right\} = \begin{bmatrix} [N_k]_i & [0] \\ [0] & [N_k]_j \end{bmatrix} \left\{ \begin{matrix} \{e\}_i \\ \{e\}_j \end{matrix} \right\} = [N_k]_{ij}\{e\}_{ij}. \tag{15.88}$$

It is known that the incremental stiffness equation of the beam element with joint panels in the global coordinates is Equation (15.8), and the relation between the elemental displacement vectors $\{D_{gy}\}$ and $\{D_{gij}\}_k$ is

$$\{D_{gy}\} = [R_g]\{D_{gij}\}_k, \tag{15.89}$$

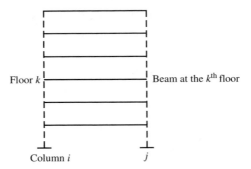

Figure 15.18 The generalized beam element

where

$$[R_g] = \begin{bmatrix} [r_g] & [0] \\ [0] & [r_g] \end{bmatrix}, \tag{15.90}$$

$$[r_g] = \begin{bmatrix} 0 & 0 & 0 & 1 & 0 & 0 & 0 & 0 \\ 0 & 0 & 0 & 0 & 1 & 0 & 0 & 0 \\ 0 & 0 & 0 & 0 & 0 & 1 & 0 & 0 \\ 0 & 0 & 0 & 0 & 0 & 0 & 1 & 0 \\ 0 & 0 & 0 & 0 & 0 & 0 & 0 & 1 \end{bmatrix}. \tag{15.91}$$

By Equations (15.89), (15.88) and (15.8), the incremental relation between the generalized forces and the generalized displacements of the beam within the span of the ith column and the jth column on the kth floor is obtained as

$$\{df_{gij}\}_k = [K_{gij}]_k \{de\}_{ij}, \tag{15.92}$$

where

$$\{df_{gij}\}_k = [N_k]_{ij}^T [R_g]^T \{dF_{g\gamma}\}, \tag{15.93}$$

$$[K_{gij}]_k = [N_k]_{ij}^T [R_g]^T [K_{g\gamma}][R_g][N_k]_{ij}. \tag{15.94}$$

Hence, the incremental stiffness equation of the generalized beam element within the span of the ith column and the jth column can be expressed as

$$\{df_{gij}\} = [K_{gij}]\{de\}_{ij}, \tag{15.95}$$

in which

$$\{df_{gij}\} = \sum_{k=1}^{n} \{df_{gij}\}_k, \tag{15.96}$$

$$[K_{gij}] = \sum_{k=1}^{n} [K_{gij}]_k, \tag{15.97}$$

where $\{df_{gij}\}$ is the generalized force vector in the generalized beam element corresponding to the generalized displacement vector $\{e\}_{ij}$ and $[K_{gij}]$ is the stiffness matrix of the generalized beam element.

15.3.2.3 *Generalized brace element*

A generalized brace element consists of all the braces between two adjacent column lines, as shown in Figure 15.19. By Equation (15.72), the relation between the end displacements of the brace on the kth storey and the generalized displacements is

$$\{D_{bij}\}_k = \left\{ \begin{array}{c} \{D_{ik-1}\} \\ \{D_{jk}\} \end{array} \right\} = \left[\begin{array}{cc} [N_{k-1}]_i & [0] \\ [0] & [N_k]_j \end{array} \right] \left\{ \begin{array}{c} \{e\}_i \\ \{e\}_j \end{array} \right\} = [N_{k-1,k}]_{ij}\{e\}_{ij}. \tag{15.98}$$

It is known that the incremental stiffness equation of the brace element with joint panels in the global coordinates is Equation (15.25), and the relation between the elemental displacement vectors $\{D_{b\gamma}\}$ and $\{D_{bij}\}_k$ is

$$\{D_{b\gamma}\} = [R_b]\{D_{bij}\}_k, \tag{15.99}$$

where

$$[R_b] = \left[\begin{array}{cc} [r_b] & [0] \\ [0] & [r_b] \end{array} \right], \tag{15.100}$$

$$[r_g] = \left[\begin{array}{cccccccc} 1 & 0 & 0 & 0 & 0 & 0 & 0 & 0 \\ 0 & 1 & 0 & 0 & 0 & 0 & 0 & 0 \\ 0 & 0 & 0 & 1 & 0 & 0 & 0 & 0 \\ 0 & 0 & 0 & 0 & 1 & 0 & 0 & 0 \\ 0 & 0 & 0 & 0 & 0 & 1 & 0 & 0 \\ 0 & 0 & 0 & 0 & 0 & 0 & 1 & 0 \\ 0 & 0 & 0 & 0 & 0 & 0 & 0 & 1 \end{array} \right]. \tag{15.101}$$

Similar to the derivation of the generalized column and beam element, the incremental stiffness equation of the generalized brace element within the span of the ith column and the jth column is obtained as

$$\{df_{bij}\} = [K_{bij}]\{de\}_{ij}, \tag{15.102}$$

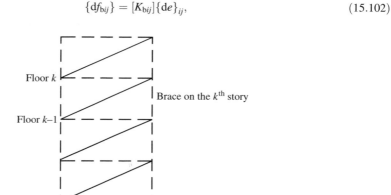

Figure 15.19   The generalized brace element

in which

$$\{df_{bij}\} = \sum_{k=1}^{n} \{df_{bij}\}_k, \tag{15.103}$$

$$[K_{bij}] = \sum_{k=1}^{n} [K_{bij}]_k, \tag{15.104}$$

$$\{df_{bij}\}_k = [N_{k-1,k}]_{ij}^{\mathrm{T}}[R_b]^{\mathrm{T}}\{dF_{b\gamma}\}, \tag{15.105}$$

$$[K_{bij}]_k = [N_{k-1,k}]_{ij}^{\mathrm{T}}[R_b]^{\mathrm{T}}[K_{b\gamma}][R_b][N_{k-1,k}]_{ij}, \tag{15.106}$$

where $\{df_{bij}\}$ is the generalized force vector in the generalized brace element corresponding to the generalized displacement vector $\{e\}_{ij}$ and $[K_{bij}]$ is the stiffness matrix of the generalized brace element.

15.3.2.4 *Generalized joint-panel element*

A generalized joint-panel element consists of all the joint panels along one column line, as shown in Figure 15.20. By Equation (15.72), the relation between the joint-panel displacements and the generalized displacements is

$$\{D_{\gamma i}\}_k = \{D_{ik}\} = [N_k]_i\{e\}_i. \tag{15.107}$$

It is known that the incremental stiffness equation of the joint-panel element in the global coordinates is Equation (15.29), and the relation between the elemental displacement vectors $\{D_{\gamma}\}$ and $\{D_{\gamma i}\}_k$ is

$$\{D_{\gamma}\} = [R_{\gamma}]\{D_{\gamma i}\}_k, \tag{15.108}$$

where

$$[R_{\gamma}] = \begin{bmatrix} 0 & 0 & 0 & 0 & 0 & 0 & 1 & 0 \\ 0 & 0 & 0 & 0 & 0 & 0 & 0 & 1 \end{bmatrix}. \tag{15.109}$$

Similar to the derivation above, the incremental stiffness equation of the generalized joint-panel element along the ith column can be obtained as

$$\{df_{\gamma i}\} = [K_{\gamma i}]\{de\}_i, \tag{15.110}$$

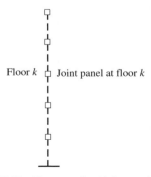

Floor k Joint panel at floor k

Figure 15.20 The generalized joint-panel element

in which

$$\{df_{\gamma i}\} = \sum_{k=1}^{n} \{df_{\gamma i}\}_k, \tag{15.111}$$

$$[K_{\gamma i}] = \sum_{k=1}^{n} [K_{\gamma i}]_k, \tag{15.112}$$

$$\{df_{\gamma i}\}_k = [N_k]_i^{\mathrm{T}} [R_\gamma]^{\mathrm{T}} \{dF_\gamma\}, \tag{15.113}$$

$$[K_{\gamma i}]_k = [N_k]_i^{\mathrm{T}} [R_\gamma]^{\mathrm{T}} [K_\gamma] [R_\gamma] [N_k]_i, \tag{15.114}$$

where $\{df_{\gamma i}\}$ is the generalized force vector in the generalized joint-panel element corresponding to the generalized displacement vector $\{e\}_i$ and $[K_{\gamma i}]$ is the stiffness matrix of the generalized joint-panel element.

15.3.3 Stiffness Equation of Generalized Structure

Assume that the space frame considered has n storeys and s column lines. Assembly of the incremental stiffness equations of all the generalized column, beam, brace and joint-panel elements in the frame yields the incremental stiffness equation for the structural system as

$$\left\{ \begin{array}{c} \{df_{\mathrm{H}}\} \\ \{df_{\mathrm{R}}\} \end{array} \right\} = \left[\begin{array}{cc} [K_{\mathrm{HH}}] & [K_{\mathrm{HR}}] \\ [K_{\mathrm{RH}}] & [K_{\mathrm{RR}}] \end{array} \right] \left\{ \begin{array}{c} \{de_{\mathrm{H}}\} \\ \{de_{\mathrm{R}}\} \end{array} \right\}, \tag{15.115}$$

in which

$$\{e_{\mathrm{H}}\} = [\{a_{u0}\}^{\mathrm{T}}, \quad \{a_{v0}\}^{\mathrm{T}}, \quad \{b_{w0}\}^{\mathrm{T}}]^{\mathrm{T}}, \tag{15.116}$$

$$\{a_{u0}\} = [a_{u01}, \quad a_{u02}, \quad \ldots, \quad a_{u0\gamma}]^{\mathrm{T}},$$

$$\{a_{v0}\} = [a_{v01}, \quad a_{v02}, \quad \ldots, \quad a_{v0\gamma}]^{\mathrm{T}},$$

$$\{b_{w0}\} = [b_{w01}, \quad b_{w02}, \ldots, \quad b_{w0\gamma}]^{\mathrm{T}},$$

$$\{e_{\mathrm{R}}\} = [\{e_{\mathrm{R}1}\}^{\mathrm{T}}, \quad \{e_{\mathrm{R}2}\}^{\mathrm{T}}, \quad \ldots, \quad \{e_{\mathrm{R}s}\}^{\mathrm{T}}]^{\mathrm{T}}, \tag{15.117}$$

$$\{e_{\mathrm{R}i}\} = [\{a_{wi}\}^{\mathrm{T}}, \quad \{b_{vi}\}^{\mathrm{T}}, \quad \{b_{ui}\}^{\mathrm{T}}, \quad \{c_{vi}\}^{\mathrm{T}}, \quad \{c_{ui}\}^{\mathrm{T}}]^{\mathrm{T}},$$

$$\{a_{wi}\} = [a_{wi1}, \quad a_{wi2}, \quad \ldots, \quad a_{wi\gamma}]^{\mathrm{T}},$$

$$\{b_{vi}\} = [b_{vi1}, \quad b_{vi2}, \quad \ldots, \quad b_{vi\gamma}]^{\mathrm{T}},$$

$$\{b_{ui}\} = [b_{ui1}, \quad b_{ui2}, \quad \ldots, \quad b_{ui\gamma}]^{\mathrm{T}},$$

$$\{c_{vi}\} = [c_{vi1}, \quad c_{vi2}, \quad \ldots, \quad c_{vi\gamma}]^{\mathrm{T}},$$

$$\{c_{ui}\} = [c_{ui1}, \quad c_{ui2}, \quad \ldots, \quad c_{ui\gamma}]^{\mathrm{T}},$$

where $\{df_{\mathrm{H}}\}$ and $\{df_{\mathrm{R}}\}$ are the incremental force vectors corresponding to the generalized displacement vectors $\{de_{\mathrm{H}}\}$ and $\{de_{\mathrm{R}}\}$, respectively.

The number of degrees of freedom in the original frame is equal to $n(3 + 5s)$ whereas that using the component mode synthesis method is $r(3 + 5s)$. Generally, the number of terms, r, in the displacement functions along the frame height may be 3–5, which indicates that for the multi-storey frame $(n > r)$, especially for the high-rise frame $(n \gg r)$, the structural degrees of freedom are largely reduced.

If the frame is subjected to horizontal loads (for example, horizontal earthquakes), $\{df_R\} = 0$. Then the stiffness equation of the structural system after static condensation becomes

$$\{df_H\} = [K_H]\{de_H\}, \tag{15.118}$$

where

$$[K_H] = [K_{HH}] - [K_{HR}][K_{RR}]^{-1}[K_{RH}]. \tag{15.119}$$

And $\{de_R\}$ can be obtained with

$$[de_R] = -[K_{RR}]^{-1}[K_{RH}]\{de_H\}. \tag{15.120}$$

15.3.4 Structural Analysis Procedure

This subsection describes the structural analysis procedure using the component mode synthesis method, with the example of frames subjected to horizontal loads.

First, denote the vector of the horizontal displacements at the origin of the floor global coordinate system as

$$\{D_H\} = [\{D_{u0}\}^T, \quad \{D_{v0}\}^T, \quad \{\theta_{w0}\}^T]^T, \tag{15.121}$$

where

$$\begin{aligned}
\{D_{u0}\} &= [D_{u01}, \quad D_{u02}, \quad \ldots, \quad D_{u0n}]^T, \\
\{D_{v0}\} &= [D_{v01}, \quad D_{v02}, \quad \ldots, \quad D_{v0n}]^T, \\
\{\theta_{w0}\} &= [\theta_{w01}, \quad \theta_{w02}, \quad \ldots, \quad \theta_{w0n}]^T.
\end{aligned} \tag{15.122}$$

And then express the floor displacements with the generalized displacements as

$$\begin{aligned}
\{D_{u0}\} &= [H_r]\{a_{u0}\}, \\
\{D_{v0}\} &= [H_r]\{a_{v0}\}, \\
\{\theta_{w0}\} &= [H_r]\{\theta_{w0}\},
\end{aligned} \tag{15.123}$$

where

$$[H_r] = \begin{bmatrix} f_1(w_1) & f_2(w_1) & \ldots & f_r(w_1) \\ f_1(w_2) & f_2(w_2) & \ldots & f_r(w_2) \\ \ldots & \ldots & \ldots & \ldots \\ f_1(w_n) & f_2(w_n) & \ldots & f_r(w_n) \end{bmatrix}. \tag{15.124}$$

By Equations (15.121), (15.123) and (15.116), the relation between the floor displacements and the generalized displacements is obtained as

$$\{D_H\} = [R_H]\{e_H\}, \tag{15.125}$$

where

$$[R_H] = \begin{bmatrix} [H_r] & & \\ & [H_r] & \\ & & [H_r] \end{bmatrix}. \tag{15.126}$$

By the energy principle, the work by the incremental generalized horizontal forces should be equal to that by the realistic incremental horizontal forces applied at the frame floors, namely

$$\{de_H\}^T\{df_H\} = \{dD_H\}^T\{dF_H\}. \tag{15.127}$$

Substituting Equation (15.125) into the above equation yields

$$\{de_H\}^T\{df_H\} = \{de_H\}^T\{R_H\}^T\{dF_H\}, \tag{15.128}$$

which leads to

$$\{df_H\} = \{R_H\}^T\{dF_H\}, \tag{15.129}$$

where $\{dF_H\}$ is the incremental vector consisting of two horizontal orthotropic forces at the origins of all the floor global coordinates and torque moments about the vertical axis through these origins, and $\{df_H\}$ is the generalized horizontal force vector.

Substituting Equation (15.129) into Equation (15.118), one has

$$[K_H]\{de_H\} = [R_H]^T\{dF_H\}. \tag{15.130}$$

As a summary, the structural analysis procedures using the component mode synthesis method can be concluded as follows:

(1) Calculate the generalized displacement incremental $\{de_H\}$ by Equation (15.130).

(2) Calculate the generalized displacement incremental $\{de_R\}$ by Equation (15.120).

(3) With $\{de_H\}$ and $\{de_R\}$, calculate the displacement incremental of structural nodes in the global coordinate system, $\{dD_{ik}\}$, by Equation (15.72).

(4) With $\{dD_{ik}\}$, calculate the elemental displacement incremental in the local coordinate system for all the beam, column, brace and joint-panel elements, respectively, by Equations (15.3), (15.12), (15.22) and (15.28).

(5) Substitute the elemental displacement incremental into the corresponding elemental stiffness equations of beam, column, brace and joint-panel elements to obtain the elemental forces (or the member resultants).

15.3.5 Numerical Example

Three space steel frame specimens were tested by Ding (1990), the geometries and loading distribution of which are given in Figure 15.21. One of the specimens is a symmetric structure, and the other two are asymmetric (the section size of the columns in axis A and axis B are different). All the beams and columns of the specimens are H sections, sizes of which are listed in Tables 15.1 and 15.2. The yielding strength of the steel material for the specimens $\sigma_s = 310\,\text{N/mm}^2$ and the elastic modulus $E = 2.075 \times 10^5\,\text{N/mm}^2$. In the experiments, vertical loads were applied at first and then horizontal loads were applied. Nonsymmetric horizontal forces were applied according to $H_A : H_B = 2 : 1$. Some measures were adopted to simulate the rigid floor.

Curves of the horizontal loads versus horizontal displacement at the top of the columns in axis A, obtained by both the component mode synthesis method and the FEM with the space bar model, are compared and given in Figure 15.22. Meanwhile, the corresponding horizontal ultimate load-carrying capacities of the frames are listed in Table 15.3. Obviously, the component mode synthesis method has good accuracy.

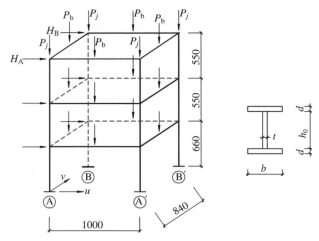

Figure 15.21   Space frame specimen

Table 15.1   Member sizes of specimen 1

Components	Section sizes (mm)			
	b	h_0	t	d
Beams	30	30	4	4.2
Columns	25	25	4	4.3

Table 15.2   Member sizes of the specimens 2 and 3

Components	Section sizes (mm)			
	b	h_0	t	d
Beams	25	25	4	4.3
Columns in axis A	30	30	4	4.2
Columns in axis B	25	25	4	4.3

Table 15.3   Comparison of ultimate horizontal loads obtained by numerical analyses and tests

Specimen member	By tests, H_t(kN)	By the space bar model, H_f(kN)	By the component mode synthesis method, H_s(kN)	$\dfrac{H_f - H_t}{H_t} \times 100\%$	$\dfrac{H_s - H_t}{H_t} \times 100\%$
1	2.10	2.16	2.23	2.86	6.31
2	1.95	2.04	2.06	4.61	5.47
3	2.40	2.48	2.56	3.33	6.67

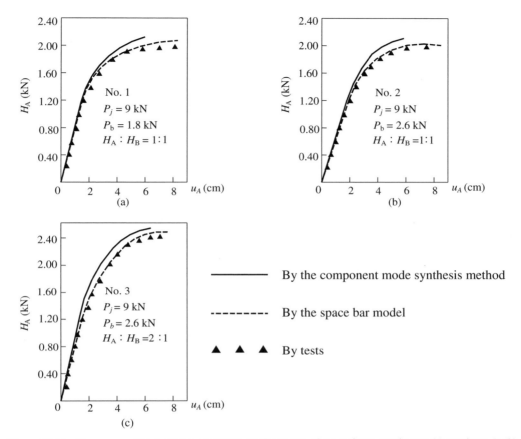

Figure 15.22   The curves of horizontal loads versus displacements of space frame specimens: (a) specimen 1; (b) specimen 2; (c) specimen 3

Because the global frame displacement is enforced in the component mode synthesis method, relatively large structural stiffness (or relatively small structural displacements) and relatively high ultimate load-carrying capacities are detected in Figure 15.22 and Table 15.3.

Part Two

Advanced Design of Steel Frames

16 Development of Structural Design Approach

The structural design approaches can be divided according to reliability measurement into deterministic approach, reliability approach based on limit states of structural members and reliability approach based on limit states of the structural system. Historically, the deterministic design approach is the earliest used, and the reliability design approach based on the limit states of structural members prevails in current practice. However, the reliability design approach will be the evolutional aim in future.

16.1 DETERMINISTIC DESIGN APPROACH

16.1.1 Allowable Stress Design (ASD) (AISC, 1989)

The allowable stress design approach was the earliest proposed after the establishment of structural analysis theory, the design principle of which is that the stress at arbitrary structural position, σ, should not be greater than the allowable stress $[\sigma]$, namely

$$\sigma \leq [\sigma], \tag{16.1}$$

in which $[\sigma]$ is determined by

$$[\sigma] = \frac{\sigma_s}{k_e}, \tag{16.2}$$

where σ_s is the material yielding strength and k_e is the safety factor.

It is assumed to ensure the structure designed keeping in elastic state in ASD so that structural analysis and design is relatively simple. However, the following drawbacks exist:

(1) The choice of the safety factor is empirical, and the structural design lacks explicit reliability significance. Although in theory the greater the safety factor the safer the structure designed, in practice with the same safety factor, different load patterns (e.g. dead or live load is dominant) or different structural materials (e.g. steel or concrete is used) will evidently result in different structural reliability levels. This cannot be reflected in ASD.

(2) Even if structural material is same, different sectional shapes of structural members or different structural forms with the same stress level can also result in different structural reliability levels.

Advanced Analysis and Design of Steel Frames Guo-Qiang Li and Jin-Jun Li
© 2007 John Wiley & Sons, Ltd

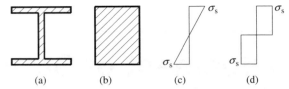

Figure 16.1   Sections of the steel member and stress distribution: (a) H section; (b) rectangular section; (c) elastic limit stress distribution; (d) plastic limit stress distribution

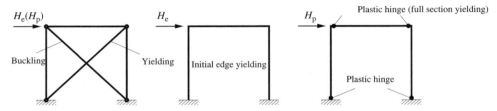

Figure 16.2   Elastic and plastic limit load-bearing capacities

For example, the steel H-shaped and rectangular sections shown in Figure 16.1 can have the same elastic limit bending moments, but their plastic limit capacities may differ from each other up to 1.4 times. Again, it can be found from comparison of the steel portal frame and the braced frame shown in Figure 16.2 that the elastic capacity of the braced frame can be same as that of the portal frame, but its plastic capacity is less because its plastic capacity is equal to its elastic one, whereas the portal frame has much capacity reservation in plasticity after elastic capacity.

16.1.2   Plastic Design (PD) (AISC, 1978)

The plastic design approach was proposed to overcome the drawback of ASD, the design principle of which is that the structural load effect S, with a consideration of safety reservation, should not be greater than the corresponding structural plastic strength R_p, namely

$$S \leq R_p \tag{16.3a}$$

or

$$k_p S_0 \leq R_p, \tag{16.3b}$$

where S_0 is the nominal value of the structural load effect and k_p is the load factor used to consider safety reservation in PD.

PD is a more rational design approach over ASD because it includes the effects of plastic development on structural limit load-bearing capacity. But the failure criterion used in PD is more dangerous than that in ASD so that generally the load factor k_p in PD is greater than the safety factor k_e used in ASD.

16.2   RELIABILITY DESIGN APPROACH BASED ON LIMIT STATES OF STRUCTURAL MEMBERS

After ASD and PD, the reliability design approach based on limit states of structural members prevails in the design of steel frame structures worldwide. This design format is also termed as load and resistance factor design (LRFD). LRFD is adopted in the codes for the design of steel structures in China (GB50017-2003), United States (AISC, 1994), United Kingdom (BS5950, 1990), European Union (Eurocode3, 1992) and Australia (AS4100, 1990). The common procedures of LRFD include the following: (1) to conduct linearly elastic structural analysis to determine the resultants of structural members under critical load and load combinations (with load factor and load combination factor); (2) to check member resultants against the limit states of structural members specified in the codes with resistance factor. If all of the limit state checks pass, the structural design is satisfactory to code requirement.

The design formula of LRFD can be expressed as

$$\gamma_{\mathrm{R}} R \geq \gamma_0 \sum \gamma_i S_{,i}, \qquad (16.4)$$

where R is the member nominal resistance, being generally the limit load-bearing capacity, calculated with material and geometric sizes of the member and according to the load form (e.g. bending, axial compression and eccentric compression) applied on the member, S_i is the nominal load effect (resultant) of the member, γ_0 is the structural importance factor, γ_{R} is the resistance factor and γ_i is the load factor.

In Equation (16.4), i indicates different type of loads, and factors γ_0, γ_{R} and γ_i are determined with structural reliability analysis so that the structural design by Equation (16.4) has the reliability level approximately close to the target one pre-assigned. LRFD is therefore a reliability-based structural design approach.

However, LRFD still has incompatibility in the following aspects:

(1) The member resultants used in design procedures 1 and 2 are incompatible (Chen, 1998). In procedure 1, linearly elastic analysis is used to obtain structural member resultants, and material nonlinearity is not taken into account. But in procedure 2, elasto-plastic limit states of structural members are considered, where material nonlinearity is accounted. Resultants will be generally redistributed in structural members after yielding occurs because a structure resists loads as an integrity rather than individual components. This incompatibility indicates that the resultants of structural members used in the limit state check are not realistic in LRFD.

(2) Instability mode assumed for structures is incompatible with the global instability mode occurring really (Liew, White and Chen, 1991). The basic assumption in the determination of buckling modes of steel frames in LRFD is 'sway or non-sway instabilities happen simultaneously to all the columns on the same frame storey', which is not true when compared to the fact that 'only individual or some columns of a frame occur elasto-plastic instability at first'. In engineering practice, the effect of structural instability is considered with an empirical effective length for frame columns in their limit state check in LRFD. However, the effective length cannot reflect the true inelastic restrain relationship among structural members.

(3) System reliability level of structures designed is incompatible with the target reliability aimed (Galambos, 1990). What reliability level ensured in LRFD is for individual structural members, not for the structural system because LRFD is based on the limit state check of structural members. There is an evident gap between the structural system and member reliability levels because the structural capacity relates not only to capacities of individual members, but also to structural ductility and redundancy, structural form and the correlation between loads and structural resistance (Hendawi and Frangopol, 1994).

The above incompatibilities may result in design irrationalities. The first is that the limit load-bearing capacity of the structural system designed is uncertain, and in majority of the cases, the design value is less than the realistic one. An examination was done where 16 planar steel frames were studied with the LRFD code method (member check) and the inelastic system method (structural check) (Ziemian, McGuire and Deierlein, 1992a). The results obtained are that the limit load-bearing capacities of the overall frames designed by the current LRFD code method are 1–58 % (average 20 %) greater than those by the inelastic system method, and the steel consumption of the frames designed by the LRFD code method is 1.5–19 % (average 12.4 %) greater than that by the inelastic system method. A similar comparison was also made for a 22-storey steel space frame by Ziemian, McGuire and Deierlein (1992b), where it was found that the overall limit load-bearing capacities of the frame designed by the LRFD code method are 14.6, 7.7 and 6.7 % greater than those by the inelastic system method under three typical load patterns, whereas the steel consumption is 13.4 % greater. Other similar results can be found for the braced and unbraced steel frames investigated by Kim and Chen (1996a, 1996b), the planar steel frames loaded in the weak axis by Ziemian and Miller (1997) and the steel portal frames by King (1991). Similar results can also be found in the Chinese code based on the LRFD method (Zhou, Duan and Chen, 1991).

The second irrationality is that the system reliability level of structures designed is ambiguous, and in majority of the cases, the design value is less than the target one (Galambos, 1990). Although previous research show that the structural reliability by LRFD is more uniform than that by ASD and PD, it is not satisfactory yet. The realistic limit load-bearing capacities of 300 concrete–steel composite columns loaded with concentrated and eccentric compression are examined with experimental tests (Lundberg and Galambos, 1996), which is not consistent with the values in design. Additionally, the reliability level calculated is also not consistent with the target one in LRFD, and note that those of steel-encased columns are smaller and those of concrete-filled tubes are greater. The realistic member reliability of three planar steel frames was checked with the stochastic finite element method (Mahadevan and Haldar, 1991). It is found that the reliabilities of the members in the same portal frame are not equal to each other, and in majority of the loading cases, the member reliability is 6–150 % greater than the target reliability. Effects of structural plasticity on member reliability are studied in reference, and results show that the failure probability of structural member may be 4.6 times that by linearly elastic analysis if material nonlinearity is considered (Xiao and Mahadevan, 1994). Similar results were also found by Lin and Corotis (1985). An example was given that the structural system reliability index reaches 3.5 though the target reliability index is 2.6 when designing a steel portal frame with LRFD for structural members (Ellingwood, 1994). As the objective is the simply supported individual member in the derivation of load factor, resistance factor and target reliability level in current LRFD, the current LRFD code method cannot reflect the realistic reliability of the members in a structural system and also cannot evaluate the structural system reliability (Moses, 1990).

16.3   STRUCTURAL SYSTEM RELIABILITY DESIGN APPROACH

With rapid development of computer technology and research progresses in the fields of structural analysis and system reliability assessment, it is possible to propose and establish a more advanced structural design approach than the current one. The advanced structural design approach should be based on the limit load-bearing capacity of the structural system and system reliability evaluation. It is a development tendency in structural engineering to combine advanced structural analysis with system reliability assessment into reliability-based advanced design (RAD) for the structural system.

The design format of RAD can be

$$\phi \cdot R_{\mathrm{n}} \geq \gamma_0 \cdot \sum \gamma_i \cdot S_{\mathrm{n}i}, \tag{16.5}$$

where S_{n} and R_{n} are the nominal values of a load distribution and the corresponding structural system resistance (limit load-bearing capacity), respectively, ϕ and γ_i are factor of structural system resistance and load factor, respectively, obtained with system reliability assessment and γ_0 is the factor relating to structural ductility, redundancy and importance. Although the design formula of RAD has the same form as that of LRFD, it is a totally new design approach based on the limit states of the structural system, no longer based on the limit states of structural members as LRFD.

RAD is an evolution of limit state design approach from member level to structural system level. It can be foreseen that RAD can not only overcome the deficient problems of the current LRFD method based on the limit states of structural members, but also produce more rational structural solutions. Furthermore, RAD can facilitate the design process and give benefits to both owner and practitioner, by waiving limit state equation check member by member and determination of the ambiguous effective length factor for frame columns.

Thereafter, the second part of this book will introduce at first the calculation of structural system reliability (Chapter 17), and then the system reliability assessment method of steel frames (Chapter 18) and finally the reliability-based advanced design approach for steel frames (Chapter 19).

17 Structural System Reliability Calculation

17.1 FUNDAMENTALS OF STRUCTURAL RELIABILITY THEORY

17.1.1 Performance Requirements of Structures

Four basic performance requirements should be satisfied for building structures:

(1) capability of resisting all kinds of loads in the phase of normal construction and utility;

(2) serviceability in normal utility;

(3) durability with normal maintenance;

(4) structural integrity during and after accidents (e.g. earthquakes and fires).

The first and fourth requirements above relate to structural safety, the second to structural serviceability and the third to durability. Only when the structure satisfies all of the four requirements, it is reliable. The structural reliability is therefore a synthesis of structural safety, serviceability and durability.

17.1.2 Performance Function of Structures

A general equation to govern the reliability of a structure can be expressed as

$$Z = g(R, S) = R - S, \tag{17.1}$$

where R and S denote structural resistance and load effect, respectively. Due to the uncertainties of structures and loads, R, S and therefore Z are random variables. Three events will happen to the structure as follows:

(1) The structure is reliable when $Z > 0$.

(2) The structure fails when $Z < 0$.

(3) The structure is in limit state when $Z = 0$.

Advanced Analysis and Design of Steel Frames   Guo-Qiang Li and Jin-Jun Li
© 2007 John Wiley & Sons, Ltd

The value of Z can be used to justify whether the structure satisfies the performance requirements or not. So Z is termed as the performance function and call the following equation,

$$Z = R - S = 0, \tag{17.2}$$

as the limit state equation of the structure.

As R and S are dependent on more basic random variables (e.g. geometric sizes of structures, sectional dimensions of structural members, material properties and so on), the more general expression of structural performance function becomes

$$Z = g(X_1, X_2, \ldots, X_n), \tag{17.3}$$

where $X_1, X_2, \ldots, X_n$ are the basic random variables.

17.1.3 Limit State of Structures

The limit state is the critical state in which the structures become disabled from reliable. When the whole structure or part of the structure in a certain state cannot satisfy a performance requirement specified in the design for structural reliability, this state is the limit state corresponding to the performance requirements.

There are two types of structural limit states. One is for load-carrying capacity and the other for serviceability.

17.1.3.1 Load-carrying capacity limit state

Load-carrying capacity limit states for structural system or structural members includes.

(1) losing equilibrium for the overall structure or part of the structure as a rigid body (e.g. overturning);

(2) exceeding material strength of structural members or connections due to static and dynamic loads;

(3) becoming a mechanism from the overall structure or part of the structure;

(4) losing stability as the overall structure or individual members (e.g. buckling in compression).

17.1.3.2 Serviceability limit state

Serviceability limit states for structural system or structural members includes

(1) the structural deformation affecting the normal utility of the structure;

(2) the local damage (e.g. cracking) affecting the normal utility and durability of the structure;

(3) the structural vibration affecting the normal utility of the structure;

(4) the other states specified affecting the normal utility of the structure.

17.1.4 Structural Reliability

Structural reliability is the probabilistic measurement of structural integrity and serviceability, which can be defined more precisely as the probability that the structure satisfies performance requirements specified in specified time and under specified conditions.

In the above definition, the specified time generally refers to the benchmark period for structural design. In the codes of many countries for structural design, the benchmark period or the specified time is taken as 50 years for normal structures. Moreover, the specified conditions actually mean the conditions of normal design, normal construction and normal use, excluding the man-made errors in the design, construction and use of the structure affecting the structural reliability.

If the probabilistic density function of Z is $f_Z(Z)$, the structural reliability p_s can be calculated as

$$p_s = P\{Z \geq 0\} = \int_0^\infty f_Z(Z)\,dZ. \tag{17.4}$$

Let p_f denote the structural failure probability, which can be expressed as

$$p_f = P\{Z < 0\} = \int_{-\infty}^0 f_Z(Z)\,dZ. \tag{17.5}$$

As event $\{Z < 0\}$ and event $\{Z \geq 0\}$ are completely opposite, therefore

$$p_s + p_f = 1, \tag{17.6}$$

or

$$p_s = 1 - p_f. \tag{17.7}$$

Structural failure probability is more straightforward and concerned in structural design so that it is calculated generally in reliability assessment of engineering structures.

If the probabilistic density functions of structural resistance, R, and load effect, S, are $f_R(R)$ and $f_S(S)$, respectively, and R, S are independent of each other, $f_Z(Z)$ becomes

$$f_Z(Z) = f_Z(R, S) = f_R(R) \cdot f_S(S), \tag{17.8}$$

and the failure probability is obtained as

$$p_f = P\{Z < 0\} = P\{R - S < 0\} = \iint_{R-S<0} f_R(R) \cdot f_S(S)\,dRdS. \tag{17.9}$$

In Equation (17.7), performing integration along R at first leads to

$$\begin{aligned}
p_f &= \int_{-\infty}^{+\infty} \left[\int_R^{+\infty} f_S(S)\,dS \right] f_R(R)\,dR \\
&= \int_{-\infty}^{+\infty} \left[1 - \int_{-\infty}^R f_S(S)\,dS \right] f_R(R)\,dR \\
&= \int_{-\infty}^{+\infty} [1 - F_S(R)] f_R(R)\,dR.
\end{aligned} \tag{17.10}$$

Performing integration along S at first leads to

$$\begin{aligned}
p_f &= \int_{-\infty}^{+\infty} \left[\int_{-\infty}^S f_R(R)\,dR \right] f_S(S)\,dS \\
&= \int_{-\infty}^{+\infty} F_R(S) f_S(S)\,dS,
\end{aligned} \tag{17.11}$$

where $F_R(\cdot)$ and $F_S(\cdot)$ are the probabilistic distribution functions of random variables, R and S, respectively.

It can be found from Equation (17.9) that as structural resistance R and load effect S are random variables, the absolutely reliable structure ($p_f = 0$ or $p_s = 1$) does not exist. From the viewpoint of probability, the aim of structural design is to make p_s sufficiently large or make p_f sufficiently small, to the extent which can be accepted.

17.1.5 Reliability Index

Assume that in the structural performance function $Z = R - S$, R and S are normal random variables uncorrelated. The mean values and standard variances of R and S are μ_R, μ_S and σ_R, σ_S, respectively. Based on the probabilistic theory knowledge, Z is also a normal random variable and its mean and standard variance are

$$\mu_Z = \mu_R - \mu_S, \tag{17.12}$$

$$\sigma_Z = \sqrt{\sigma_R^2 + \sigma_S^2}. \tag{17.13}$$

Then the structural failure probability is

$$p_f = P\{Z < 0\} = P\left\{\frac{Z}{\sigma_Z} < 0\right\} = P\left\{\frac{Z - \mu_Z}{\sigma_Z} < -\frac{\mu_Z}{\sigma_Z}\right\}. \tag{17.14}$$

If

$$\beta = \frac{\mu_Z}{\sigma_Z}, \tag{17.15}$$

$$Y = \frac{Z - \mu_Z}{\sigma_Z}, \tag{17.16}$$

one has

$$p_f = P\{Y < -\beta\} = \Phi(-\beta), \tag{17.17}$$

where Y is the standardized normal random variable and Φ is the standard normal distribution function.

Substituting Equation (17.15) into Equation (17.14) yields

$$p_f = P\{Z < \mu_Z - \beta\sigma_Z\}. \tag{17.18}$$

Equation (17.18) can be illustrated with Figure 17.1. When β reduces, the area of the hatched part in Figure 17.1 and therefore the failure probability, p_f, increases, whereas this area and p_f reduce if β increases. This indicates that β is a quantitative index of the structural reliability and is defined as the structural reliability index.

Substituting Equations (17.12) and (17.13) into Equation (17.14) results in the expression of β for the case that R and S are normal random variables uncorrelated, given by

$$\beta = \frac{\mu_R - \mu_S}{\sqrt{\sigma_R^2 + \sigma_S^2}}. \tag{17.19}$$

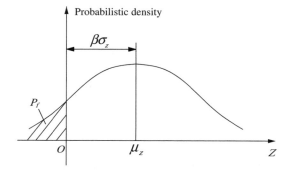

Figure 17.1   Reliability index versus failure probability

Table 17.1 Corresponding values of reliability index β with failure probability p_f

β	1.0	1.5	2.0	2.5	3.0	3.5	4.0	4.5
p_f	1.59×10^{-1}	6.68×10^{-2}	2.28×10^{-2}	6.21×10^{-3}	1.35×10^{-3}	2.33×10^{-4}	3.17×10^{-5}	3.40×10^{-6}

If R and S are log-normal random variables uncorrelated, Equation (17.14) becomes

$$
\begin{aligned}
p_f &= P\{Z < 0\} = P\{R - S < 0\} = P\{R < S\} \\
&= P\left\{\frac{R}{S} < 1\right\} = P\left\{\ln\frac{R}{S} < \ln 1\right\} \\
&= P\{\ln R - \ln S < 0\}.
\end{aligned}
\tag{17.20}
$$

As $\ln R$ and $\ln S$ are normal random variables, the structural reliability index in this case can be obtained with

$$
\beta = \frac{\mu_{\ln R} - \mu_{\ln S}}{\sqrt{\sigma_{\ln R}^2 + \sigma_{\ln S}^2}},
\tag{17.21}
$$

where $\mu_{\ln R}$, $\mu_{\ln S}$ are mean values and $\sigma_{\ln R}$, $\sigma_{\ln S}$ are standard variances of $\ln R$ and $\ln S$, respectively. It has been proved that the statistics of $\ln X$ (assuming that X is the log-normal random variable) are

$$
\mu_{\ln X} = \ln \mu_X - \frac{1}{2}\ln(1 + \delta_X^2),
\tag{17.22}
$$

$$
\sigma_{\ln X} = \sqrt{\ln(1 + \delta_X^2)},
\tag{17.23}
$$

where δ_X is the coefficient of variation of X.

With Equations (17.22) and (17.23), the expression of β for the case that R and S are log-normal random variables uncorrelated is obtained as

$$
\beta = \frac{\ln \dfrac{\mu_R \sqrt{1 + \delta_S^2}}{\mu_S \sqrt{1 + \delta_R^2}}}{\sqrt{\ln\left[(1 + \delta_R^2)(1 + \delta_S^2)\right]}}.
\tag{17.24}
$$

When the basic random variables are not normal or log-normal distribution or the performance function is not linear, it is difficult to explicitly express the reliability index with statistics of basic variables. In such a condition, β can be calculated by using the structural failure probability with Equation (17.17), i.e.

$$
\beta = -\Phi^{-1}(p_f),
\tag{17.25}
$$

where $\Phi^{-1}(\cdot)$ is the inverse distribution function of the standard normal variable.

A group of corresponding values between β and p_f are listed in Table 17.1.

17.2 THE FIRST-ORDER SECOND-MOMENT (FOSM) METHODS FOR STRUCTURAL RELIABILITY ASSESSMENT

The definition of the structural reliability index is based on the distribution function of the basic random variables. If the variables are normally distributed and the performance function of the structure is linear, then it has a simple expression with only the statistics of the variables. In reality, however, the problem becomes much complicated. For example, the structural performance function is usually nonlinear and in many cases the variables are non-normally distributed. In such conditions, as mentioned above, the exact reliability index cannot be calculated directly. However, if the reliability index can be approximately estimated with the statistics of basic random variables, it is useful in engineering practice. The FOSM method is such a method relying on only the first and second moments of the basic random variables to estimate the reliability index. Two FOSM methods, the central point method and the design point method, are to be introduced in this section.

17.2.1 Central Point Method

The central point method was proposed in the early stage of structural reliability study. Its principle is to expand the structural performance function at the mean values of basic variables, called as the central point, as Taylor's series and retain only the first term at first and then calculate the mean and variance of the performance function approximately. Finally, the reliability index is expressed directly with the mean and variance of the performance function.

17.2.1.1 Linear performance function

Assume the performance function as

$$Z = a_0 + \sum_{i=1}^{n} a_i X_i, \tag{17.26}$$

where a_0, a_i $(i = 1, 2, \ldots, n)$ are known coefficients and X_i is the random variables in the performance function.

The mean and standard variance of the performance function are

$$\mu_Z = a_0 + \sum_{i=1}^{n} a_i \mu_{X_i}, \tag{17.27}$$

$$\sigma_Z = \sqrt{\sum_{i=1}^{n} (a_i \sigma_{X_i})^2}. \tag{17.28}$$

According to the central limit principle, the distribution of the performance function approaches to the normal distribution with the increase in the number of the random variables in the performance function. So, when n is a large number, the reliability index can be approximately calculated with

$$\beta = \frac{\mu_Z}{\sigma_Z} = \frac{a_0 + \sum_{i=1}^{n} a_i \mu_{X_i}}{\sqrt{\sum_{i=1}^{n} (a_i \sigma_{X_i})^2}}. \tag{17.29}$$

And the structural failure probability can be determined with Equation (17.17).

17.2.1.2 Nonlinear performance function

Generally, the structural performance function is nonlinear. Assume

$$Z = g(X_1, X_2, \ldots, X_n). \tag{17.30}$$

Expand the performance function at the mean values of the basic variables as the Taylor's series and retain only the first (linear) term as

$$Z \approx g(\mu_{X_1}, \mu_{X_2}, \ldots, \mu_{X_n}) + \sum_{i=1}^{n} \frac{\partial g}{\partial X_i}\bigg|_{\mu_X} (X_i - \mu_{X_i}), \tag{17.31}$$

where the subscribe, μ_X, indicates the assignment of the basic variables with the mean value.

So, the mean and standard variance of performance function become

$$\mu_Z = g(\mu_{X_1}, \mu_{X_2}, \ldots, \mu_{X_n}), \tag{17.32}$$

$$\sigma_Z = \sqrt{\sum_{i=1}^{n} \left(\frac{\partial g}{\partial X_i}\bigg|_{\mu_X} \sigma_{X_i} \right)^2}. \tag{17.33}$$

And the reliability index is approximately obtained with

$$\beta = \frac{\mu_Z}{\sigma_Z} = \frac{g(\mu_{X_1}, \mu_{X_2}, \ldots, \mu_{X_n})}{\sqrt{\sum_{i=1}^{n} \left(\frac{\partial g}{\partial X_i}\bigg|_{\mu_X} \sigma_{X_i} \right)^2}}. \tag{17.34}$$

In the central point method, the reliability index can be expressed directly and no iteration is needed. But the following two disadvantages are evident in this method:

(1) It does not consider the distribution type of random variables. Because the central point method is based on variables normally distributed, for non-normal variables it makes error in predicting the reliability index.

(2) For the nonlinear structural performance function, only the first linear term in Taylor's expansion is retained in the central point method so that its accuracy depends on the extent of discrepancy between the linearized and the realistic limit state function. Although generally the central point is not at the realistic limit state surface governed by the limit state equation with the performance function, the error cannot be avoided.

17.2.2 Design Point Method

The design point method is an improvement over the central point method. The principles are as follows:

(1) For the nonlinear structural performance function, namely the limit state equation, $g(X) = 0$, being a nonlinear surface, the linearized approximation is not expanded at the central point but at one point $X^* = [X_1^*, X_2^*, \ldots, X_n^*]^T$ satisfying $g(X) = 0$.

(2) Non-normal variable, X_i, is transformed into an equivalent normal variable at point X^*, and the effect of distribution type of the variable is therefore considered.

The point X^* is called the design point.

Let

$$Z = g(X) = g(X_1, X_2, \ldots, X_n). \tag{17.35}$$

Transform X space to $\hat{X}$ space by

$$\hat{X}_i = \frac{X_i - \mu_{X_i}}{\sigma_{X_i}}, \tag{17.36}$$

and the performance function in $\hat{X}$ space is then

$$\hat{Z} = \hat{g}(\hat{X}) = \hat{g}(\hat{X}_1, \hat{X}_2, \dots, \hat{X}_n). \tag{17.37}$$

In $\hat{X}$ space, the tangent plane equation of $\hat{Z} = 0$ at the point $\hat{X}^* = [\hat{X}_1^*, \hat{X}_2^*, \dots \hat{X}_n^*]^T$ corresponding to the design point X^* is

$$\hat{g}(\hat{X}_1^*, \hat{X}_2^*, \dots \hat{X}_n^*) + \sum_{i=1}^n \frac{\partial \hat{g}}{\partial \hat{X}_i}\bigg|_{\hat{X}^*} (\hat{X}_i - \hat{X}_i^*) = 0. \tag{17.38}$$

As X^* is one of points on the surface of $\hat{Z} = 0$, i.e.

$$\hat{g}(\hat{X}_1^*, \hat{X}_2^*, \dots \hat{X}_n^*) = 0, \tag{17.39}$$

the tangent plane equation is simplified to

$$\sum_{i=1}^n \frac{\partial \hat{g}}{\partial \hat{X}_i}\bigg|_{\hat{X}^*} (\hat{X}_i - \hat{X}_i^*) = 0. \tag{17.40}$$

Actually the distance from the origin of the coordinates in $\hat{X}$ space to the tangent plane defined by Equation (17.40) is the reliability index desired (Li, 1985), i.e.

$$\beta = \frac{-\sum_{i=1}^n \frac{\partial \hat{g}}{\partial \hat{X}_i}\big|_{\hat{X}^*} \hat{X}_i^*}{\sqrt{\sum_{i=1}^n \left(\frac{\partial \hat{g}}{\partial \hat{X}_i}\big|_{\hat{X}^*}\right)^2}}. \tag{17.41}$$

Let

$$\alpha_i = \frac{-\frac{\partial \hat{g}}{\partial \hat{X}_i}\big|_{\hat{X}^*}}{\sqrt{\sum_{i=1}^n \left(\frac{\partial \hat{g}}{\partial \hat{X}_i}\big|_{\hat{X}^*}\right)^2}}. \tag{17.42}$$

It can be proved that α_i is the directional cosine from the origin to the design point X^* in $\hat{X}$ space. Hence, one has

$$\hat{X}_i^* = \alpha_i \beta. \tag{17.43}$$

Returning to X space, one has

$$X_i^* = \mu_{X_i} + \alpha_i \beta \sigma_{X_i}, \tag{17.44}$$

because

$$\frac{\partial \hat{g}}{\partial \hat{X}_i}\bigg|_{\hat{X}^*} = \frac{\partial g}{\partial X_i}\bigg|_{X^*} \sigma_{X_i}. \tag{17.45}$$

Substituting Equation (17.45) into Equation (17.42) leads to

$$\alpha_i = \frac{-\sum_{i=1}^n \frac{\partial g}{\partial X_i}\big|_{X^*} \sigma_{X_i}}{\sqrt{\sum_{i=1}^n \left(\frac{\partial g}{\partial X_i}\big|_{X^*} \sigma_{X_i}\right)^2}}. \tag{17.46}$$

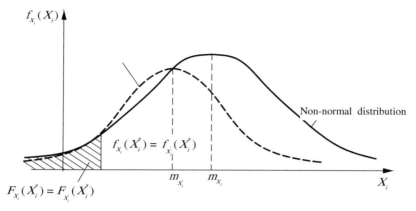

Figure 17.2   Conditions of equivalent normal distribution

In addition, the design point X^* satisfies

$$g(X_1^*, X_2^*, \ldots X_n^*) = 0. \tag{17.47}$$

There are $2n + 1$ simultaneous equations provided by Equations (17.44), (17.46) and (17.47), which can be used to solve X_i^*, α_i $(i = 1, 2, \ldots, n)$ and β, a total of [Pallavi1]$2n + 1$ unknowns. As $g(\cdot) = 0$ is usually nonlinear, iterative technique is needed to obtain the final results of the unknowns.

If any variable X_i is non-normal, it should be transformed to the equivalent normal variable X_i' at the design point X^* (see Figure 17.2) to estimate the reliability index with Equation (17.41).

According to the condition for the same value of probabilistic distribution functions at the design point X^*, one has

$$F_i(X_i^*) = \Phi\left[\frac{X_i^* - \mu_{X_i'}}{\sigma_{X_i'}}\right], \tag{17.48}$$

from which it can be derived that

$$\mu_{X_i'} = X_i^* - \Phi^{-1}[F_i(X_i^*)]\sigma_{X_i'}. \tag{17.49}$$

According to the condition for the same value of probabilistic density functions at the design point X^*, one has

$$f_i(X_i^*) = \frac{1}{\sigma_{X_i'}}\phi\left[\frac{X_i^* - \mu_{X_i'}}{\sigma_{X_i'}}\right], \tag{17.50}$$

from which it can be derived that

$$\sigma_{X_i'} = \phi\left[\frac{X_i^* - \mu_{X_i'}}{\sigma_{X_i'}}\right]\bigg/ f_i(X_i^*) = \phi\{\Phi^{-1}[F_i(X_i^*)]\}/f_i(X_i^*), \tag{17.51}$$

where $\Phi(\cdot)$ and $\Phi^{-1}(\cdot)$ are respectively the distribution function of the standard normal variable and its inverse function; and $\phi(\cdot)$ is the density function of the standard normal variable.

The general calculation procedure of the design point method for determining the structural reliability index can be summarized as follows:

(1) Express the performance function in the form of $g(X_1, X_2, \ldots, X_n) = 0$, and determine the distribution type and statistics of all the basic random variables.

(2) Assume the initial values of X_i^* and β. Generally, the mean value of X_i is taken as the initial value of X_i^*.

(3) For non-normal random variable X_i, calculate the mean, $\mu_{X_i'}$, and standard variance, $\sigma_{X_i'}$, of the equivalent normal variable, X_i', according to Equations (17.49) and (17.51), and replace the corresponding values, μ_{X_i} and σ_{X_i}, of the original variable, X_i.

(4) Calculate the directional cosine, α_i, according to Equation (17.46).

(5) Solve the reliability index, β, using Equation (17.47) with Equation (17.44).

(6) Update X_i^* with Equation (17.44).

(7) Repeat steps (3)–(6) until the error of the updated X_i^* is allowable.

17.3   EFFECTS OF CORRELATION AMONG RANDOM VARIABLES

The reliability calculation mentioned above is based on the independence among random variables in the structural performance function considered. In reality, those random variables affecting the structural reliability are possibly correlated. For example, correlation exists between the earthquake effect and the gravity effect on internal forces in the structure, and also between structural member sizes and material properties. It is therefore necessary to consider variable correlation in structural reliability calculation.

Let the structural performance function be

$$Z = g(X_1, X_2, \cdots, X_n),$$

and assume the correlation factor between the two arbitrary variables X_i and X_j to be ρ_{ij} (when $i \neq j$, $|\rho_{ij}| \leq 1$; when $i = j$, $\rho_{ij} = 1$). The reliability index can be approximately determined with

$$\beta \approx \frac{\mu_Z}{\sigma_Z} = \frac{g(\mu_{X_1}, \mu_{X_2}, \ldots, \mu_{X_n})}{\sqrt{\sum_{i=1}^{n}\sum_{j=1}^{n}\left(\frac{\partial g}{\partial x_i}\bigg|_{x=\mu}\frac{\partial g}{\partial x_j}\bigg|_{x=\mu}\rho_{ij}\sigma_{x_i}\sigma_{x_j}\right)}}. \tag{17.52}$$

It can be proven that only when $g(\cdot)$ is a linear expression and all random variables are normal distributed, Equation (17.52) is the exact expression for β, otherwise it is an approximate one.

17.4   STRUCTURAL SYSTEM RELIABILITY AND BOUNDARY THEORY

As the load-bearing limit state of structures corresponds to structural safety and is generally more important than the limit state of structural serviceability, in the remaining part of this book, the structural reliability is mainly referred to as the load-bearing limit state.

Structural reliability comprises structural component or member (including connections) reliability and structural system reliability. Structural member reliability is based on the limit state of member failure, and structural system reliability is based on the limit state of overall structure failure. As the overall structure failure must be initiated from member failure, structural system reliability is necessarily related to structural member reliability. This section will introduce the basic concepts of the system reliability and the upper–lower boundary method.

17.4.1   Basic Concepts

17.4.1.1   Failure feature of structural member

Due to the material property and feature of load applied, the members (connection can be taken as a particular member) comprising the structure can be divided into brittle or ductile types.

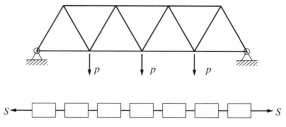

Figure 17.3   Static-determinant truss and serial model

A brittle member is the one that loses its function completely once failure occurs. For example, a reinforced concrete column in compression loses its load-bearing capacity once failure occurs.

A ductile member is the one that still maintains its function even when failure occurs. For example, a member made of steel with yielding plateau in tension or in bending can still retain load-bearing capacity even when yielding achieves.

Brittle or ductile members have different effects on structural system reliability.

17.4.1.2 *Failure model of structural system*

A structure is made of individual members. The different way for the members to form the structure and the different failure feature that the members possess will lead to different structural failure modes. Three basic failure models of structural system can be concluded as serial model, parallel model and serial–parallel model.

(1) Serial model

If any one of the structural members fails, the entire structure fails; the structure can be logically represented by the serial model.

All of the static-determinant structures can be represented with the serial model. For a static-determinant truss shown in Figure 17.3, each member can be taken as one component of the serial model and the system fails once any one of the components consisting of the system fails. No matter whether the members in the static-determinant structures are brittle or ductile, the structural system reliability is not influenced.

(2) Parallel model

If after one or more than one member failures, the structural function can still be retained by the members left alone or together with the members failed with ductile failure feature, this structure can be logically represented with the parallel model. Failure of redundant structures can be represented with the parallel model. Each column in the multispan bent frame, shown in Figure 17.4, is one component in the parallel model, and only after all of the columns fail, the structural system fails. As for the steel beam with fixed ends, shown in Figure 17.5, only after three plastic hinges at the mid-span and the two ends have produced, the beam fails. If a plastic hinge is regarded as one component in a parallel system, the steel beam can also be modelled with the parallel model.

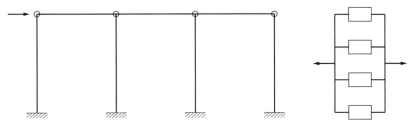

Figure 17.4   Bent frame structure and parallel model

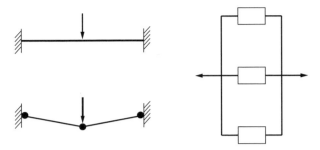

Figure 17.5   Steel beam with fixed ends and parallel model

The structural reliability of a parallel system is influenced by the failure feature (brittle or ductile) of the members consisting of the structural system. Although brittle members have no contribution after failure, failure sequence of members should be taken into account in calculation of structural system reliability. Ductile members, however, sustain their function after failure so that only the final failure mode of the structure need to be considered in the structural system reliability determination.

(3) Serial–parallel model

If the final failure mode of a redundant structure comprising ductile members is not unique, the structure can be represented with the serial–parallel model.

Plastic hinge mechanism is the failure mode of the single-span single-storey steel frame shown in Figure 17.6. There are a total of [Pallavi2]three plastic hinge modes possible (see Figure 17.6), and once any one of the three modes occurs, the frame collapses. So, the frame can be actually represented with a serial model comprising three subparallel models, namely the serial–parallel model.

In weakly redundant structures comprising brittle members, when one member collapses, the failure probability of the other members will increase rapidly and the structural redundancy hardly improves the structural system reliability. So, for this kind of structures, the subparallel components can be simplified to only one component and structures can be represented with the serial model.

17.4.1.3 Correlation among member failure and failure modes

The reliability of a structural member depends on the load effect on the member and the corresponding resistance of the member. Within one structure, the load effect of each member is from the same load so that

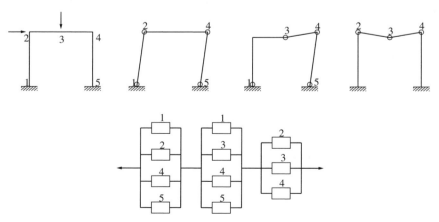

Figure 17.6   Plastic hinge mechanism of frame and serial–parallel model

load effects of different structural members are highly correlated. Additionally, all or parts of structural members are probably made of the same batch of steel so that resistances of different structural members are also correlated. So, failures of different structural members correlate each other.

As shown in Figure 17.6, different failure modes of a redundant structure generally include the same member failure. It is therefore necessary to consider the correlation among failures of different members in the assessment of structural system reliability.

Due to the correlations discussed above, evaluation of structural system reliability becomes very complicated and difficult.

17.4.2 Upper–Lower Boundary Method

In particular cases, the structural system reliability can be calculated with the structural member reliability based on probabilistic theory. Denote the safe state of a structural member with X_i, the failure state of the member with $\bar{X}_i$, the failure probability of the member with $P_{\mathrm{f}i}$ and the failure probability of the structural system with P_{f} for the following discussion.

17.4.2.1 Serial system

As for a serial structural system, when safe states of components are independent, one has

$$P_{\mathrm{f}} = 1 - P\left(\prod_{i=1}^{n} X_i\right) = 1 - \prod_{i=1}^{n}(1 - P_{\mathrm{f}i}). \tag{17.53}$$

When the safe states of components are fully and positively correlated, the failure probability of the system is

$$P_{\mathrm{f}} = 1 - P(\min_{i \in 1,n} X_i) = 1 - \min_{i \in 1,n}(1 - P_{\mathrm{f}i}) = \max_{i \in 1,n} P_{\mathrm{f}i}, \tag{17.54}$$

where n is the number of the components consisting of the system.

Generally, a realistic structural system falls within the above two extreme cases. The failure probability of a real serial system is therefore between the above two results, namely

$$\max_{i \in 1,n} P_{\mathrm{f}i} \le P_{\mathrm{f}} \le 1 - \prod_{i=1}^{n}(1 - P_{\mathrm{f}i}). \tag{17.55}$$

It can be observed that the reliability of static-determinant structures is always less than or equal to the minimum reliability of structural members.

17.4.2.2 Parallel system

For a parallel structural system, when the safe states of components are independent, one has

$$P_{\mathrm{f}} = P\left(\prod_{i=1}^{n} \bar{X}_i\right) = \prod_{i=1}^{n} P_{\mathrm{f}i}. \tag{17.56}$$

When they are fully and positively correlated, the failure probability of the system becomes

$$P_{\mathrm{f}} = P(\min_{i \in 1,n} \bar{X}_i) = \min_{i \in 1,n} P_{\mathrm{f}i}. \tag{17.57}$$

So, generally

$$\prod_{i=1}^{n} P_{fi} \le P_f \le \min_{i \in 1,n} P_{fi}. \tag{17.58}$$

It can be found from the above analysis that the system reliability of a redundant structure is always greater than or equal to the maximum member reliability if the failure mode of the structure is unique. However, if the structure has several failure modes, the reliability of each failure mode is always greater than or equal to the maximum member reliability whereas the system reliability of the structure is always less than or equal to the minimum reliability of various structural failure modes.

17.5 SEMI-ANALYTICAL SIMULATION METHOD FOR SYSTEM RELIABILITY

By the above discussion, there is some relationship between structural system reliability and member reliability. In the past, many efforts have been made and some methods have been proposed to calculate the system reliability with the reliabilities of the member consisting of the system. However, such methods are difficult in practical application, the reasons of which are as follows:

(1) Generally, a realistic structure has many members so that it is hard to correlate all the failure modes of the structural system with the failure of the structural member;

(2) System reliability is influenced by the correlation among member failure and among failure modes of the structural system. The calculation of the system reliability may become too complicated to consider the above correlation;

(3) When the effect of geometric nonlinearity on structural system failure is not negligible (e.g. steel frames are such systems), the relationship between the system failure mode and the failure of members is hard to be established explicitly.

To overcome the above difficulties, a semi-analytical simulation method is proposed in this section for calculating structural system reliability, where the performance function of a structural system is expressed with the simplest form of load and system resistance, and the random simulation technique, approximate probability fitting and the design point method are employed.

17.5.1 General Principle

The performance function for the reliability assessment of a structural system can be generally written as

$$G = R - S, \tag{17.59}$$

where R and S represent the structural resistance and the corresponding load on the structure, respectively. As the probabilistic statistics of the load S can be found in the load codes, FOSM methods can be applied to Equation (17.59), and the failure probability or reliability index of the structural system can be obtained, provided that the probabilistic feature and statistics of the structural resistance are determined. So, the structural system reliability assessment can be reduced to sufficiently simulate the probabilistic characteristics of the structural resistance, R, under the corresponding load S applied.

In this section, the probabilistic characteristics of the structural resistance, R, is determined with random sampling approach. The Monte Carlo sampling strategy is used to determine the strength of the material and the geometrical sizes of a sample structure according to the probabilistic characteristics of the strength of the material and the geometrical sizes of the structure considered, which are normally available. By implementing analysis of the ultimate load-bearing capacity of the sample structure subjected to the load, S, using the knowledge introduced in Part One of this book, a sample of the structural resistance, R, is obtained. Many samples of R may result in the statistics of R, which may further lead to the probabilistic density function of R approximately.

In the following sections are discussed various techniques of random sampling for random variables and approximation of the probabilistic density function of random variables using their statistics.

17.5.2 Random Sampling

17.5.2.1 Systematic sampling (SS)

Let $\{x_i\}$, $i = 1, 2, \ldots, N$, denote a set of independent realizations of the random variable, X. Fractile constraints employed as sampling rule in systematic sampling hold

$$F(x_i) = \frac{2i - 1}{2N}, \quad (i = 1, 2, \ldots, N), \tag{17.60}$$

where $F(\cdot)$ is the cumulative distribution function of the random variable, X (Melchers, 1987).

For a random vector $\bar{X} = \{X_1, X_2, \ldots, X_K\}$, the components of which are independent, let x_{jk} be the jth simulated value of the kth component of $\bar{X}$. Define $\bar{P} = \{p_{jk}\}$ to be an $N \times K$ matrix, each column of which is an independent random permutation of $\{1, 2, \ldots, N\}$. Then x_{jk} is obtained by

$$F_k(x_{jk}) = \frac{2p_{jk} - 1}{2N}, \quad (j = 1, 2, \ldots, N; k = 1, 2, \ldots, K), \tag{17.61}$$

where $F_k(\cdot)$ is the cumulative distribution function of X_k.

17.5.2.2 Updated system sampling (USS)

Matrix $\bar{P}$ in system sampling is produced randomly, which makes it possible to introduce statistic correlation among each column of $\bar{P}$ and reduce the efficiency of sampling. A method to improve statistic correlation was proposed by Florian (1992), where Spearman parameter is used to describe such column correlation of $\bar{P}$. The definition is

$$T_{ij} = 1 - \frac{6 \sum_k (R_{ki} - R_{kj})^2}{N(N - 1)(N + 1)}, \quad (k = 1, 2, \ldots, N; i, j = 1, 2, \ldots, K), \tag{17.62}$$

where T_{ij} is Spearman parameter between the input variables i and j, the value of which is within $[-1, 1]$, and R_{ki} and R_{kj} are the random numbers in column i and column j of row k in matrix R.

Let R be the random number matrix in sampling and T the column correlation matrix of R. The element T_{ij} ($i, j = 1, 2, \ldots, K$) is the Spearman parameter of column i and column j of R. Evidently, T is symmetric and is equal to unit matrix when each column is fully independent. When T is a positive determinant, its Cholesky partition is

$$T = Q \cdot Q^{\mathrm{T}}, \tag{17.63}$$

where Q is the lower triangular matrix. Let

$$S = Q^{-1}, \tag{17.64}$$

and make transformation

$$R_s = R \cdot S^{\mathrm{T}}. \tag{17.65}$$

The correlation between various columns of the matrix, R_s, can be described with T_s. It can be proven that T_s is closer to unit matrix I than T. In other words, the difference between the corresponding elements of T_s and I is less than that between the corresponding elements of T and I. The column correlation of the random number produced according to R_s is therefore improved.

An example with random number and sample size is given in Table 17.2. The matrix of T for the random numbers sampled as shown in Table 17.2 is listed in Table 17.3, which shows improvement of the column correlation after modifications.

Table 17.2   Random number matrix for $K = 5$ and $N = 10$

Number of samples	Non-modified Variables					Modified, 1st iteration Variables					Modified, 2nd iteration Variables				
	1	2	3	4	5	1	2	3	4	5	1	2	3	4	5
1	4	1	2	5	8	4	1	3	2	10	4	1	3	2	10
2	1	9	10	1	9	1	10	8	3	7	1	10	8	3	7
3	5	6	9	3	7	5	6	9	5	2	5	6	9	5	2
4	7	3	4	7	1	7	2	6	6	1	7	2	6	6	1
5	10	5	8	4	10	10	4	10	4	9	10	4	10	4	9
6	6	8	6	10	4	6	8	5	10	8	6	8	5	10	8
7	9	10	3	2	6	9	9	2	1	4	9	9	2	1	3
8	2	2	7	6	5	2	3	7	7	3	2	3	7	7	4
9	8	7	5	9	3	8	7	4	9	6	8	7	4	9	6
10	3	4	1	8	2	3	5	1	8	5	3	5	1	8	5

17.5.2.3 Antithetic variates (AV)

In AV, negative correlation between different cycles of simulation is induced in order to improve the simulation efficiency (Ayyub and Haldar, 1984). If r is the random number uniformly distributed in the interval [0, 1] and is used to determine the estimator, Z_1, the random number $1 - r$ can be used in another run to obtain Z_2 and an improved estimator for the random variable, Z, could be given as AV by

$$Z = \frac{1}{2}(Z_1 + Z_2).$$ (17.66)

17.5.2.4 Combined sampling technique

Simulation of AV simulation is very easy to combine with direct Monte Carlo simulation or other variance-reduction techniques. In this chapter, the Monte Carlo simulation by combining updated systemic sampling with AV is employed to sample the structural resistance of structural systems. Numerical examples will show that this combination can improve the efficiency of the Monte Carlo simulation.

17.5.2.5 Treatment of correlated random variables

The effects of possible correlation between random variables in structural reliability analysis should be considered in certain cases. Generally, the correlated variables can be transformed to a set of uncorrelated variables by using an orthogonal matrix consisting of eigenvectors of the covariance matrix of the basic random vectors. In this method, series of independent random numbers need be converted to dependent ones in Monte Carlo simulation, and the transformation for a standard normal vector with covariance was provided by Melchers (1987).

Table 17.3   Correlation matrix of random numbers10

Variables	Non-modified Variables					Modified, 1st iteration Variables					Modified, 2nd iteration Variables				
	1	2	3	4	5	1	2	3	4	5	1	2	3	4	5
1	1.00	0.26	−0.14	0.10	−0.02	1.00	0.02	0.02	−0.07	0.05	1.00	0.02	0.02	−0.07	−0.03
2	0.26	1.00	0.32	−0.27	0.15	0.02	1.00	−0.02	0.01	0.04	0.02	1.00	−0.02	0.01	−0.03
3	−0.14	0.32	1.00	−0.42	0.54	0.02	−0.02	1.00	−0.08	−0.05	0.02	−0.02	1.00	−0.08	0.01
4	0.10	−0.27	−0.42	1.00	−0.73	−0.07	0.01	−0.08	1.00	−0.13	−0.07	0.01	−0.08	1.00	−0.05
5	−0.02	0.15	0.54	−0.73	1.00	0.05	0.04	−0.05	−0.13	1.00	−0.03	−0.03	0.01	−0.05	1.00

EXAMPLE   **309**

If the desired dependent variables are non-normal, the treatment of approximate transformation to equivalent variables with normal distribution (Hohenbichler and Rackwitz, 1981) can be made at first and then the linear transformation aforementioned can be done.

17.5.3 Exponential Polynomial Method (EPM)

The EPM (Er, 1998) is employed in this chapter to fit the probability density function (PDF) of the structural resistance, R. EPM is, in fact, a method based on the principle of approximating a numerically specified probability distribution with maximum entropy probability distribution (Kapur, 1994).

For a standardized random variable X, its PDF, $f(x)$, could be written by the truncated form of exponential polynomial as

$$f(x) = c \cdot e^{Q_n(x)}, \tag{17.67}$$

where

$$Q_n(x) = \begin{cases} \sum_{i=1}^{n} a_i x^i & (x \in [\alpha, \beta]) \\ -\infty & (x \notin [\alpha, \beta]) \end{cases},$$

$$c = \frac{1}{\int_{\alpha}^{\beta} e^{Q_n(x)} \cdot \mathrm{d}x}.$$

Experience shows that the contribution of $e^{Q_n(x)}$ to the normalizing constant, c, becomes negligible as the value of x is greater than 4σ or less than -4σ, where σ denotes the standard deviation of X. Hence, the values of α and β can be set to -4σ and 4σ, respectively, in the numerical practice.

In Equation (17.67), a_i ($i = 1, 2, \ldots, n$) are progressive coefficients and can be determined by using the first $2(n-1)$th moments of the random variable by

$$\begin{bmatrix} \tilde{\mu}_0 & 2\tilde{\mu}_1 & \cdots & n\tilde{\mu}_{n-1} \\ \tilde{\mu}_1 & 2\tilde{\mu}_2 & \cdots & n\tilde{\mu}_n \\ \vdots & \vdots & \ddots & \vdots \\ \tilde{\mu}_{n-1} & 2\tilde{\mu}_n & \cdots & n\tilde{\mu}_{2(n-1)} \end{bmatrix} \begin{Bmatrix} a_1 \\ a_2 \\ \vdots \\ a_n \end{Bmatrix} = \begin{Bmatrix} 0 \\ -1 \\ \vdots \\ -(n-1)\tilde{\mu}_{n-2} \end{Bmatrix}, \tag{17.68}$$

where $\tilde{\mu}_i$ denotes the ith moment of X and $\tilde{\mu}_0 = 1$. Equation (17.71) indicates that the number of moments of X needed for unique determination of the n unknown parameters a_i is $2(n-1)$.

Numerical examples (Er, 1998) showed that the EPM-estimated PDF with $n = 4$ has sufficient accuracy. Those examples demonstrated as well that EPM-estimated PDF has satisfactory tail behaviour, which is very important in structural reliability assessment due to the small failure probability of structural systems.

17.6 EXAMPLE

17.6.1 A Steel Beam Section

Consider the elastic limit state in bending of a steel beam to study the efficiency of different sampling techniques. The structural performance function is

$$Z = f_y S - M, \tag{17.69}$$

where f_y is the steel yielding strength, S is the elastic section modulus and M is the bending moment applied. The statistics of each random variable are listed in Table 17.4.

The failure probability of the beam for distribution type I is 1.183×10^{-3} by direct Monte Carlo simulation (DMCS) with sample size of 2×10^5, and that for the distribution type II is 3.196×10^{-3} by DMCS with sample size of 1×10^5. The DMCS results will be used as exact solution in this chapter.

Table 17.4 Statistics of random variables of the steel beam

Variables	Mean	Coefficient of variation	Distribution type I	Distribution type II
Yielding strength (f_y)	275.52 (MPa)	0.125	Normal	Log-normal
Elastic section module (S)	8.19×10^{-4} (m³)	0.05	Normal	Log-normal
Moment (M)	113 (kN m)	0.20	Normal	Extreme type I

Four sampling techniques are used to compare the efficiency of sampling. They are SS, USS, SS+AV and USS+AV. Ten times of calculations are conducted for each sampling technique with a sampling size of 500 for distribution type I and 300 for distribution type II, the results of which are given in Tables 17.5 and 17.6, respectively. With the comparison listed, it can be found that combination of variance-reduction techniques (the last three sampling methods) can improve the sampling efficiency and the best one is USS+AV.

Generally, the accuracy of sampling simulation increases with the sampling size. The results of sampling size study for distribution types I and II, with the USS+AV sampling technique, are plotted in Figures 17.7 and 17.8. It can be observed that when sampling size achieves 300, the failure probability of the steel beam is

Table 17.5 Failure probability of the steel beam with sampling size of 500 for distribution type I case ($\times 10^{-3}$)

Number of calculation	SS method	USS method	SS + AV method	USS + AV method
1	0.898	0.970	1.159	1.172
2	1.174	0.895	1.166	1.042*
3	1.276	1.043	1.041	1.185
4	1.186	1.138	1.169	1.187
5	1.700	1.275	1.144	1.173
6	1.253	0.994	1.158	1.182
7	1.023	1.236	1.251	1.326*
8	1.297	1.071	1.211	1.196
9	1.053	1.133	1.003	1.203
10	1.053	1.108	0.999	1.170
Mean	1.191	1.086	1.130	1.184
Standard variance	0.220	0.117	0.086	0.068
Coefficient of variation	0.184	0.108	0.076	0.057

Note: The values marked with * are maximum or minimum failure probability of the steel beam numerically obtained.

Table 17.6 Failure probability of the steel beam with sampling size of 300 for distribution type II case ($\times 10^{-3}$)

Number of calculation	SS method	USS method	SS + AV method	USS + AV method
1	3.076	3.131	3.165	3.081
2	3.212	3.111	3.032	3.052
3	3.133	3.057	3.008	3.061
4	2.985	3.081	3.190	3.150*
5	3.075	3.040	3.055	3.109
6	2.995	3.092	3.014	3.037
7	3.048	3.153	3.150	3.031*
8	3.048	3.055	3.058	3.115
9	3.267	3.030	3.051	3.082
10	3.113	3.086	3.213	3.073
Mean	3.095	3.083	3.094	3.079
Standard variance	0.090	0.040	0.077	0.037
Coefficient of variation	0.029	0.013	0.025	0.012

Note: The values marked with * are maximum or minimum failure probability of the steel beam numerically obtained.

EXAMPLE   **311**

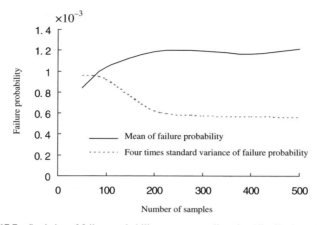

Figure 17.7   Statistics of failure probability versus sampling size (distribution type I case)

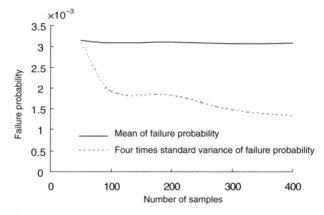

Figure 17.8   Statistics of failure probability versus sampling size (distribution type II case)

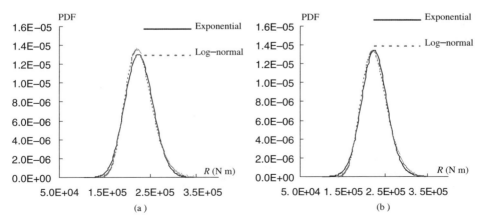

Figure 17.9   Probabilistic density function, PDF, of moment resistance, R, of the steel beam (distribution type I case):
(a) for minimum failure probability; (b) for maximum failure probability

Table 17.7 The mean value, variance and parameters in EPM for distribution type I case

Parameters	For minimum failure probability	For maximum failure probability
μ_R ($\times 10^5$ N m)	2.2561	2.2568
σ_R ($\times 10^5$ N m)	0.3007	0.3057
a_1	$-0.047\ 36$	$-0.048\ 03$
a_2	-0.4607	-0.5566
a_3	$0.016\ 55$	$0.014\ 59$
a_4	$0.004\ 412$	$0.008\ 139$
C	2.5541	2.4311

Table 17.8 The mean value, variance and parameters in EPM for distribution type II case

Parameters	For minimum failure probability	For maximum failure probability
μ_R ($\times 10^5$ N m)	2.2560	2.2566
σ_R ($\times 10^5$ N m)	0.2998	0.3041
a_1	-0.2202	-0.2376
a_2	-0.4732	-0.4194
a_3	$0.079\ 24$	$0.089\ 30$
a_4	$-0.011\ 48$	$-0.021\ 60$
C	2.5266	2.5959

to be convergent. So, it is possible to use relatively small sampling size to achieve satisfactory accuracy in estimating structural reliability, when the suitable sampling technique, such as USS + AV, is used.

The PDF of the moment resistance of the beam, $R = f_y S$, is given in Figures 17.9 and 17.10 for distribution types I and II, respectively, where the maximum and minimum failure probabilities are corresponding to the values in Tables 17.5 and 17.6 marked with asterisk. At the same time, the PDF of log-normal distribution with the same mean value and variance as those by EPM are also plotted in Figures 17.9 and 17.10. It can be observed that PDF by EMP is very close to that of log-normal distribution, which agrees with the general assumption that the distribution of structural member resistance satisfies log-normal type. The mean value and variance, and undeterminated parameters and integration (normalizing) constant in EPM are listed in Tables 17.7 and 17.8.

The effect of correlation between material property and elastic section modulus on the failure probability of the beam is demonstrated in Figure 17.11. It can be observed that positive correlation increases the failure probability whereas negative correlation reduces failure probability in this study.

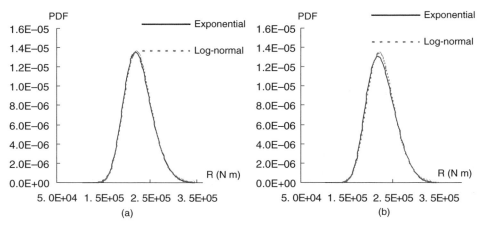

Figure 17.10 Probabilistic density function, PDF, of moment resistance, R, of the steel beam (distribution type II case): (a) for minimum failure probability; (b) for maximum failure probability

EXAMPLE   **313**

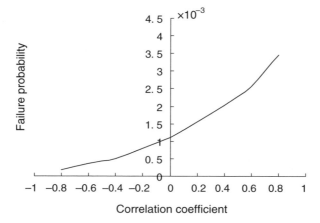

Figure 17.11   Effect of correlation of variables on failure probability of the steel beam

17.6.2   A Steel Portal Frame

In this example a steel portal frame, as shown in Figure 17.12, is investigated. Uncertainties of the frame are represented by the limit yielding moments M_c and M_b respectively for the columns and the beam employed in the portal frame. The statistics of M_c and M_b and those of the load, P, are given in Table 17.9, with assumed normal distribution.

The limit state function of the portal frame can be expressed as

$$G = R - S = \lambda \cdot S - S, \tag{17.70}$$

where $R = \lambda \cdot S$ is the structural resistance (i.e. limit load-bearing capacity), S is the load effect including the actions of the two concentrated loads, P and $2P$, and λ is the load factor.

Two different models for structural analysis are used to determine the structural resistance. One is the second-order inelastic structural analysis and the other is the rigid-plastic structural analysis. Failure probabilities of this frame with DMCS are $P_{f_i} = 2.548 \times 10^{-3}$ (by second-order inelastic analysis model,

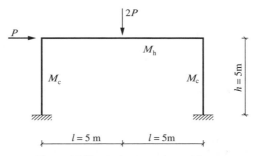

Figure 17.12   A simple steel portal frame

Table 17.9   Statistics of the variable in the portal frame

Variables	M_c	M_b	P
Mean value	75 kN m	150 kN m	20 kN
Coefficient of variation	0.05	0.05	0.3

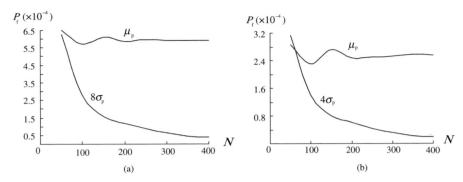

(a)　　　　　　　　　　　　　　　　　　　(b)

Figure 17.13 Statistics of failure probability versus sampling size of the portal frame: (a) rigid-plastic analysis; (b) second-order inelastic analysis

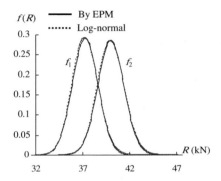

Figure 17.14 PDF of structural resistance of the portal frame

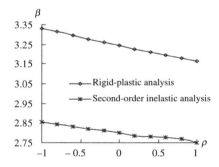

Figure 17.15 Effect of correlation on the system reliability of the portal frame

Table 17.10 The mean, standard deviation and parameters in EPM-estimated PDF for structural resistance of the portal frame

Parameters	f_1	f_2
μ_R (kN)	37.25	40.00
σ_R (kN)	1.40	1.41
a_1	0.015 90	0.0000
a_2	−0.5182	−0.5001
a_3	−0.0053	0.0000
a_4	0.0007	0.0029
C	2.5007	2.4833

EXAMPLE **315**

with 10^6 samples) and $P_{f_2} = 5.920 \times 10^{-4}$ (by rigid-plastic analysis model, with 10^7 samples). The reliability index corresponding to P_{f_1} and P_{f_2} are $\beta_1 = 2.800$ and $\beta_2 = 3.245$, respectively. By comparing the results of these two analysis models, one can find that large error may result from rigid-plastic analysis (the failure probability is one power smaller than second-order inelastic analysis in this case) because it does not consider material inelasticity and geometric nonlinearity. For steel frames, limit load-bearing capacities are generally influenced by material and geometric nonlinearities. It is therefore necessary to adopt second-order inelastic analysis model in reliability calculation of steel frames.

The variations of the mean and standard deviation of the failure probability of the portal frame with the number of trials, obtained by simulations of SS with AV, are shown in Figure 17.13. It can be found that the calculation is to be convergent from sampling size 400.

The EPM-based PDF curves of the structural resistance of the portal frame obtained by second-order inelastic analysis (f_1) and rigid-plastic analysis (f_2) are given in Figure 17.14. The mean and standard deviation of the structural resistance simulation and the parameters in the expression of EPM-estimated PDF for the portal frame are listed in Table 17.10, and the corresponding curves f_1 and f_2 in Figure 17.14.

The effects of correlation, ρ, between M_c and M_b on the reliability index of the portal frame for the above-mentioned two models of structural analysis are shown in Figure 17.15.

18 System Reliability Assessment of Steel Frames

Significant random variables with existing statistics are used in this chapter to evaluate the system reliability of steel frames, by the semi-analytical simulation method given in the last chapter, under two typical load cases.

18.1 RANDOMNESS OF STEEL FRAME RESISTANCE

The randomness of the sectional resistance in the structural component is due to the randomness of material properties, geometric dimension and analysis model. As for the randomness of the structural resistance for steel frames, it becomes very complicated. In addition to the above three factors, structural imperfection (both geometrical and mechanical imperfections) in the process of manufacture and instalment also produces uncertainties to the structural resistance. For steel frame structures, it is recognized that the uncertainty of the yielding strength of steel is the most principal factor contributing to the randomness of the structural resistance. In this section, uncertainties of the sectional dimension and analysis model will also be involved and other factors are ignored because little knowledge can be used in a practical assessment currently.

To clarify the basis of investigation, the following assumptions are made:

(1) The uncertainties of material properties are simplified and represented with no more than the uncertainty of the yielding strength of steel. Other property parameters are treated as deterministic.

(2) The uncertainties of sectional dimensions are expressed only with that of section resisting modulus. In other words, the uncertainties of sectional dimensions are considered to affect only bending stiffness, not axial and shear stiffness, of frame members.

(3) A special random factor is introduced to consider the effect of the uncertainty of the analysis model on the system reliability of steel frames.

(4) All random factors mentioned above, i.e. yielding strength of steel, section resisting modulus and analysis model factor, are assumed to be satisfied with normal distribution.

By the initial and ultimate yielding equations introduced in Chapter 5, the initial yielding moment M_{sN} and ultimate yielding moment M_{pN} of frame members coupled with the axial force depend on those moments in pure bending, M_s and M_p, provided that the axial capacity of this member is deterministic. Furthermore,

Advanced Analysis and Design of Steel Frames   Guo-Qiang Li and Jin-Jun Li
© 2007 John Wiley & Sons, Ltd

Table 18.1   Statistics of the fundamental variables influencing structural resistance of the steel frame

Variables			Mean/normal values	Coefficient of variation
Geometric size		Breadth	1.000	0.0135
		Thickness	1.000	0.0350
Yielding strength	Q235 steel	$t \leq 16\,\text{mm}$	1.070	0.081
		$16\,\text{mm} < t \leq 40\,\text{mm}$	1.074	0.077
		$40\,\text{mm} < t \leq 60\,\text{mm}$	1.118	0.066
		$60\,\text{mm} < t \leq 100\,\text{mm}$	1.087	0.066
		$t \leq 16\,\text{mm}$	1.040	0.066
	Q345 steel	$16\,\text{mm} < t \leq 35\,\text{mm}$	1.025	0.076
		$35\,\text{mm} < t \leq 50\,\text{mm}$	1.125	0.057
		$50\,\text{mm} < t \leq 100\,\text{mm}$	1.184	0.083
Factor of structural analysis model		Vertical load	1.000	0.075
		Vertical plus horizontal loads	1.000	0.075

precluding randomness of the sectional plastic factor, χ_p, indicates that the uncertainty of M_p is actually same as that of M_s because $M_p = \chi_p M_s$. As $M_s = W_e \cdot \sigma_s$, it comprises two random factors, sectional resisting modulus and yielding strength of steel, and becomes the unique random variable in the sampling of the structural resistance for steel frames according to the assumption mentioned above.

The elastic resisting modulus W_e is a function of sectional dimensions, including breadth and thickness of the section. The statistics of the dimensions of steel sections are listed in Table 18.1, and all of these random variables are assumed to be satisfied with normal distribution. Based on the information in Table 18.1, the mean value and standard variance of elastic resisting modulus W_e can be calculated with the function method. And the distribution of W_e can also be assumed to be normal because

(1) by χ^2 verification, W_e does not refuse the normal distribution and the D-value, which actually indicates the discrepancy between realistic and analytical frequencies, is very small;

(2) the mean value and standard variance of W_e obtained with the quasi-normalization principle are very close to those obtained with the function method.

In the reliability evaluation of steel frames with tapered members, a beam or column within one taper ratio is assumed to be one 'member', and the randomness of each 'member' depends on elastic section modulus at the two ends. For the tapered member generally used, with linearly varying web height and constant flange breadth, the distribution of elastic section modulus along member length is a second-order function. If more than one element is subdivided within one such 'member', the elastic section modulus of internal elements can be calculated with the second-order interpolation.

The statistics of Q235 and Q345 steel are given in Table 18.1 according to different plate thicknesses (Chen, Li and Xia, 1985; Dai et al., 2000). The yielding strength is assumed to be normally distributed.

The uncertainty parameters of the analysis model factor are also listed in Table 18.1 (Val, Bljuger and Yankelevsky, 1997) and assumed to be normally distributed too.

18.2   RANDOMNESS OF LOADS

Consider two load cases, respectively, for steel portal frames and multi-storey steel frames, as shown in Figure 18.1(a)–(d). In load cases (a) and (c), the incremental vertical loads are the dead and live loads; in load case (b), the constant vertical load is the dead load and the incremental vertical and horizontal load is the wind load; in load case (d), the constant vertical loads are dead and live loads, and the incremental

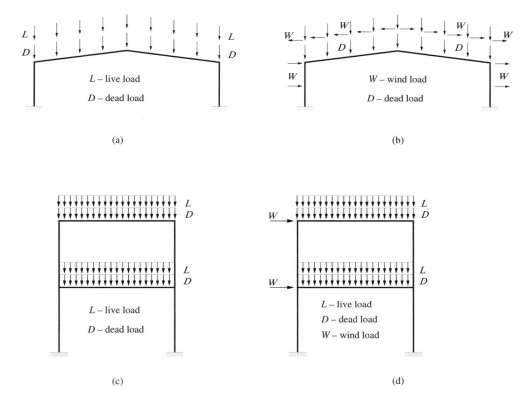

Figure 18.1 Loading cases of steel frames considered: (a) steel portal frame subjected to vertical loads; (b) steel portal frame subjected to horizontal and vertical loads; (c) multi-storey steel frame subjected to vertical loads; (d) multi-storey steel frame subjected to horizontal and vertical loads

horizontal load is the wind load. The statistics of loads considered in the above load cases are listed in Table 18.2 (Ellingwood and Galambos, 1981; Li, 1985).

18.3 SYSTEM RELIABILITY EVALUATION OF TYPICAL STEEL FRAMES

Based on the information given in Tables 18.1 and 18.2, a practical evaluation on the system reliability of steel portal frames and multi-storey frames can be made.

18.3.1 Effect of Correlation Among Random Variables

To determine the appropriate correlation parameters of random variables used in system reliability evaluation, examine the effect of correlation parameters of random variables on the system reliability of steel frames.

Table 18.2 Statistics of various loads

	Mean/normal value	Coefficient of variation	Distribution type
Deal load	1.06	0.07	Normal
Live load	1.00	0.25	Extreme, type I
Wind load	1.00	0.193	Extreme, type I

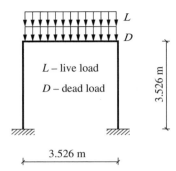

Figure 18.2   A steel frame used to study the correlation effect

Evaluate the system reliability of a steel portal frame under vertical (dead and live) loads, as shown in Figure 18.2, with the semi-analytical simulation method. The nominal values of yielding strength, dead load and live load are 235 MPa, 50 kN/m and 50 kN/m, respectively. The sections of beams and columns are all W8X31, and the elastic modulus of the material is 206 GPa. In the calculation of the structural system resistance (structural limit load-bearing capacity), vertical dead and live loads are both incremental up to system failure.

It is assumed that the yielding strength of steel is completely independent of elastic section modulus of frame members, but correlation among yielding strengths or elastic section moduli of the different frame members exists. The effects of such correlation on the system reliability of the steel portal frame are given in Table 18.3, where ρ_σ denotes correlation among yielding strength of different steel members and ρ_W denotes correlation among elastic section modulus of different members.

The results in Table 18.3 show that ρ_σ has more impact on the system reliability index of steel portal frames than ρ_W, and the system reliability reduces with the increase of the correlation of yielding strength. As in reality the material of the members for the same steel frame is generally from the same steel production, the strong correlation of yielding strength can be accepted easily. Additionally, for the purpose of conservative consideration, $\rho_\sigma = 1.00$ will be adopted in the following investigations. Considering that the deviation coefficient of the elastic section modulus of normal steel frame members ranges from 0.043 to 0.045, which is much less than that of the yielding strength of steel being 0.081, the effect of the correlation of the elastic section modulus on the structural system reliability is small. Hence, the following values are recommended for ρ_W in evaluating the system reliability of steel frames, i.e. $\rho_W = 0.0$ for steel portal frames and $\rho_W = 0.5$ for multi-storey steel frames.

18.3.2   Evaluation of Structural System Reliability Under Vertical Loads

18.3.2.1   Steel portal frame

Consider the load case as in Figure 18.1(a) and a steel portal frame with tapered beams and columns as shown in Figure 18.3. The cross-sectional dimension of the flange of all the frame members is constantly 150 mm×6 mm, and the thickness of the web is 4 mm. The web height of the frame members varies linearly

Table 18.3   Effect of correlation of fundamental variables on the system reliability index of the steel portal frame

	ρ_σ		
ρ_W	1.000	0.500	0.000
1.000	2.7231	2.8287	2.9274
0.500	2.7629	2.8728	2.9689
0.000	2.8178	2.9001	3.0150

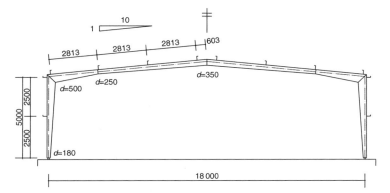

Figure 18.3 A steel portal frame

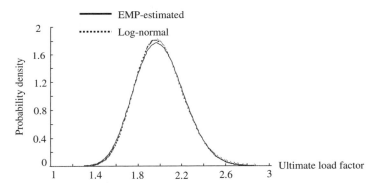

Figure 18.4 PDF of the ultimate load factor of the steel portal frame under vertical loads

along the member length, and the governing values are given in Figure 18.3. In this steel portal frame building, the column spacing is 6 m, and the nominal values of the vertical dead and live loads on the roof of the building are 0.40 and 0.30 kN/m², respectively. In the calculation of the structural system resistance (i.e. structural limit load-bearing capacity) of the frame, the vertical dead and live loads are both gradually increased up to the system failure of the frame.

Take the reference load as the sum of the nominal dead and live loads. As the ultimate load of the frame can be expressed as the ultimate load factor multiplied with the reference load, the randomness of the frame resistance can then be expressed as that of the ultimate load factor. By sampling the elastic resisting modulus and yielding strength of the members comprising the frame, the mean value and standard variance of the ultimate load factor of the frame are obtained through structural nonlinear analysis. The statistics and the parameters for the EPM-estimated PDF curve of the ultimate load factor (see Figure 18.4) are given in Table 18.4. The log-normal distribution curve with the same mean and variance values of the EPM-estimated PDF is also plotted, both of which agree well in Figure 18.4.

Table 18.4 Statistics and parameters for the PDF curve of ultimate load factor of the steel portal frame under vertical loads

M	σ	a_1	a_2	a_3	a_4	c
1.9967	0.2224	−0.1112	−0.4592	0.039 29	−0.008 880	2.5535

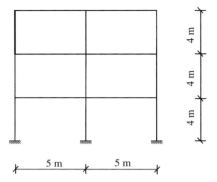

Figure 18.5 A multi-storey steel frame

Simply applying the first-order second-moment (FOSM) design point method to performance function $Z = R - D - L$, the system reliability of the steel portal frame under vertical dead and live loads can be determined. The reliability index and failure probability of the frame are equal to 3.8027 and 7.1562×10^{-5}, respectively.

18.3.2.2 Steel multi-storey frame

Check a two-span three-storey steel frame, as shown in Figure 18.5, under vertical (dead and live) loads as shown in Figure 18.1(c). The nominal value of steel yielding strength is 235 MPa. The values of the nominal dead and live loads are both 60 kN/m on the first and second floors and 30 kN/m on the top floor (roof). W16X50 is used for the frame beams and W16X67 is used for the columns. The steel elastic modulus is 206 GPa. In the calculation of the structural system resistance (structural limit load-bearing capacity) of the frame, the vertical dead and live loads are both gradually increased up to the system failure of the frame.

Take the reference load as the sum of the nominal dead and live loads. By sampling, the mean value and standard variance of the ultimate load factor of the frame are obtained to be 2.2209 and 0.2567, respectively. The statistics and the parameters for the EPM-estimated PDF curve of the ultimate load factor (see Figure 18.6) are given in Table 18.5. The log-normal distribution curve with the same mean and variance values of the EPM-estimated PDF is also plotted, both of which agree well in Figure 18.6.

As the performance function of the frame is $Z = R - D - L$, the system reliability index of the frame under the vertical dead and live loads can be easily obtained using the FOSM design point method. The

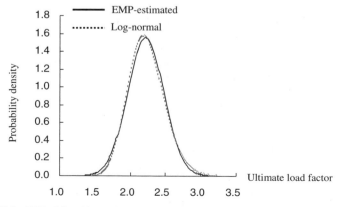

Figure 18.6 PDF of the ultimate load factor of the multi-storey steel frame under vertical loads

Table 18.5 Statistics and parameters for the PDF curve of ultimate load factor of the multi-storey steel frame under vertical loads

M	Σ	a_1	a_2	a_3	a_4	c
2.2209	0.2567	−0.0182	−0.5002	0.0061	0.0008	2.5061

reliability index and the corresponding failure probability of the frame are $\beta = 3.8069$ and $P_f = 7.0366 \times 10^{-5}$, respectively.

18.3.3 Evaluation of Structural System Reliability Under Horizontal and Vertical Loads

18.3.3.1 Steel portal frame

Consider the load case shown in Figure 18.1(b) and the steel portal frame shown in Figure 18.3. Wind load is involved in the assessment of the system reliability of the frame. The nominal value of the basic wind pressure is 0.50 kN/m². The wind shape factor is determined according to the Chinese code (CECS, 2002). In the analysis of the ultimate load-bearing capacity of the frame, the vertical dead load remains constant (equal to its nominal value) and the wind load keeps increasing up to the frame failure.

By sampling, the mean value and standard variance of the ultimate wind load factor of the frame are 4.0247 and 0.4420, respectively. The statistics and the parameters for the EPM-estimated PDF curve of the ultimate load factor (see Figure 18.7) are given in Table 18.6. The log-normal distribution curve with the same mean and variance values of the EPM-estimated PDF is also plotted, both of which agree well in Figure 18.7.

Simply applying the FOSM design point method to the performance function $Z = R - W$, the system reliability of the frame under wind loads can be determined. The reliability index and failure probability of the frame are equal to 5.3543 and 4.2944×10^{-8}, respectively.

Table 18.6 Statistics and parameters for the PDF curve of the ultimate wind load factor of the steel portal frame

M	Σ	a_1	a_2	a_3	a_4	C
4.0247	0.4420	−0.076 09	−0.4952	0.025 71	−0.001 738	2.5101

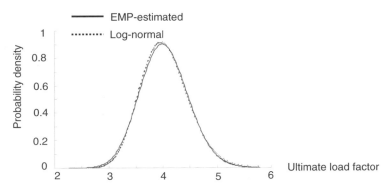

Figure 18.7 PDF of the ultimate wind load factor of the steel portal frame

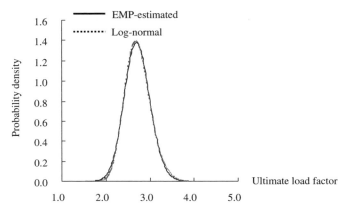

Figure 18.8 PDF of the ultimate wind load factor of the multi-storey steel frame

Table 18.7 Statistics and parameters for the PDF curve of the ultimate wind load factor of the multi-storey steel frame

μ	σ	a_1	a_2	a_3	a_4	c
2.7121	0.2903	-0.0627	-0.5019	0.0209	0.0007	2.5053

18.3.3.2 Steel multi-storey frame

Check the load case in Figure 18.1(d) and the steel frame in Figure 18.5. Wind load, which is assumed to be applied at beam–column joints of the frame, is to be introduced in the structural system reliability evaluation. The nominal value of the wind load at the level of the first and second floors is 100 and 50 kN at the third floor. In the calculation of the structural system resistance (structural limit load-bearing capacity), the vertical dead and live loads remain constant, whereas wind load is gradually increased up to the system failure of the frame.

By sampling, the mean value and standard variance of the ultimate wind load factor of the frame are 2.7121 and 0.2903, respectively. The statistics and the parameters for the EPM-estimated PDF curve of the ultimate wind load factor (see Figure 18.8) are given in Table 18.7. The log-normal distribution curve with the same mean and variance values of the EPM-estimated PDF is also plotted, both of which agree well in Figure 18.8.

As the performance function of the frame is $Z = R - W$, the system reliability index of the frame under both the horizontal and vertical loads can be obtained with the FOSM design point method. The reliability

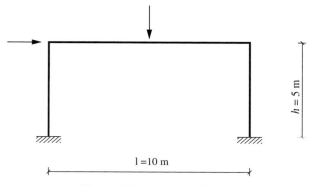

Figure 18.9 A steel portal frame

Table 18.8   Statistics of member resistance and loads of the portal frame

	Mean value	Coefficient of variation	Distribution type
Load V	80 kN	0.2	Normal
Load H	50 kN	0.3	Extreme, type I
Ultimate yielding moment of frame components	180 kN m	0.1	Log-normal

index and the corresponding failure probability of the frame are $\beta = 3.9813$ and $P_f = 3.4272 \times 10^{-5} =$, respectively.

18.4   COMPARISON OF SYSTEM RELIABILITY EVALUATION

Currently, the failure mode method is used in majority for structural system reliability evaluations. In the failure mode method, the dominant system failure modes of a structure are first identified with some particular techniques and then the failure probability of each failure mode is calculated by establishing the performance function of the failure mode based on the virtual work of mechanism. The system reliability of the structure can be obtained finally with a consideration of correlation among all of the dominant failure modes.

For the sake of comparing the failure mode method with the semi-analytical simulation method proposed in the last chapter, the system reliability of a steel portal frame (see Figure 18.9) is analysed with such two methods. The statistics of the member resistance and loads used in the analysis are given in Table 18.8.

The reliability index obtained by the failure mode method ranges from 2.91 to 2.93 (Li, 1985) and that by the semi-analytical simulation method is 2.6980. It is revealed by this comparison that as the failure mode method cannot consider the effects of material and geometric nonlinearities, it may overestimate the system reliability of steel frames.

19 Reliability-Based Advanced Design of Steel Frames

Research efforts have been made for the advanced design of steel frames recently. The advanced design of structures is the structural design based on advanced analysis. Currently, the advanced design of steel frames is proposed by simply substituting the limit load-bearing capacity of the structural system obtained with advanced analysis into the existing design formula based on structural member reliability (Kim and Chen, 1996a, 1996b), which avoids structural system reliability evaluation. This kind of advanced design is called the preliminary advanced design (PAD), which cannot produce the design of steel frames with certain structural system reliability (Buonopane and Schafer, 2006). With the viewpoint of structural reliability, the difference is tangible between the design formula based on structural member reliability and the structural system reliability as discussed in Chapter 16. The uncertainty of the design objective and the ambiguity of the target reliability of the design may be obtained if such a difference is neglected in the advanced design. It is absolutely necessary to establish the advanced design approach for steel frames based on the structural system limit state and system reliability evaluation. This design may be called the reliability-based advanced design (RAD), which is able to produce the design of steel frames with certain structural system reliability.

There are two methods in the structural reliability design, i.e. complete probabilistic method and approximate probabilistic method. The complete probabilistic method is based on direct evaluation of structural system reliability, and the system reliability is accurate. However, it is not practical or it is very difficult to present a direct evaluation of system reliability in practice, especially for engineering structures widely used. Currently, the complete probabilistic method is used only in the design of some special and costive structures such as nuclear power shells and offshore platforms. In contrast, the approximate probabilistic method is a simplified reliability design method, which has a form familiar to structural engineers with nominal values of loads and material yielding strengths, and load and resistance factors. As the values of load and resistance factors are constant, the structural reliability by the design is not accurate and uniform, but within a limited scope accepted. Generally, the approximate probabilistic design method is accepted as routine work due to its ease of application.

In this chapter, the practical design formula of the reliability-based advanced design is to be established for steel portal frames and multi-storey steel frames with the approximate probabilistic method, based on advanced analysis and evaluation of structural system reliability.

19.1 STRUCTURAL DESIGN BASED ON SYSTEM RELIABILITY

19.1.1 Target Reliability of Design

The target reliability of the structural design largely affects the results of the design. An excessively high target reliability will result in an unexpectedly strong structure and an unreasonable construction cost,

Table 19.1 Annual death rate of some events

Accident	Annual fatal rate	Accident	Annual fatal rate
Climbing, car racing	5.0×10^{-3}	Travel by car	2.5×10^{-5}
Air travel	1×10^{-4}	Swimming	3×10^{-5}
Mining	7×10^{-4}	Construction	3×10^{-5}
Building fire	2×10^{-5}	Electric shot	6×10^{-6}
Thunder	5×10^{-7}	Storm	4×10^{-6}

whereas a structure with an inadequate target reliability will not satisfy the performance requirements relating to its safety, serviceability and durability. So, the selection of the target reliability for the design of a structure should be based on the optimum balance between structural reliability and economy. Four aspects are necessarily considered for this purpose, which are (1) public psychology, (2) structural importance, (3) nature of structural damage and (4) endurance of society and economy to structural failure.

One might attempt to establish reliability targets by considering risks that arise from other exposures and human activities. Table 19.1 lists the annual death rates due to some events. General public think that climbing and car racing are somewhat dangerous, air travel is relatively safe, travel by car is safe and death due to electric or lightning shock is nearly impossible. From the analysis of public psychology, the dangerous rate accepted by bold people is about 10^{-3} per year, whereas this value for cautious people is 10^{-4} per year. Such an analysis indicates furthermore that if the annual death rate is 10^{-5} per year or less, normal people will not consider fatalness anymore, which may be adopted as the reference of the reliability target for a safe structure. So, the engineering structure with annual failure rate less than 1×10^{-4} can be regarded to have less safety, that less than 1×10^{-5} will be safe whereas that less than 1×10^{-6} will be extremely safe. As the design life for general structures is usually 50 years, the structures with failure probabilities of 5×10^{-3}, 5×10^{-4} and 5×10^{-5} in the design life are accordingly thought to be less safe, safe and extremely safe, respectively. The corresponding reliability index ranges from 2.5 to 4.0.

The target reliability of the structural design is also relevant to structural importance. For very important structures (e.g. nuclear power station and national broadcast emitting tower), the reliability of design should be targeted higher, whereas for temporary or secondary structures (e.g. temporary warehouse and bicycle shed), it may be reduced proportionally. Three classes of structural importance are employed in many countries, namely important, general and less important. The fundamental target reliability is often based on general structures, and then the target reliability of important structures decreases by one power and that of less important structures increases by one power.

As the damage of brittle structures (e.g. masonry structures) is nearly unpredictable, its sequence is more serious than that of ductile structures (e.g. steel structures). The general treatment on the design of brittle structures is to raise the target reliability properly.

One increasingly effective factor for the selection of the target reliability for structural design is the endurance of society and economy to structural failure. Generally speaking, the more developed the economy, the more the concern paid by public to structural failure which leads to higher target structural reliability.

Calibration is also a compromised method to determine the target reliability used in the probability-based structural design, especially in the early stage from the empirical design to the probabilistic design. The

Table 19.2 Target reliability indices in the current Chinese code for components of building structures

	Structural importance		
	Important	General	Less important
Ductile structure	3.7	3.2	2.7
Brittle structure	4.2	3.7	3.2

Table 19.3 Target reliability indices and corresponding failure reliability for the reliability-based advanced design of steel frames

Important structure		General structure		Less important structure	
β	P_f	β	P_f	β	P_f
4.2	1.3×10^{-5}	3.7	1.1×10^{-4}	3.2	6.9×10^{-4}

reliability in the traditional structural design is calibrated and succeeded in the new probabilistic design. As an example, the target reliability of structural members adopted in the current probabilistic design code for building structures in China is calibrated from that in the previous semi-empirical and semi-probabilistic design methods. The target reliability indices are listed in Table 19.2.

The reliability-based advanced design aims at the limit states of the structural system, and its target reliability index can be rationally higher than that aiming at the limit states of structural components because of the failure severity of the structural system. The values of target reliability indices proposed for the reliability-based advanced design of steel frames are listed in Table 19.3.

19.1.2 Load and Load Combination

As discussed in Chapter 18, three types of loads (dead, live and wind loads) and two load combinations are considered for steel portal frames and multi-storey steel frames. The statistics of such loads have been given in Chapter 18.

For multi-storey steel frames, denote load case I as a combination of dead and live loads and load case II as that of dead plus live loads and wind loads. As for steel portal frames, it is not necessary to combine live and wind loads because roof suction produced by wind load is offset by live load. So for steel portal frames, denote load case II as a combination of dead and wind loads. In low-rise steel portal frames, load case I (vertical load case) generally controls the structural design, as the effect of the horizontal action induced by wind is less dominant.

19.1.3 Practical Design Formula

The practical design formula of RAD can be written as

$$\phi \cdot R_n \geq \gamma_0 \cdot \sum \gamma_i \cdot S_{ni}, \tag{19.1}$$

where R_n and S_n are the nominal values of structural system resistance (limit load-bearing capacity) and load effects, respectively, ϕ and γ_i are, respectively, the factor of the structural system resistance and the load factor, obtained with system reliability assessment, and γ_0 is the factor relating to structural ductility, redundancy and importance. Although the design formula of RAD has the same form as that of LRFD (load and resistance factor design), RAD is a totally new design approach based on the limit states of structural systems, no longer based on the limit states of structural components as LRFD.

19.1.3.1 Factor of structural importance

Referring the following values of structural importance factor for the member reliability design (GB 50068, 2001):

(1) not less than 1.1, for important structures;

(2) not less than 1.0, for general structures;

(3) not less than 0.9, for less important structures,

the values of structural importance factor for the system reliability design or RAD are recommended as

(1) not less than 1.1, for important structures;

(2) not less than 1.0, for general structures;

(3) not less than 0.9, for less important structures.

19.1.3.2 Load and resistance factors

The principle to calculate the load and resistance factors is to make the nominal system resistance R_K^* determined by satisfying the design target reliability index as close as possible to the nominal system resistance R_K by the following design formula:

$$R_K = \gamma_R(\gamma_G G_K + \gamma_L L_K). \tag{19.2}$$

With the above principle, the realistic reliability index of the structure designed by the above formula may be close to the target one. Such a principle can therefore be transferred to minimize ε, given by

$$\varepsilon = \sum_{i=1}^{m}\sum_{j=1}^{n}\left(1 - \frac{R_{Kij}}{R_{Kij}^*}\right)^2, \tag{19.3}$$

where R_{Kij}^* is the nominal system resistance by the complete probabilistic method aiming at the design target reliability index, under the ith group of values of γ_G and γ_L, and the jth dead/live load ratio, R_{Kij} is the nominal system resistance by design formula (19.2), under the same conditions as above, m is the total number of groups of γ_G and γ_L, and n is the number of dead/live load ratios.

The flow chart to calculate the load and resistance factors is given in Figure 19.1, which is similar to that used for the member reliability design except the statistics herein are from system reliability evaluation. The main processes are assigning the values of resistance factors, satisfying Equation (19.3), and finally obtaining load factors.

The statistics used for determining the design formula of RAD of steel portal frames are tabulated in Table 19.4 and those for RAD of multi-storey steel frames in Table 19.5.

19.1.3.3 Design formula

Aiming at the target reliability index to be 3.7, the design formulae can be determined, respectively, for steel portal frames and multi-storey steel frames made from Q235 or Q345 steel as follows.

(1) Steel portal frames

– Under load case I:

$$0.80R_n \geq 1.00G_n + 1.80L_n \quad \text{for Q235 steel,} \tag{19.4a}$$

$$0.80R_n \geq 1.05G_n + 2.20L_n \quad \text{for Q345 steel,} \tag{19.4b}$$

where R_n is the nominal value of structural resistance under load case I determined by advanced analysis, and G_n and L_n are, respectively, the nominal values of the vertical dead load and live load.

– Under load case II:

$$0.80R_n(G = 1.00G_n) \geq 1.60W_n \quad \text{for Q235 steel,} \tag{19.5a}$$

$$0.80R_n(G = 1.00G_n) \geq 1.85W_n \quad \text{for Q345 steel,} \tag{19.5b}$$

where R_n is the nominal value of structural resistance under load case II determined by advanced analysis when the dead load adopts its nominal values and W_n is the nominal value of the wind load.

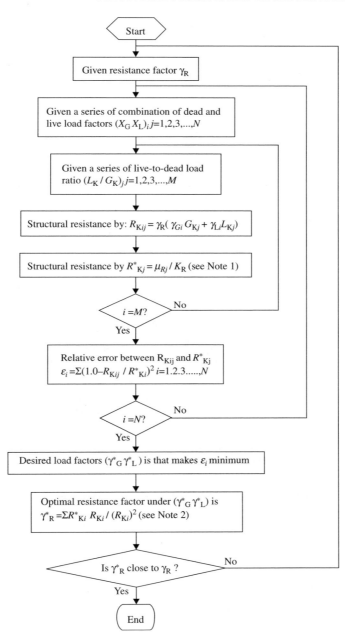

Figure 19.1   Flow chart for calculation of resistance and load factors

Note 1. μ_{Rj} is determined by the FOSM design point method with the statistics listed in Table 19.3.

Note 2. R_{Kj} is obtained by $R_{Kj} = \gamma_R(\gamma_G^* G_{Kj} + \gamma_L L_{Kj})$.

Note 3. Subscript K means the nominal value.

Table 19.4   Statistics for the design formula of RAD of steel portal frames

Steel grade	Load case	Resistance factor (given)	Load factor (to be determined)		Load ratio	Statistics of resistance	
Q235	I	1.25	Deal load 1.00–1.40[b]	Live load 1.00–2.50[b]	Live load/dead load 0.50–1.10[c]	COV 0.115	K_R^a 1.20
	II	1.25	Dead load[d] 1.0	Wind load 1.00–2.00[b]	—	COV 0.100	K_R^a 1.25
Q345	I	1.25	Dead load 1.00–1.40[b]	Live load 1.40–2.95[b]	Live load/dead load 0.50–1.10[c]	COV 0.110	K_R^a 1.03
	II	1.25	Dead load[d] 1.0	Wind load 1.10–2.65[b]	—	COV 0.112	K_R^a 1.08

[a] K_R denotes the ratio of mean value to nominal value.
[b] Let the incremental be 0.05.
[c] Let the incremental be 0.15.
[d] The value of dead load is assumed to be constant, which is equal to the nominal value in load case II.

Table 19.5   Statistics for the design formula of RAD of multi-storey steel frames

Steel grade	Load case	Resistance factor (given)	Load factor (to be determined)		Load ratio	Statistics of resistance	
Q235	I	1.20	Deal load 1.00–1.60[b]	Live load 1.40–3.00[b]	Live load/dead load 0.50–1.10[c]	COV 0.110	K_R^a 1.035
	II	1.20	Dead load[d] 1.0	Wind load 1.10–2.65[b]	—	COV 0.108	K_R^a 1.065
Q345	I	1.20	Dead load 1.00–1.60[b]	Live load 1.40–3.00[b]	Live load/dead load 0.50–1.10[c]	COV 0.110	K_R^a 1.046
	II	1.20	Dead load[d] 1.0	Wind load 1.10–2.65[b]	—	COV 0.121	K_R^a 1.103

[a] K_R denotes the ratio of mean value to nominal value.
[b] Let the incremental be 0.05.
[c] Let the incremental be 0.15.
[d] The value of dead load is assumed to be constant, which is equal to the nominal value in load case II.

The comparison of the relative error given by Equation (19.3) among different load factors and target reliability indices is plotted in Figure 19.2, and the numerical results are also given in Table 19.6.

(2) Multi-storey steel frames

– Under load case I:

$$0.83R_n \geq 1.10G_n + 2.25L_n \quad \text{for Q235 steel,} \tag{19.6a}$$

$$0.83R_n \geq 1.10G_n + 2.20L_n \quad \text{for Q345 steel,} \tag{19.6b}$$

where R_n is the nominal value of structural resistance under load case I determined by advanced analysis, and G_n and L_n are, respectively, the nominal values of the vertical dead load and live load.

-   Under load case II:

$$0.83R_n((G+L) = 1.00(G_n + L_n)) \geq 1.95W_n \quad \text{for Q235 steel,} \tag{19.7a}$$

$$0.83R_n((G+L) = 1.00(G_n + L_n)) \geq 1.90W_n \quad \text{for Q345 steel,} \tag{19.7b}$$

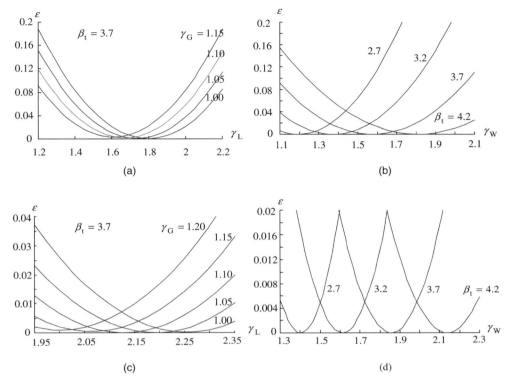

Figure 19.2 Relative error of structural system resistance of steel portal frames with fixed $\gamma_R = 1.25$: (a) variation of relative error with dead and live load factors for Q235 steel under load case I; (b) variation of relative error with target reliability and wind load factors for Q235 steel under load case II; (c) variation of relative error with dead and live load factors for Q345 steel under load case I; (d) variation of relative error with target reliability and wind load factors for Q345 steel under load case II

where R_n is the nominal value of structural resistance under load case II determined by advanced analysis when dead and live loads adopt their nominal values and W_n is the nominal value of the wind load.

The comparison of the relative error given by Equation. (19.3) among different load factors and target reliability indices is plotted in Figure 19.3, and the numerical results are also given in Table 19.7.

Table 19.6 Numerical results of ε varying with target reliability and load factors for the steel portal frame (given $\gamma_R = 1.25$)

Steel grade		Load case I			Load case II	
	β_t	γ_G	γ_L	ε	γ_W	ε
Q235	2.7	1.00	1.20	0.1612×10^{-2}	1.20	0.2141×10^{-4}
	3.2	1.00	1.50	0.7238×10^{-3}	1.35	0.2353×10^{-3}
	3.7	1.00	1.80	0.2835×10^{-3}	1.60	0.2672×10^{-3}
	4.2	1.00	2.15	0.1022×10^{-3}	1.80	0.3340×10^{-4}
Q345	2.7	1.05	1.55	0.7999×10^{-3}	1.40	0.1038×10^{-5}
	3.2	1.05	1.85	0.7703×10^{-4}	1.60	0.5415×10^{-4}
	3.7	1.05	2.20	0.4822×10^{-4}	1.85	0.1268×10^{-4}
	4.2	1.05	2.60	0.7879×10^{-4}	2.15	0.3049×10^{-4}

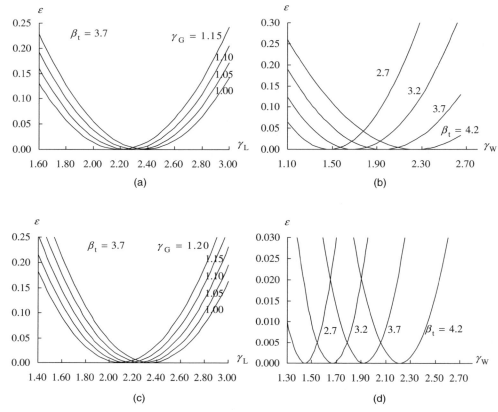

Figure 19.3 Relative error of structural system resistance of multi-storey steel frames with fixed $\gamma_R = 1.20$: (a) variation of relative error with dead and live load factors for Q235 steel under load case I; (b) variation of relative error with target reliability and wind load factors for Q235 steel under load case II; (c) variation of relative error with dead and live load factors for Q345 steel under load case I; (d) variation of relative error with target reliability and wind load factors for Q345 steel under load case II

Table 19.7 Numerical results of ε varying with target reliability and load factors for multi-storey steel frames (given $\gamma_R = 1.20$)

| Steel grade | β_t | Load case I | | | Load case II | |
		γ_G	γ_L	ε	γ_W	ε
Q235	2.7	1.10	1.60	$5.083\,98 \times 10^{-4}$	1.45	$2.536\,47 \times 10^{-4}$
	3.2	1.10	1.90	$7.090\,22 \times 10^{-5}$	1.70	$1.460\,03 \times 10^{-5}$
	3.7	1.10	2.25	$2.552\,64 \times 10^{-5}$	1.95	$1.614\,55 \times 10^{-7}$
	4.2	1.10	2.70	$2.217\,63 \times 10^{-4}$	2.25	$9.661\,20 \times 10^{-6}$
Q345	2.7	1.10	1.55	$2.807\,23 \times 10^{-4}$	1.45	$1.070\,29 \times 10^{-5}$
	3.2	1.10	1.85	$1.023\,53 \times 10^{-4}$	1.65	$8.154\,08 \times 10^{-5}$
	3.7	1.10	2.20	$1.402\,13 \times 10^{-4}$	1.90	$1.222\,49 \times 10^{-4}$
	4.2	1.10	2.65	$1.130\,74 \times 10^{-4}$	2.20	$5.762\,73 \times 10^{-5}$

19.2 EFFECT OF CORRELATION ON LOAD AND RESISTANCE FACTORS

If correlation exists among the random variables in the structural performance function, it will influence the nominal values of the structural system resistance R_K^* determined by targeting a definite reliability index and then influence the calculation of the load and resistance factors. In this section, such an effect is examined with an example of a multi-storey frame using Q345 steel under load case I and aiming at a target reliability index of 3.7. The effects of correlation on the relative error of the structural system resistance are given in Figures 19.4–19.6, and the numerical values are listed in Tables 19.8–19.10.

From the observation of Figures 19.4 and 19.5 and Tables 19.8 and 19.9, it can be found that at the level of the same target reliability index, the load factor decreases with the increase of positive correlation between the system resistance R and the dead load G or the live load L, whereas the load factor increases with the

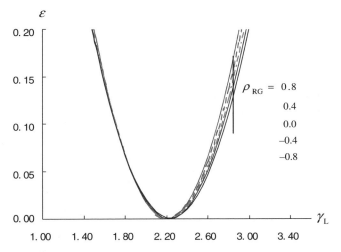

Figure 19.4 Effect of correlation of structural system resistance and dead load on relative error of the structural system resistance with fixed $\gamma_R = 1.20$

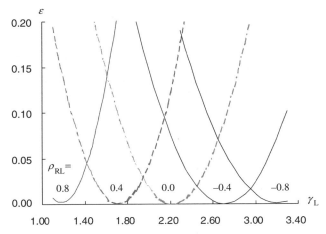

Figure 19.5 Effect of correlation of structural system resistance and live load on relative error of the structural system resistance with fixed $\gamma_R = 1.20$

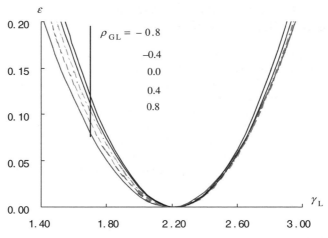

Figure 19.6   Effect of correlation of dead and live loads on relative error of the structural system resistance with fixed $\gamma_R = 1.20$

Table 19.8   Effect of correlation of structural system resistance and dead load on load factors with fixed $\gamma_R = 1.20$

ρ_{RG}	γ_G	γ_L	γ_R	ϵ
0.8	1.00	2.20	1.2009	2.38192×10^{-4}
0.4	1.05	2.20	1.2034	1.17839×10^{-4}
0.0	1.10	2.20	1.2061	1.40213×10^{-4}
−0.4	1.15	2.20	1.2092	3.19424×10^{-4}
−0.8	1.20	2.20	1.2125	6.63761×10^{-4}

Table 19.9   Effect of correlation of structural system resistance and live load on load factors with fixed $\gamma_R = 1.20$

ρ_{RL}	γ_G	γ_L	γ'_R	ε
0.8	1.10	1.20	1.1923	2.33222×10^{-3}
0.4	1.10	1.70	1.2042	8.43900×10^{-4}
0.0	1.10	2.20	1.2061	1.40213×10^{-4}
−0.4	1.10	2.70	1.2030	2.71523×10^{-4}
−0.8	1.10	3.20	1.2001	7.97594×10^{-4}

Table 19.10   Effect of correlation of dead and live loads on load factors with fixed $\gamma_R = 1.20$

ρ_{GL}	γ_G	γ_L	γ'_R	ε
−0.8	0.90	2.20	1.2092	3.00753×10^{-4}
−0.4	1.00	2.20	1.2090	2.86100×10^{-4}
0.0	1.10	2.20	1.2061	1.40213×10^{-4}
0.4	1.20	2.20	1.2014	5.40652×10^{-5}
0.8	1.30	2.20	1.1952	2.21674×10^{-4}

increase of negative correlation between the system resistance R and the dead load G or the live load L. In other words, such positive correlation reduces structural reliability, and the negative one increases it. As the coefficient of variation of live loads is greater than that of dead loads, correlation between R and L is more significant.

From Figure 19.6 and Table 19.10, it can be seen that at the level of the same target reliability index, the load factor decreases with the increase of negative correlation between G and L, whereas it increases with the increase of positive correlation between G and L. In other words, such negative correlation reduces structural reliability, and the positive one increases it.

19.3 COMPARISON OF DIFFERENT DESIGN METHODS

19.3.1 For Steel Portal Frames

19.3.1.1 Example frames

Three typical steel portal frames with tapered members, as shown in Figures 19.7 (frame I), 19.8 (frame II) and 19.9 (frame III), are used to compare the design results of RAD, PAD and LRFD, respectively.

The nominal values of loads are given in Figures 19.7–19.9. The dead/live load ratios of the three frames are selected as 0.60, 0.80 and 1.00, respectively, to consider the effects of different load ratios on the design results of the frames. The building shape factor to calculate wind pressure in load case II is determined according to MBMA (1996).

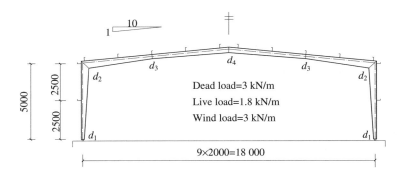

Figure 19.7 Steel portal frame I

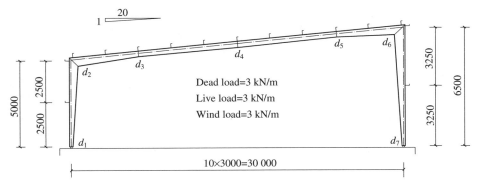

Figure 19.8 Steel portal frame II

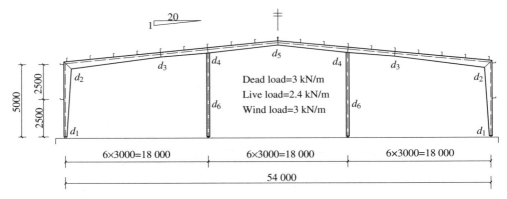

Figure 19.9 Steel portal frame III

If the load and resistance factors in AISC (1994) are used for PAD, the design formulae of PAD are

- Under load case I:

$$0.90R_n \geq 1.20G_n + 1.60L_n \quad \text{for Q235 and Q345 steel,} \tag{19.8a}$$

where R_n is the nominal value of the structural resistance under load case I determined by advanced analysis, and G_n and L_n are, respectively, the nominal values of the dead load and the live load.

- Under load case II:

$$0.90R_n(G = 1.20G_n) \geq 1.30W_n \quad \text{for Q235 and Q345 steel,} \tag{19.8b}$$

where R_n is the nominal value of the structural resistance under load case II determined by advanced analysis when the dead load adopts its nominal values and W_n is the nominal value of the wind load.

19.3.1.2 Design results

The sectional sizes and steel consumptions of the example frames designed, respectively, by RAD, PAD and LRFD are listed in Table 19.11, where Q345 steel is used for all cases. For the sake of comparison, all structural members in the same frame have the same flange and identical web thickness, but different web heights.

The frames are designed by checking member by member in LRFD for both load case I and load case II. In RAD and PAD, the limit load-bearing capacity of the frames is first obtained with advanced analysis for load case I, and it is used to check Equation (19.4b) in RAD and Equation (19.8a) in PAD. If they are

Table 19.11 Results of the example frames designed with various methods

Design method		Flange (mm²)	Web thickness (mm)	Web height (mm)							Steel consumption (10³ kg)
				d_1	d_2	d_3	d_4	d_5	d_6	d_7	
Frame I	LRFD	135 × 6.5	4.0	200	700	200	305				0.724
	PAD	120 × 5.6	3.2	200	600	200	252				0.537
	RAD	125 × 6.0	3.6	200	600	200	252				0.601
Frame II	LRFD	210 × 9.8	5.4	300	900	300	600	300	1080	300	2.344
	PAD	170 × 8.4	4.8	300	800	300	600	300	900	300	1.757
	RAD	170 × 8.6	5.4	300	900	300	600	300	1080	300	1.952
Frame III	LRFD	165 × 7.8	4.4	200	600	200	630	300	400		2.640
	PAD	130 × 6.6	3.4	200	600	200	630	300	400		1.879
	RAD	155 × 7.4	3.8	200	600	200	630	300	400		2.325

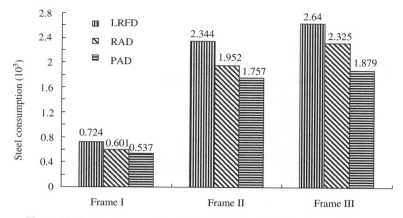

Figure 19.10   Steel consumption of LRFD, RAD and PAD for the portal frames

satisfied, the same procedures are further used to check Equation (19.5b) in RAD and Equation (19.8b) in PAD for load case II.

Due to asymmetry of frame II, structural analysis and structural safety check of load case II should consider left and right winds.

The comparison of steel consumptions of different designs for the frames is also illustrated in Figure 19.10.

19.3.1.3 Reliability calibration

To further study the difference among RAD, PAD and LRFD, the structural system reliability of the above designs is calibrated. The reliability indices and failure probabilities are listed in Table 19.12. Two values of those in frame II under load case II correspond to the wind loads from left and right sides, respectively.

By reliability calibration, it can be seen that the design of low-rise steel portal frames is controlled by vertical loads (load case I) rather than horizontal loads (load case II). In load case I, the system reliability of the frames designed with LRFD can reach about 4.6 and is greater than the target one. The system reliability index of the frames by PAD is about 2.7, which is expected to be relatively low. For the frames designed with RAD, the system reliability index is about 3.7, which is consistent with the target one.

Table 19.12   Structural system reliability of the example portal frames by LRFD, PAD and RAD

		Load case I		Load case II	
Design method		β	P_f	β	P_f
Frame I	LRFD	4.5941	0.2173×10^{-5}	7.7788	0.3662×10^{-14}
	PAD	2.5356	0.5613×10^{-2}	5.5898	0.1137×10^{-7}
	RAD	3.5500	0.1928×10^{-3}	5.9339	0.1480×10^{-8}
Frame II	LRFD	4.9795	0.3187×10^{-6}	7.9186	0.1201×10^{-14}
				8.1890	0.1317×10^{-15}
	PAD	2.8938	0.1903×10^{-2}	6.8842	0.2907×10^{-11}
				7.1897	0.3246×10^{-12}
	RAD	3.9914	0.3284×10^{-4}	7.2965	0.1477×10^{-12}
				7.4158	0.6045×10^{-13}
Frame III	LRFD	4.4110	0.5144×10^{-5}	7.3979	0.6918×10^{-13}
	PAD	2.8339	0.2299×10^{-2}	6.1091	0.5009×10^{-9}
	RAD	3.9389	0.4094×10^{-4}	6.8728	0.3148×10^{-11}

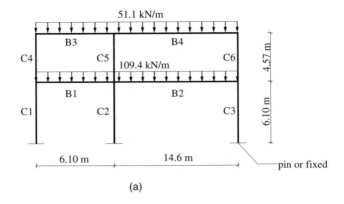

(a)

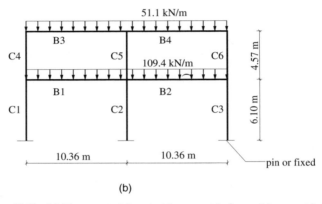

(b)

Figure 19.11   Multi-storey steel frames: (a) asymmetric frame; (b) symmetric frame

19.3.2   For Multi-Storey Steel Frames

19.3.2.1 Introduction of the frames studied

Four two-storey two-span steel frames (Zhou, 2000) are given in Figure 19.11, which are an asymmetric frame with hinge base (U-P), a symmetric frame with hinge base (S-P), an asymmetric frame with clamp base (U-F) and a symmetric frame with clamp base (S-F). The sectional sizes of the four frames are given in Table 19.13 and the statistics for structural reliability evaluation are listed in Table 19.14. The nominal value of steel yielding strength is 248 MPa and the elastic modulus is 206 GPa. The loads given in Figure 19.11 are the sum of nominal dead and live loads.

19.3.2.2   Structural reliability evaluation

It is assumed in structural reliability evaluation that

(1) all members are bending about the strong axis;

(2) out-of-plane restrain of all members is sufficient;

(3) the dead to live load ratio is 1.0.

Table 19.13 Cross section of the multi-storey steel frames

Component	Frame			
	U-P	U-F	S-P	S-F
C1	W12 × 19	W12 × 14	W14 × 53	W14 × 53
C2	W14 × 159	W14 × 145	W14 × 99	W14 × 74
C3	W14 × 145	W14 × 145	W14 × 53	W14 × 53
C4	W6 × 9	W6 × 9	W14 × 43	W14 × 53
C5	W14 × 145	W14 × 145	W14 × 26	W12 × 22
C6	W14 × 145	W14 × 145	W14 × 43	W14 × 53
B1	W30 × 116	W33 × 118	W36 × 135	W33 × 130
B2	W36 × 182	W36 × 182	W36 × 135	W33 × 130
B3	W24 × 55	W24 × 55	W27 × 84	W24 × 76
B4	W30 × 116	W30 × 108	W27 × 84	W24 × 76

Table 19.14 Statistics of the multi-storey steel frames

	Mean/standard deviation	Coefficient of variation	Distribution type
Dead load	1.00	0.08	Normal
Live load	1.00	0.25	Extreme, type I
Yield strength	1.05	0.10	Log-normal

The four steel frames are designed with AISC (1994) and load case I with combination of $1.2D + 1.6L$ governs the design results. The reliability of each member, based on the load combination of $1.2D + 1.6L$, is listed in Table 19.15. At the same time, the system reliability of these four frames is evaluated with the semi-analytical simulation method presented in Chapter 17 and the statistics given in Table 19.14, the results of which are given in Table 19.16. A comparison of member and system reliability is listed in Table 19.17.

The results in Tables 19.15–19.17 show that the system reliability of a steel frame is always greater than the reliability of the critical member in the frame because the steel frame is generally redundant and the force redistribution effect is considered in the system reliability evaluation but ignored in the member reliability

Table 19.15 Reliability of each member in the multi-storey steel frames

Component	Frame							
	U-P		U-F		S-P		S-F	
	β	P_f	β	P_f	β	P_f	β	P_f
C1	4.28	9.4×10^{-6}	4.15	1.7×10^{-5}	3.31	4.6×10^{-4}	3.33	4.3×10^{-4}
C2	3.21	6.6×10^{-4}	3.33	4.4×10^{-4}	3.67	1.2×10^{-4}	3.26	5.6×10^{-4}
C3	4.09	2.1×10^{-5}	4.00	3.2×10^{-5}	3.28	5.2×10^{-4}	3.33	4.4×10^{-4}
C4	5.36	4.1×10^{-8}	5.48	2.2×10^{-8}	2.97	1.5×10^{-3}	2.63	4.3×10^{-3}
C5	3.00	1.4×10^{-3}	3.14	8.4×10^{-4}	3.83	6.5×10^{-5}	3.20	6.8×10^{-4}
C6	3.07	1.1×10^{-3}	3.28	5.3×10^{-4}	2.98	1.5×10^{-3}	2.63	4.3×10^{-3}
B1	2.59	4.8×10^{-3}	2.81	2.5×10^{-3}	2.77	2.9×10^{-3}	2.53	5.8×10^{-3}
B2	2.60	4.7×10^{-3}	2.53	5.7×10^{-3}	2.77	2.9×10^{-3}	2.53	5.8×10^{-3}
B3	2.71	3.4×10^{-3}	2.62	4.4×10^{-3}	3.11	9.3×10^{-4}	2.41	8.0×10^{-3}
B4	2.96	1.5×10^{-3}	2.51	6.0×10^{-3}	3.11	9.3×10^{-4}	2.41	8.0×10^{-3}

Table 19.16 Structural system reliability of the multi-storey steel frames

U-P		U-F		S-P		S-F	
β	$P_{\mathrm f}$	β	$P_{\mathrm f}$	β	$P_{\mathrm f}$	β	$P_{\mathrm f}$
3.2047	6.7599×10^{-4}	3.4121	3.2232×10^{-4}	3.3794	3.6322×10^{-4}	3.3026	4.7896×10^{-4}

Table 19.17 Comparison of structural system reliability with member reliability for the multi-storey steel frames

	U-P	U-F	S-P	S-F
Critical component(s)	B1	B4	B1 and B2	B3 and B4
Failure probability of member P_{fm}	4.8×10^{-3}	6.0×10^{-3}	2.9×10^{-3}	8.0×10^{-3}
Failure probability of system P_{fs}	6.76×10^{-4}	3.22×10^{-4}	3.63×10^{-4}	4.79×10^{-4}
$P_{\mathrm{fm}}/P_{\mathrm{fs}}$	7	19	8	17

calculation. In addition, the ratio of the failure probability of the critical member against the structural system, $P_{\mathrm{fm}}/P_{\mathrm{fs}}$, relates to the redundancy of steel frames and those values of $P_{\mathrm{fm}}/P_{\mathrm{fs}}$ for the two clamped frames are greater than those of the two hinged ones. In other words, the higher the structural redundancy, the more evident the effect of force redistribution on the structural system reliability.

The member reliability is not uniformly distributed in frame, as illustrated in Table 19.15, and ranges from 2.41 to 5.48, which indicates that the structural design with only member by member checking cannot retain certain system reliability. However, the structural design based on the limit state of the structural system can provide the structure a definite reliability at system level, as shown in Table 19.16.

19.3.2.3 Effect of dead/live load ratios

The member and system reliabilities of the frames above are obtained under the condition that the dead/live load ratio be 1.0. In this subsection, the dead/live load ratios 0.75 and 1.25 are used to analyse the effect of this load ratio on the reliability results of the U-P frame. The member reliability and system reliability results of the U-P frame are given in Tables 19.18 and 19.19, respectively. The ratio $P_{\mathrm{fm}}/P_{\mathrm{fs}}$ is listed in Table 19.20.

As the variance of live load is generally greater than that of dead load, member and system reliability increases with the dead/live load ratio, which can be seen from Tables 19.18 and 19.19. From Table 19.20, it can be found that the system reliability is more sensitive to the dead/live load ratio.

Table 19.18 Variation of member reliabilities of the U-P frame with dead/live load ratio

	Dead/live load ratio					
	0.75		1.0		1.25	
Component	β	$P_{\mathrm f}$	β	$P_{\mathrm f}$	β	$P_{\mathrm f}$
C1	4.08	2.2×10^{-5}	4.28	9.4×10^{-6}	4.46	4.1×10^{-6}
C2	3.08	1.0×10^{-3}	3.21	6.6×10^{-4}	3.33	4.4×10^{-4}
C3	3.92	4.5×10^{-5}	4.09	2.1×10^{-5}	4.26	1.0×10^{-5}
C4	5.13	1.5×10^{-7}	5.36	4.1×10^{-8}	5.58	1.2×10^{-8}
C5	2.89	1.9×10^{-3}	3.00	1.4×10^{-3}	3.09	1.0×10^{-3}
C6	2.96	1.5×10^{-3}	3.07	1.1×10^{-3}	3.17	7.8×10^{-4}
B1	2.52	5.8×10^{-3}	2.59	4.8×10^{-3}	2.66	4.0×10^{-3}
B2	2.53	5.6×10^{-3}	2.60	4.7×10^{-3}	2.67	3.8×10^{-3}
B3	2.63	4.3×10^{-3}	2.71	3.4×10^{-3}	2.79	2.7×10^{-3}
B4	2.86	2.1×10^{-3}	2.96	1.5×10^{-3}	3.05	1.2×10^{-3}

Table 19.19   Variation of system reliability of the U-P frames with dead/live load ratio

Dead/live load ratio	0.75		1.0		1.25	
	β	P_f	β	P_f	β	P_f
Reliability index and failure probability	2.9681	1.4981×10^{-3}	3.2047	6.7599×10^{-4}	3.4264	3.0588×10^{-4}

Table 19.20   Comparison of system reliability with reliability of the critical member in the U-P frame

Dead/live load ratio	0.75	1.0	1.25
Critical member	B1	B1	B1
Failure probability of critical member P_{fm}	5.8×10^{-3}	4.8×10^{-3}	4.0×10^{-3}
System reliability P_{fs}	1.50×10^{-3}	6.76×10^{-4}	3.06×10^{-4}
P_{fm}/P_{fs}	4	7	13

19.3.2.4 *Comparison of different design approaches*

Select the U-P frame as an example to compare the results by different design approaches, RAD, PAD and LRFD, as above for steel portal frames. The design formula for load case I in PAD, adopting the load and resistance factors as in AISC LRFD, can be

$$0.90R_n \geq 1.20G_n + 1.60L_n \quad \text{for Q235 and Q345 steel,} \tag{19.9}$$

where R_n is the nominal value of the structural resistance under load case I determined by advanced analysis, and G_n and L_n are, respectively, the nominal values of the dead load and live load.

Assume that Q345 steel is used for the U-P frame and a comparison of the steel consumption is given in Table 19.21, for design results by RAD, PAD and LRFD.

Table 19.21   Sectional sizes and steel consumption of the U-P frame

	Design method		
	LRFD	PAD	RAD
C1	W12 × 19	W12 × 14	W12 × 16
C2	W14 × 132	W14 × 99	W14 × 109
C3	W14 × 109	W14 × 82	W14 × 99
C4	W10 × 12	W10 × 12	W10 × 12
C5	W14 × 109	W14 × 99	W14 × 109
C6	W14 × 109	W14 × 99	W14 × 109
B1	W27 × 84	W27 × 84	W27 × 84
B2	W36 × 135	W30 × 108	W30 × 108
B3	W18 × 40	W18 × 40	W18 × 40
B4	W27 × 94	W27 × 84	W27 × 94
Steel consumption (× 10^3 kg)	9.59	8.51	9.12

Table 19.22   Structural system reliability calibration of the U-P frame

Design method	β	P_f
LRFD	4.4591	4.1148×10^{-6}
PAD	3.2033	6.7923×10^{-4}
RAD	3.8427	6.0856×10^{-5}

In the above three approaches, the frame is designed by checking member by member in LRFD. In RAD and PAD, the limit load-bearing capacity of the frame is first obtained with advanced analysis, and then it is used to check the design requirement, Equation (19.6b) in RAD or Equation (19.9) in PAD.

Reliability of the frames designed by the three different approaches is calibrated, as given in Table 19.22. It is noted that the system reliability index of the frame designed with LRFD is about 4.46, which is higher than the target one. However, the system reliability index of the frame by PAD is about 3.2, which is expected to be relatively low. For the frame designed with RAD, the system reliability index is about 3.8427, which is consistent with the target one.

References/Bibliography

Abdel-Jaber, M., Beale, R. G., and Godley, M. H. R. (2005). 'Numerical study on semi-rigid racking frames under sway,' *Computers and Structures*, **83**(28–30), 2463–75.

Ackroyd, M. H. (1987). 'Design of flexibly connected unbraced steel building frames,' *Journal of Constructional Steel Research*, **8**, 261–86.

Ackroyd, M. H. and Gerstle, K. H. (1982). 'Behavior of type 2 steel frames,' *Journal of the Structural Engineering Division, ASCE*, **108**(ST7), 1541–56.

Ain, A. K., Redwood, R. G., and Feng, L. (1993). 'Seismic response of concentrically braced dual steel frames,' *Canadian Journal of Civil Engineering*, **20**(4), 672–87.

AISC. (1978). *Specification for the Design, Fabrication and Erection of Structural Steel for Buildings*, American Institute of Steel Construction, Chicago, IL.

AISC. (1980). *Manual of Steel Construction*, American Institute of Steel Construction, Chicago, IL.

AISC. (1986). *Load and Resistance Factor Design Specification for Structural Steel Buildings*, American Institute of Steel Construction, Chicago, IL.

AISC. (1989). *Allowable Stress Design Specification for Structural Steel Buildings*, American Institute of Steel Construction, Chicago, IL.

AISC. (1994). *Manual of Steel Construction, Load and Resistance Factor Design*, 2nd edn, American Institute of Steel Construction, Chicago, IL.

Al-Gahtani, H. J. (1996). 'Exact stiffness for tapered members,' *Journal of Structural Engineering, ASCE*, **122**(10), 1234–9.

Al-Gahtani, H. J. (1998). 'Exact analysis of non-prismatic beams,' *Journal of Engineering Mechanics, ASCE*, **124**(11), 1291–3.

Almusalllam, T. H. and Richard, R. M. (1993). 'Steel frame analysis with flexible joints exhibiting a strain-softning behavior,' *Computers and Structures*, **46**(1), 55–63.

Alvarez, R. J. and Birnstiel, C. (1969). 'Inelastic analysis of multistory multibay frames,' *Journal of the Structural Division, ASCE*, **95**(ST11), 2477–503.

Amadio, C., Fedrigo, C., Fragiacomo, M., and Macorini, L. (2004). 'Experimental evaluation of effective width in steel-concrete composite beams,' *Journal of Constructional Steel Research*, **60**(2), 199–220.

Amadio, C. and Fragiacomo, M. (2003). 'Analysis of rigid and semi-rigid steel-concrete composite joints under monotonic loading Part I: Finite element modelling and validation,' *Steel and Composite Structures*, **3**(5), 349–69.

Argyris, J. H., Boni, B., Hindenlang, U., and Leiber, M. (1982). 'Finite element analysis of two- and three-dimensional elasto-plastic frames — the Natural Approach,' *Computer Methods in Applied Mechanics and Engineering*, **35**(2), 221–48.

Aristizabal-Ochoa, J. D. (1993). 'Static, stability and vibration of non-prismatic beams and columns,' *Journal of Sound and Vibration*, **162**(3), 441–55.

Aristizabal-Ochoa, J. D. (1994). 'K-factor for columns in any type of construction: Nonparadoxical approach,' *Journal of Structural Engineering, ASCE*, **120**(4), 1273–90.

Aristizabal-Ochoa, J. D. (1997a). 'Elastic stability of beam-columns with flexural connections under various conservative end axial forces,' *Journal of Structural Engineering, ASCE*, **123**(9), 1194–200.

Aristizabal-Ochoa, J. D. (1997b). 'Story stability of braced, partially braced, and unbraced frames: Classical approach,' *Journal of Structural Engineering, ASCE*, **123**(6), 799–807.

Aristizabal-Ochoa, J. D. (1998). 'Minimum stiffness of bracing for multi-column framed structures,' *Structural Engineering and Mechanics*, **6**(3), 305–25.

Arlekar, J. N. and Murty, C. V. R. (2004). 'Improved truss model for design of welded steel moment-resisting frame connections,' *Journal of Structural Engineering, ASCE*, **130**(3), 498–510.

AS4100. (1990). *Steel Structures*, Standards Australia, Sydney.

Ashraf, M., Nethercot, D. A., and Ahmed, B. (2004). 'Sway of semi-rigid steel frames Part 1: Regular frames,' *Engineering Structures*, **26**(12), 1809–19.

Attalla, M. R., Deierlein, G. G., and McGuire, W. (1994). 'Spread of plasticity: Quasi-plastic-hinge approach,' *Journal of Structural Engineering, ASCE*, **120**(8), 2451–73.

Au, S. K. and Beck, J. L. (1999). 'A new adaptive importance sampling scheme for reliability calculation,' *Structural Safety*, **21**(2), 135–58.

Avery, P. and Mahendran, M. (1998). 'Pseudo plastic zone analysis of steel frame structures comprising non-compact sections,' *Research Monograph 98-7*, Physical Infrastructure Center, School of Civil Engineering, QUT, Brisbane.

Avery, P. and Mahendran, M. (2000a). 'Analytical benchmark solutions for steel frame structures subjected to local buckling effects,' *Advances in Structural Engineering*, **3**(3), 215–29.

Avery, P. and Mahendran, M. (2000b). 'Distributed plasticity analysis of steel frame structures comprising non-compact sections,' *Engineering Structures*, **22**(8), 901–19.

Avery, P. and Mahendran, M. (2000c). 'Large-scale testing of steel frame structures comprising non-compact sections,' *Engineering Structures*, **22**(8), 920–36.

Avery, P. and Mahendran, M. (2000d). 'Pseudo plastic zone analysis of steel frame structures comprising non-compact sections,' *Structural Engineering and Mechanics*, **10**(4), 371–92.

Ayoub, A. (2005). 'A force-based model for composite steel–concrete beams with partial interaction,' *Journal of Constructional Steel Research*, **61**(3), 387–414.

Ayyub, B. M. and Chia, C. Y. (1992). 'Generalized conditional expectation for structural reliability assessment,' *Structural Safety*, **11**(2), 131–46.

Ayyub, B. M. and Haldar, A. (1984). 'Practical structural reliability techniques,' *Journal of Structural Engineering, ASCE*, **110**(8), 1701–24.

Azizinamini, A., Bradburn, J., H., and Radziminski, J. B. (1987). 'Initial stiffness of semi-rigid steel beam-to-column connections,' *Journal of Constructional Steel Research*, **8**, 71–90.

Bai, Y. and Terndrup Pedersen, P. (1993). 'Elastic-plastic behaviour of offshore steel structures under impact loads,' *International Journal of Impact Engineering*, **13**(1), 99–115.

Baldassino, N. and Bernuzzi, C. (2000). 'Analysis and behaviour of steel storage pallet racks,' *Thin-Walled Structures*, **37**(4), 177–304.

Balendra, T. and Huang, X. (2003). 'Overstrength and ductility factors for steel frames designed according to BS 5950,' *Journal of Structural Engineering, ASCE*, **129**(8), 1019–35.

Banerjee, J. R. (1985). 'Exact Bernoulli-Euler dynamic stiffness matrix for a range of tapered beams,' *International Journal for Numerical Method in Engineering*, **21**(11), 2289–302.

Banerjee, J. R. (1986). 'Exact Bernoulli-Euler static stiffness matrix for a range of tapered beam-columns,' *International Journal for Numerical Method in Engineering*, **23**(9), 1615–28.

Baotou Steel & Iron Design and Research Institute (2000). 'Analysis and design of steel structures,' *Mechanical Industry Press*, Beijing (in Chinese).

Baron, F. and Venkatesan, M. S. (1971). 'Nonlinear formulations of beam-column effects,' *Journal of the Structural Division, ASCE*, **97**(4), 1305–40.

Barsoum, R. S. and Gallagher, R. H. (1970). 'Finite element analysis of torsional and torsional- flexural problems,' *International Journal for Numerical Methods in Engineering*, **2**(3), 335–52.

Bath, K. J. (1996). *Finite Element Procedures*, Prentice Hall, Englewood Cliff, NJ.

Bathe, K. J. and Bolourchi, S. (1979). 'Large displacement analysis of three- dimensional beam structures,' *International Journal for Numerical Methods in Engineering*, **14**(7), 961–86.

Bayo, E., Cabrero, J. M., and Gil, B. (2005). 'An effective component-based method to model semi-rigid connections for the global analysis of steel and composite structures,' *Engineering Structures*, **28**(1), 97–108.

Bayo, E. and Loureiro, A. (2001). 'An efficient and direct method for buckling analysis of steel frame structures,' *Journal of Constructional Steel Research*, **57**(12), 1321–36.

Bennett, R. M. (1990). 'Structural analysis methods for system reliability,' *Structural Safety*, **7**(2–4), 109–14.

Berczynski, S. and Wroblewski, T. (2005). 'Vibration of steel-concrete composite beams using the Timoshenko beam model,' *Journal of Vibration and Control*, **11**(6), 829–48.

Berman, J. W., Celik, O. C., and Bruneau, M. (2005). 'Comparing hysteretic behavior of light-gauge, steel plate shear walls and braced frames,' *Engineering Structures*, **27**(3), 475–85.

Bjerager, P. (1988). 'Probability integration by directional simulation,' *Journal of Engineering Mechanics, ASCE*, **114**(8), 1285–302.

Blandford, G. E. and Glass, G. C. (1987). 'Static/dynamic analysis of locally buckling frames,' *Journal of Structural Engineering, ASCE*, **113**(2), 363–80.

Bogosian, D. D., Dunn, B. W., and Chrostowski, J. D. (1999). 'Blast analysis of complex structures using physics-based fast-running models,' *Computers and Structures*, **72**(1–3), 81–92.

Boswell, L. F. and O'Conner, M. A. (1988). 'Moment rotation characteristics of bolted extended end-Plate connections for use in semi-rigid analysis,' *Proceeding of the 4th International Conference on Tall Buildings*, Vol. I, Hongkong.

Bradford, M. A. and Azhari, M. (1995). 'Inelastic local buckling of plates and plate assemblies using bubble functions,' *Engineering Structures*, **17**(2), 95–103.

Brown, N. D., Hughes, A. F., and Anderson, D. (2001). 'Prediction of the initial stiffness of ductile end-plate steel connections,' *Proceedings of the Institution of Civil Engineers-Structures and Buildings*, **146**(1), 17–29.

Bruneau, M., Engelhardt, M., Filiatrault, A., *et al.* (2005). 'Review of selected recent research on US seismic design and retrofit strategies for steel structures,' *Progress in Structural Engineering and Materials*, **7**(3), 103–14.

BS5950. (1990). *Structural Use of Steelwork in Building, Part 1, Code of Practice for Design in Simple and Continuous Construction: Hot Rolled Sections*, British Standards Institution, London.

Buonopane, S. G. and Schafer, B. W. (2006). 'Reliability of steel frames designed with advanced analysis,' *Journal of Structural Engineering, ASCE*, **132**(2), 267–76.

Buonopane, S. G., Schafer, B. W., and Igusa, T. (2003). 'Reliability implications of advanced analysis in design of steel frames,' *Advances in Structures*, Vol. 1, pp. 547–53, ASSCCA'03.

Bursi, O. S. and Gerstle, K. H. (1994). 'Analysis of flexibly connected braced steel frames,' *Journal of Constructional Steel Research*, **30**(1), 61–83.

Bursi, O. S., Sun, F. F., and Postal, S. (2005a). 'Non-linear analysis of steel–concrete composite frames with full and partial shear connection subjected to seismic loads,' *Journal of Constructional Steel Research*, **61**, 67–92.

Bursi, O. S., Sun, F. F., and Postal, S. (2005b). 'Non-linear analysis of steel-concrete composite frames with full and partial shear connection subjected to seismic loads,' *Journal of Constructional Steel Research*, **61**(1), 67–92.

Cabrero, J. M. and Bayo, E. (2005). 'Development of practical design methods for steel structures with semi-rigid connections,' *Engineering Structures*, **27**(8), 1125–37.

Calderoni, B. and Rinaldi, Z. (2002). 'Seismic performance evaluation for steel MRF: Non linear dynamic and static analyses,' *Steel and Composite Structures*, **2**(2), 113–28.

Cas, B., Bratina, S., Saje, M., and Planinc, I. (2004). 'Non-linear analysis of composite steel-concrete beams with incomplete interaction,' *Steel and Composite Structures*, **4**(6), 489–507.

Castro, J. M., Elghazouli, A. Y., and Izzuddin, B. A. (2005). 'Modelling of the panel zone in steel and composite moment frames,' *Engineering Structures*, **27**(1), 129–44.

Cavaleri, L. and Papia, M. (2003). 'A new dynamic identification technique: Application to the evaluation of the equivalent strut for infilled frames,' *Engineering Structures*, **25**(7), 889–901.

CECS (2002). *Technical Specification for Steel Structure of Light- Weight Buildings with Gabled Frames*, China Planning Press, Beijin.

Chan, S. L. and Chui, P. P. T. (1997). 'A generalized design-based elastoplastic analysis of steel frames by section assemblage concept,' *Engineering Structures*, **19**(8), 628–36.

Chan, S. L. and Chui, P. P. T. (2000). *Non-Linear Static and Cyclic Analysis of Steel Frames with Semi-Rigid Connections*, Elsevier Science, Amsterdam.

Chan, S. L. and Gu, J. X. (2000). 'Exact tangent stiffness for imperfect beam-column members,' *Journal of Structural Engineering, ASCE*, **126**(9), 1094–102.

Chan, S. L., Huang, H. Y., and Fang, L. X. (2005). 'Advanced analysis of imperfect portal frames with semirigid base connections,' *Journal of Engineering Mechanics, ASCE*, **131**(6), 633–40.

Chan, S. L., Kitipornchi, S., and Al-Berman, F. G. A. (1991). 'Elastio-plastic analysis of box-beam-columns including local buckling effects,' *Journal of Structural Engineering, ASCE*, **117**(7), 1946–62.

Chan, S. L., Liu, Y. P., and Zhou, Z. H. (2005). 'Limitation of effective length method and codified second-order analysis and design,' *Steel and Composite Structures*, **5**(2–3), 181–92.

Chan, S. L. and Zhou, Z. H. (1994). 'Pointwise equilibrating polynomial element for nonlinear analysis of frames,' *Journal of Structural Engineering, ASCE*, **120**(6), 1703–17.

Chan, S. L. and Zhou, Z. H. (1995). 'Second-order elastic analysis of frames using single imperfect element per member,' *Journal of Structural Engineering, ASCE*, **121**(6), 939–45.

Chan, S. L. and Zhou, Z. H. (2000). 'Non-linear integrated design and analysis of skeletal structures by one element per member,' *Engineering Structures*, **22**(3), 246–57.

Chen, G. X., Li, J. H., and Xia, Z. Z. (1985). 'Statistics of steel yielding strength and sectional geometry,' *Journal of Chongqing Institute of Architectural Egineering* (in Chinses).

Chen, M. (2000a). 'Nonlinear inelastic analysis of steel-concrete composite frames', *Ph.D. Thesis*, The National University of Singapore.

Chen, S. and Gu, P. (2005). 'Load carrying capacity of composite beams prestressed with external tendons under positive moment,' *Journal of Constructional Steel Research*, **61**(4), 515–30.

Chen, W. F. (1993). *Advanced Analysis in Steel Frames*, CRC Press, Boca Raton, FL.

Chen, W. F. (1998). 'Implementing advanced analysis for steel frame design,' *Progress in Structural Engineering and Materials*, **1**(3), 323–8.

Chen, W. F. (2000b). 'Structural stability: from theory to practice,' *Engineering Structures*, **22**(2), 116–22.

Chen, W. F. and Atsuta, T. (1976). *Theory of Beam-Columns, Vol.1: In-Plane Behavior and Design*, McGraw-Hill Inc., New York.

Chen, W. F. and Atsuta, T. (1977). *Theory of Beam-Columns, Vol.II: Out-Plane Behavior and Design*, McGraw-Hill, Inc., New York.

Chen, W. F. and Chan, S. L. (1995). 'Second-order inelastic analysis of steel frames using element with midspan and end springs,' *Journal of Structural Engineering, ASCE*, **121**(3), 530–41.

Chen, W. F. and Kim, S.-E. (1997). *LRFD Steel Design Using Advanced Analysis*, CRC Press, Boca Raton, FL.

Cheong-Siat-Moy, F. (1999). 'An improved K-factor formula,' *Journal of Structural Engineering, ASCE*, **125**(2), 169–74.

Chikho, A. H. and Kirby, P. A. (1995). 'An approximate method for estimation of bending moments in continuous and semirigid frames,' *Canadian Journal of Civil Engineering*, **22**(6), 1120–32.

Chiou, Y.-J. and Hsiao, P.-A. (2005). 'Large displacement analysis of cyclically loaded inelastic structures,' *Journal of Structural Engineering, ASCE*, **131**(12), 1803–10.

Choi, S. H. and Kim, S. E. (2002). 'Optimal design of steel frame using practical nonlinear inelastic analysis,' *Engineering Structures*, **24**(9), 1189–201.

Chopra, A. K., Goel, R. K., and Chintanapakdee, C. (2003). 'Statistics of single-degree-of-freedom estimate of displacement for pushover analysis of buildings,' *Journal of Structural Engineering, ASCE*, **129**(4), 459–69.

Chrysanthakopoulos, C., Bazeos, N., and Beskos, D. E. (2006). 'Approximate formulae for natural periods of plane steel frames,' *Journal of Constructional Steel Research*, **62**(6), 592–604.

Chui, P. P. T. and Chan, S. L. (1997). 'Vibration and deflection characteristics of semi-rigid jointed frames,' *Engineering Structures*, **19**(2), 1001–10.

Clarke, M. and Bridge, R. (1996). 'The design of steel frames using the notional load approach,' *Proceedings of the 3rd International Colloquium on Stability of Metal Structures (SSRC): 33–42*, Lehigt University, Bethlehem, PA.

Clarke, M. J. (1992). 'Advanced analysis of steel building frames,' *Journal of Constructional Steel Research*, **23**(1–3), 1–29.

Cleghorn, W. L. and Tabarrok, B. (1992). 'Finite element formulation of a tapered Timoshenko beam for free vibration analysis,' *Journal of Sound and Vibration*, **152**(3), 461–70.

Clement, D. E. and Williams, M. S. (2004). 'Seismic design and analysis of a knee braced frame building,' *Journal of Earthquake Engineering*, **8**(4), 523–43.

Clough, R. W. and Benuska, K. L. (1967). 'Nonlinear earthquake behavior of tall buildings,' *Journal of Engineering Mechanics, ASCE*, **93**(EM3), 129–46.

Clough, R. W., Benuska, K. L., and Wilson, E. L. (1965). 'Inelastic earthquake response of tall buildings,' *Proceedings of the 3rd World Conference of Earthquake Engineering (WCEE)*.

Connor, J., Logcher, R. D., and Chan, S. C. (1968). 'Nonlinear analysis of elastic framed structures,' *Journal of the Structural Division, ASCE*, **94**(ST6), 1525–47.

Corradi, L., Poggi, C., and Setti, P. (1990). 'Interaction domains for steel beam-column in fire conditions,' *Journal of Constructional Steel Research*, **17**(2), 217–35.

Council on Tall Buildings (1979). 'Structure design of tall steel buildings,' Volume SB of *Monograph on Planning and Design of Tall Building*, ASCE, New York.

Crandall, S. H. (1980). 'Non-Gaussian closure for random vibration of non-linear oscillators,' *International Journal for Non-Linear Mechanics*, **15**(2), 303–13.

Dai, G. X., Li, L. C., Xia, Z. Z., and Huang, Y. M. (2000). 'Statistics and analysis of new property parameters of structural building steel,' *Building Structures* , **30**(4), 31–2 (in Chinese).

Dall'Asta, A. and Zona, A. (2004a). 'Comparison and validation of displacement and mixed elements for the non-linear analysis of continuous composite beams,' *Computers and Structures*, **82**(23–26), 2117–30.

Dall'Asta, A. and Zona, A. (2004b). 'Three-field mixed formulation for the non-linear analysis of composite beams with deformable shear connection,' *Finite Elements in Analysis and Design*, **40**(4), 425–48.

Das, P. K. and Zheng, Y. (2000). 'Cumulative formation of response surface and its use in reliability analysis,' *Probabilistic Engineering Mechanics*, **15**(4), 309–15.

Davies, J. M. (2006). 'Strain hardening, local buckling and lateral-torsional buckling in plastic hinges,' *Journal of Constructional Steel Research*, **62**(1–2), 27–34.

Davies, J. M., Engel, P., Liu, T. T. C., and Morris, L. J. (1990). 'Realistic modeling of steel portal frame behavior,' *The Structural Engineer*, **68**(1), 1–6.

De Lima, L. R. O., da Silva, L. S., Vellasco, P., and de Andrade, S. A. L. (2004). 'Experimental evaluation of extended endplate beam-to-column joints subjected to bending and axial force,' *Engineering Structures*, **26**(10), 1333–47.

Deierlein, G. G. (1997). 'Steel-framed structures,' *Progress in Structural Engineering and Materials*, **1**(1), 10–7.

Della Corte, G., De Matteis, G., Landolfo, R., and Mazzolani, F. M. (2002). 'Seismic analysis of MR steel frames based on refined hysteretic models of connections,' *Journal of Constructional Steel Research*, **58**(10), 1331–45.

Dey, A. and Mahadevan, S. (1998). 'Ductile structural system reliability analysis using adaptive importance sampling,' *Structural Safety*, **20**(2), 137–54.

Ding, J. M. (1987). Elasto-plastic stability and load-carrying capacity analysis of multi-story steel frames, *Master dissertation*, Department of Structural Engineering, Tongji University, Shanghai, China (in Chinese).

Ding, J. M. (1990), 'Elasto-plastic stability analysis of space steel frames', *PhD dissertation*, Department of Structural Engineering, Tongji University, Shanghai, China (in Chinese).

Dissanayake, U. I., Davison, J. B. and Burgess, I. W. (1995). 'Development of a computer program for the analysis of composite frames with partial interaction,' *Proceedings of the Fourth Pacific Structural Steel Conference*, V3, Singapore, Pergamon/Elsevier Science, 179–186.

Ditlevsen, O. and Melchers, R. E. (1990). 'General multi-dimensional probability integration by directional simulation,' *Computers and Structures*, **36**(2), 355–68.

Duan, L. and Chen, W. F. (1989). 'Design interaction equation for steel beam-columns,' *Journal of Structural Engineering, ASCE*, **115**(5), 1225–43.

Edwood, R. G., Lu, F., Bouchard, G., and Paultre, P. (1991). 'Seismic response of concentrically braced steel frames,' *Canadian Journal of Civil Engineerng*, **18**(6), 1062–77.

Eisenberger, M. (1995). 'Non-uniform torsional analysis of variable and open cross-section bars,' *Thin-Walled Structures*, **21**(2), 93–105.

El-Tawil, S. and Deierlein, G. G. (2001a). 'Nonlinear analysis of mixed steel-concrete frames. I: Element formulation,' *Journal of Structural Engineering, ASCE*, **127**(6), 647–55.

El-Tawil, S. and Deierlein, G. G. (2001b). 'Nonlinear analysis of mixed steel-concrete frames. II: Implementation and verification,' *Journal of Structural Engineering, ASCE*, **127**(6), 656–65.

El-Tawil, S., Vidarsson, E., Mikesell, T., and Kunnath, S. K. (1999). 'Inelastic behavior and design of steel panel zones,' *Journal of Structural Engineering, ASCE*, **125**(2), 183–93.

Elghazouli, A. Y., Broderick, B. M., Goggins, J. *et al.* (2005). 'Shake table testing of tubular steel bracing members,' *Proceedings of the Institution of Civil Engineers-Structures and Buildings*, **158**(4), 229–41.

Ellingwood, B. R. (1994). 'Probability-based codified design: Past accomplishments and future challenges,' *Structural Safety*, **13**(3), 159–76.

Ellingwood, B. R. (2000). 'LRFD: Implementing structural reliability in professional practice,' *Engineering Structures*, **22**(2), 106–15.

Ellingwood, B. R. and Galambos, T. V. (1981). *General Specifications for Structural Design Loads in Probabilistic Methods in Structural Engineering*, ASCE Press, Reston, VG.

Engelund, S. and Rackwitz, R. (1993). 'A benchmark study on importance sampling technique in structural reliability,' *Structural Safety*, **12**(4), 255–76.

Er, G. K. (1998). 'A method for multi-parameter PDF estimation of random variables,' *Structural Safety*, **20**(1), 25–36.

Ermoupoulos, J. C. (1985a). 'Buckling of tapered bars under stepped axial loads,' *Journal of Structural Engineering, ASCE*, **112**(6), 1346–54.

Ermoupoulos, J. C. (1985b). 'Stability of frames with tapered built-up members,' *Journal of Structural Engineering, ASCE*, **111**(9), 1978–93.

Ermoupoulos, J. C. (1988). 'Slope-deflection method and bending of tapered bars under stepped loads,' *Journal of Construction Steel Research*, **11**(2), 121–41.

Eurocode3. (1992). *ENV-1993-1-1, Design of Steel Structures*, Commission of the European Communities, European Prenorm, Brussel.

Fabbrocino, G., Manfredi, G., and Cosenza, E. (1999). 'Non-linear analysis of composite beams under positive bending,' *Computers and Structures*, **70**(1), 77–89.

Fabbrocino, G., Manfredi, G., and Cosenza, E. (2002). 'Modelling of continuous steel–concrete composite beams: Computational aspects,' *Computers and Structures*, **80**(27–30), 2241–51.

Faella, C., Martinelli, E. and Nigro, E. 'One-dimensional finite element approach for the analysis of steel concrete composite frames,' *Proceedings of the First International Conference on Steel and Composite Structures*, Pusan, Korea, 1245–52.

Faella, C., Martinelli, E., and Nigro, E. (2003). 'Shear connection nonlinearity and deflections of steel–concrete composite beams: A simplified method,' *Journal of Structural Engineering, ASCE*, **129**(1), 12–20.

Feng, Y. S. (1988). 'The theory of structural redundancy and its effect on structural design,' *Computers and Structures*, **28**(1), 15–24.

Florian, A. (1992). 'An efficient sampling scheme: Updated latin hypercube sampling,' *Probabilistic Engineering Mechanics*, **7**(2), 123–30.

Foley, C. M. and Schinler, D. (2003). 'Automated design of steel frames using advanced analysis and object-oriented evolutionary computation,' *Journal of Structural Engineering, ASCE*, **129**(5), 648–60.

Foley, C. M. and Vinnakota, S. (1999). 'Inelastic behavior of multistory partially restrained steel frames: Part II,' *Journal of Structural Engineering, ASCE*, **125**(8), 862–69.

Foliente, G. C., Leicester, R. H., and Pham, L. (1998). 'Development of the CIB proactive program on performance based building codes and standards', *CIB Report*.

Foschi, R. O., Li, H., and Zhang, J. (2002). 'Reliability and performance-based design: A computational approach and applications,' *Structural Safety*, **24**(2–4), 205–18.

Fragiacomo, M., Amadio, C., and Macorini, L. (2004). 'Finite-element model for collapse and long-term analysis of steel–concrete composite beams,' *Journal of Structural Engineering, ASCE*, **130**(3), 489–97.

Frye, M. J. and Morris, G. A. (1975). 'Analysis of flexibly connected steel frames,' *Canadian Journal of Civil Engineers*, **2**(2), 280–91.

Funk, R. R. and Wang, K. T. (1988). 'Stiffnesses of nonprismatic member,' *Journal of Structural Engineering, ASCE*, **114**(2), 489–95.

Galambos, T. V. (1988). *Guide to Stability Design Critical for Metal Structures*, 4th edn, John Wiley & Sons, Inc, New York.

Galambos, T. V. (1990). 'System reliability and structural design,' *Structural Safety*, **7**(2–4), 101–8.

Gallagher, R. H. (1970). 'Matrix dynamic and instability analysis with non-uniform elements,' *International Journal for Numerical Method in Engineering*, **2**(1), 265–75.

Gao, L. (1995). 'Safety evaluation of frames with P. R. connection,' *Journal of Structural Engineering, ASCE*, **121**(7), 1101–9.

GB50017-2003. (2003). *Design Code of Steel Structures* (in Chinese).

Ghobarah, A. and Abou Elfath, H. (2001). 'Rehabilitation of a reinforced concrete frame using eccentric steel bracing,' *Engineering Structures*, **23**(7), 745–55.

Ghobarah, A. and Ramadan, T. (1991). 'Seimic analysis of links of various lengths in ecentrically braced frames,' *Canadian Journal of Civiel Engineering*, **18**(1), 140–8.

Giberson, M. F. (1969). 'Two nonlinear beams with definitions of ductility,' *Journal of the Structural Division, ASCE*, **95**(2), 137–57.

Girhammar, U. A. and Gopu, V. K. A. (1993). 'Composite beam-columns with Interlayer slip – exact analysis,' *Journal of Structural Engineering, ASCE*, **119**(4), 1265–82.

Gizejowski, M. A., Barszcz, A. M., Branicki, C. J., and Uzoegbo, H. C. (2006). 'Review of analysis methods for inelastic design of steel semi-continuous frames,' *Journal of Constructional Steel Research*, **62** (1–2), 81–92.

Gizejowskia, M. A., Barszcz, A. M., Branicki, C. J., and Uzoegbo, H. C. (2006). 'Review of analysis methods for inelastic design of steel semi-continuous frames,' *Journal of Constructional Steel Research*, **62**(1–2), 81–92.

Goda, K. and Hong, H. P. (2006). 'Optimal seismic design for limited planning time horizon with detailed seismic hazard information,' *Structural Safety*, **28**(3), 247–60.

Gong, B. and Shahrooz, B. M. (2000). 'Steel-concrete composite coupling beams — behavior and design,' *Engineering Structures*, **23**(11), 1480–90.

Gong, B. and Shahrooz, B. M. (2001). 'Concrete-steel composite coupling beams. I: Component testing,' *Journal of Structural Engineering, ASCE*, **127**(6), 625–31.

Gong, Y. L., Xu, L., and Grierson, D. E. (2005). 'Performance-based design sensitivity analysis of steel moment frames under earthquake loading,' *International Journal for Numerical Methods in Engineering*, **63**(9), 1229–49.

Gong, Y. L., Xu, L., and Grierson, D. E. (2006). 'Sensitivity analysis of steel moment frames accounting for geometric and material nonlinearity,' *Computers and Structures*, **84**(7), 462–75.

Gorman, M. (1984). 'Structural resistance moments by quadrature,' *Structural Safety*, **2**(1), 73–81.

Goto, Y. and Chen, W. F. (1987). 'Second-order elastic analysis for frame design,' *Journal of Structural Engineering, ASCE*, **113**(7), 1501–19.

Grigoriu, M. (1983). 'Approximate analysis of complex reliability problems,' *Structural Safety*, **1**(2), 277–88.

Gu, J. X. and Chan, S. L. (2005). 'Second-order analysis and design of steel structures allowing for member and frame imperfections,' *International Journal for Numerical Methods in Engineering*, **62**(5), 601–15.

Guan, X. L. and Melchers, R. E. (2001). 'Effect of response surface parameter variation on structural reliability estimates,' *Structural Safety*, **23**(4), 429–44.

Guo, Y. L. and Chen, S. F. (1991). 'Elasto-plastic interactive buckling of cold-formed channel columns,' *Journal of Structural Engineering, ASCE*, **117**(8), 2278–98.

Gupta, A. and Krawinkler, H. (2000). 'Behavior of ductile SMRFs at various seismic hazard levels,' *Journal of Structural Engineering, ASCE*, **126**(1), 98–107.

Gupta, A. K. (1986). 'Frequent-dependent matrices for tapered beams,' *Journal of Structural Engineering, ASCE*, **112**(1), 85–103.

Gupta, R. S. and Rao, S. S. (1978). 'Finite element eigenvalue analysis of tapered and twisted Timoshenko beams,' *Journal of Sound and Vibration*, **56**(2), 187–200.

Gurung, N. and Mahendran, M. (2002). 'Comparative life cycle costs for new steel portal frame building systems,' *Building Research and Information*, **30**(1), 35–46.

Hadianfard, M. A. and Razani, R. (2003). 'Effects of semi-rigid behavior of connections in the reliability of steel frames,' *Structural Safety*, **25**(2), 123–38.

Han, L. H. and Yang, Y. F. (2005). 'Cyclic performance of concrete-filled steel CHS columns under flexural loading,' *Journal of Constructional Steel Research*, **61**(4), 423–52.

Hancock, G. J. (1981). 'Interaction buckling in I-section columns,' *Journal of the Structural Division, ASCE*, **107**(1), 165–79.

Hancock, G. J. and Rasmussen, K. J. R. (1998). 'Recent research on thin-walled beam-columns,' *Thin-Walled Structures*, **32**(1–3), 3–18.

Harbitz, A. (1986). 'An efficient sampling method for probability of failure calculation,' *Structural Safety*, **3**(2), 109–15.

Hasham, A. S. and Rasmussen, K. J. R. (1998). 'Section capacity of thin-walled I-sections in combined compression and major axis bending,' *Journal of Structural Engineering, ASCE*, **124**(4), 351–9.

Heger, F. J. (1993). 'Public safety-is it compromised by new LRFD design standards,' *Journal of Structural Engineering, ASCE*, **119**(4), 1251–64.

Heldt, T. J. and Mahendran, M. (1998). 'Full scale experiments of a steel portal frame building,' *Journal of the Australian Institute of Steel Construction*, **32**(4), 1–21.

Hellesland, J. and Bjorhovde, R. (1996). 'Improved frame stability analysis with effective lengths,' *Journal of Structural Engineering, ASCE*, **122**(11), 1275–83.

Hendawi, S. and Frangopol, D. M. (1994). 'System reliability and redundancy in structural design and evaluation,' *Structural Safety*, **16**(1–2), 47–71.

Higginbotham, A. B. and Hanson, R. D. (1976). 'Axial hysteretic behavior of steel members,' *Journal of the Structural Engineering Division, ASCE*, **106**(ST7), 1365–81.

Higgins, C. and Newell, J. (2004). 'Confined steel brace for earthquake resistant design,' *Engineering Journal, AISC*, **41**(4), 187–202.

Hoenderkamp, J. C. D. and Bakker, M. C. M. (2003). 'Analysis of high-rise braced frames with outriggers,' *Structural Design of Tall and Special Buildings*, **12**(4), 335–50.

Hohenbichler, M. and Rackwitz, R. (1981). 'Non-normal dependent vectors in structural safety,' *Journal of Engineering Mechanics, ASCE*, **107**(6), 1227–38.

Hong, H. P. and Lind, N. C. (1996). 'Approximate reliability analysis using normal polynomial and simulation results,' *Structural Safety*, **18**(4), 329–39.

Igarashi, S., Inoue, K., Asano, M., and Ogawa, K. (1973). 'Restoring force characteristics of steel diagonal bracing,' *Proceedings of the 5th World Conference of Earthquake Engineering (WCEE)*.

Institute, B. S. I. D. a. R. (2000). *Design and Calculation of Steel Structures*, Press of Mechanical Industry, Beijing.

Isenberg, J., Pereyra, V., and Lawver, D. (2002). 'Optimal design of steel frame structures,' *Applied Numerical Mathematics*, **40**(1–2), 59–71.

Iu, C. K., Chan, S. L., and Zha, X. X. (2005). 'Nonlinear pre-fire and post-fire analysis of steel frames,' *Engineering Structures*, **27**(11), 1689–702.

Izzuddin, B. A., Elnashai, S., So, A. K. W., and Chan, S. L. (1995). 'Buckling and geometrically nonlinear analysis of frames using one element member,' *Journal of Constructional Steel Research*, **32**(2), 227–30.

Izzuddin, B. A. and Elnashi, A. S. (1994). 'Buckling and geometrically nonlinear analysis of frames using one element/ member,' *Journal of Constructional Steel Research*, **28**(3), 321–2.

Izzuddin, B. A. and Smith, D. L. (2000). 'Efficient nonlinear analysis of elasto-plastic 3D R/C frames using adaptive techniques,' *Computers and Structures*, **78**(4), 549–73.

Jain, A. K. and Goel, S. C. (1980). 'Hysteretic model of steel member for earthquake response of braced frames,' *Proceedings of the 7th World Conference of Earthquake Engineering (WCEE)*.

Jain, A. K. and Goel, S. C. (1978). 'Hysteresis models for steel members subjected to cyclic buckling or cyclic end moment and buckling,' *Rep. No. UMEE 78R6*, Department of Civil Engineering, University of Michigan.

Jang, J. J. (1999). 'Application of linear structural analysis program for steel frame ultimate limit state analysis,' *Structural Design of Tall Buildings*, **8**(3), 205–14.

JGJ99-98. (1998). *Technique Specification for High-rise Civilian Steel Buildings*, China Architectural Industry Press, Beijing (in Chinese).

Jiang, J. J. (1998). *Theory of Concrete Structural Engineering*, Chinese Architectural Industrial Press, Beijing (in Chinese).

Jiang, X. M., Chen, H., and Liew, J. Y. R. (2002). 'Spread-of-plasticity analysis of three-dimentional steel frames,' *Journal of Constructional Steel Research*, **58**(2), 193–212.

Jurkiewiez, B., Buzon, S., and Sieffert, J. G. (2005). 'Incremental viscoelastic analysis of composite beams with partial interaction,' *Computers and Structures*, **83**(21–22), 1780–91.

Jurkiewiez, B. and Hottier, J. M. (2005). 'Static behaviour of a steel–concrete composite beam with an innovative horizontal connection,' *Journal of Constructional Steel Research*, **61**(9), 1286–300.

Just, D. J. (1977). 'Plane frameworks of tapering box and I-section,' *Journal of the Structural Division, ASCE,* **103**(1), 71–86.

Kahn, L. F. and Hanson, R. D. (1976). 'Inelastic cycle of axially loaded steel members,' *Journal of the Structural Engineering Division, ASCE,* **102**(ST5), 947–59.

Kameshki, E. S. and Saka, M. P. (2001). 'Optimum design of nonlinear steel frames with semi-rigid connections using a genetic algorithm,' *Computers and Structures,* **79**(17), 1593–604.

Kameshki, E. S. and Saka, M. P. (2003). 'Genetic algorithm based optimum design of nonlinear planar steel frames with various semirigid connections,' *Journal of Constructional Steel Research,* **59**(1), 109–34.

Kanchanalai, T. (1977). 'The design and behavior of beam-columns in unbraced steel frames,' *AISI Project No. 189, Report No. 2,* Civil Engineering/Structures Research Lab., University of Texas at Austin.

Kapur, J. N. (1994). 'Approximating a Given Probability Distribution by a Maximum Entropy Distribution', *Measures of Information and their Applications, Part C, 49: 446–457,* John Wiley & Sons, New Delhi, India.

Karabalis, D. L. (1983). 'Static, dynamic and stability analysis of structures composed of tapered beams,' *Computers and Structures,* **16**(6), 731–48.

Karamchandani, A. and Cornell, C. A. (1991). 'Adaptive hybrid conditional expectation approaches for reliability estimation,' *Structural Safety,* **11**(1), 59–76.

Karavasilis, T. L., Bazeos, N., and Beskos, D. E. (2006). 'Maximum displacement profiles for the performance based seismic design of plane steel moment resisting frames,' *Engineering Structures,* **28**(1), 9–22.

Kasai, K. and Popov, E. P. (1986). 'General behavior of WF steel shear link beams,' *Journal of the Structural Engineering Division, ASCE,* **112**(ST2), 362–82.

Kassimali, A. (1983). 'Large deformation analysis of elastic-plastic frames,' *Journal of Structural Engineering, ASCE,* **109**(8), 1869–86.

Katsuki, S. and Frangopol, D. M. (1994). 'Hyperspace division method for structural reliability,' *Journal of Engineering Mechanics, ASCE,* **120**(11), 2405–27.

Kattner, M. and Crisinel, M. (2000). 'Finite element modelling of semi-rigid composite joints,' *Computers and Structures,* **78**(1–3), 341–53.

Kennedy, D. J. L., Albert, C., and MacCrimmon, R. A. (2000). 'Inelastic incremental analysis of an industrial pratt truss,' *Engineering Structures,* **22**(2), 146–54.

Khudada, A. E. and Geschwindner, L. F. (1997). 'Nonlinear dynamic analysis of steel frames by modal superposition,' *Journal of Structural Engineering, ASCE,* **123**(11), 1519–27.

Khulief, Y. and Bazoune, A. (1992). 'Frequencies of rotating tapered beams with different boundary conditions,' *Computers and Structures,* **42**(5), 781–95.

Kim, M. C., Lee, G. C., and Chang, K. C. (1995). 'Inelastic buckling of tapered members with accumulated strain,' *Structural Engineering and Mechanics,* **3**(6), 611–22.

Kim, S. E. (1998). 'Direct design of truss bridges using advanced analysis,' *Structural Engineering and Mechanics,* **6**(8), 871–82.

Kim, S. E. and Chen, W. F. (1996a). 'Practical advanced analysis for braced steel frame design,' *Journal of Structural Engineering, ASCE,* **122**(11), 1266–74.

Kim, S. E. and Chen, W. F. (1996b). 'Practical advanced analysis for semi-rigid frame design,' *Engineering Journal, AISC,* **33**(4), 129–41.

Kim, S. E. and Chen, W. F. (1996c). 'Practical advanced analysis for unbraced steel frames design,' *Journal of Structural Engineering, ASCE,* **122**(11), 1259–65.

Kim, S. E. and Chen, W. F. (1997). 'Further studies of practical advanced analysis for weak-axis bending,' *Engineering Structures,* **19**(6), 407–16.

Kim, S. E. and Chen, W. F. (1999). 'Design guide for steel frames using advanced analysis program,' *Engineering Structures,* **21**(4), 352–64.

Kim, S. E. and Choi, S. H. (2001). 'Practical advanced analysis for semi-rigid space frames,' *International Journal of Solids and Structures,* **38**(50–51), 9111–31.

Kim, S. E. and Choi, S. H. (2005). 'Practical second-order inelastic analysis for three-dimensional steel frames subjected to distributed load,' *Thin-Walled Structures,* **43**(1), 135–60.

Kim, S. E., Choi, S. H., Kim, C. S., and Ma, S. S. (2004). 'Automatic design of space steel frame using practical nonlinear analysis,' *Thin-Walled Structures,* **42**(9), 1273–91.

Kim, S. E., Cuong, N. H., and Lee, D. H. (2006). 'Second-order inelastic dynamic analysis of 3-D steel frames,' *International Journal of Solids and Structures,* **43**(6), 1693–709.

Kim, S. E. and Kang, K. W. (2004). 'Large-scale testing of 3-D steel frame accounting for local buckling,' *International Journal of Solids and Structures,* **41**(18–19), 5003–22.

Kim, S. E., Kang, K. W., and Lee, D. H. (2003). 'Full-scale testing of space steel frame subjected to proportional loads,' *Engineering Structures,* **25**(1), 69–79.

Kim, S. E., Kim, M. K., and Chen, W. F. (2000). 'Improved refined plastic hinge analysis accounting for strain reversal,' *Engineering Structures*, **22**(1), 15–25.

Kim, S. E. and Lee, J. (2002). 'Improved refined plastic-hinge analysis accounting for lateral torsional buckling,' *Journal of Constructional Steel Research*, **58**(11), 1431–53.

Kim, S. E., Lee, J., and Park, J. S. (2002). '3-D second-order plastic-hinge analysis accounting for lateral torsional buckling,' *International Journal of Solids and Structures*, **39**(8), 2109–28.

Kim, S. E., Lee, J., and Park, J. S. (2003). '3-D second-order plastic-hinge analysis accounting for local buckling,' *Engineering Structures*, **25**(1), 81–90.

Kim, S. E. and Lee, J. H. (2001). 'Improved refined plastic-hinge analysis accounting for local buckling,' *Engineering Structures*, **23**(8), 1031–42.

Kim, S. E., Lee, J. S., Choi, S. H., and Kim, C. S. (2004). 'Practical second-order inelastic analysis for steel frames subjected to distributed load,' *Engineering Structures*, **26**(1), 51–61.

Kim, S. E., Park, M. H., and Choi, S. H. (2001a). 'Direct design of three-dimensional frames using practical advanced analysis,' *Engineering Structures*, **23**(11), 1491–502.

Kim, S. E., Park, M. H., and Choi, S. H. (2001b). 'Practical advanced analysis and design of three-dimensional truss bridges,' *Journal of Constructional Steel Research*, **57**(8), 907–23.

Kim, S. E., Song, W. K., and Ma, S. S. (2004). 'Optimal design using genetic algorithm with nonlinear elastic analysis,' *Structural Engineering and Mechanics*, **17**(5), 707–25.

Kim, W. and Lü, L. W. (1992). 'Cyclic lateral load analysis of composite frames,' *Composite Construction in Steel and Concrete II: Proceedings of an Engineering Foundation Conference*, 366–81.

King, W. S. (1991). 'On second-order inelastic methods for steel frame design,' *Journal of Structural Engineering, ASCE*, **118**(2), 408–28.

King, W. S. and Chen, W. F. (1994). 'Practical second-order inelastic analysis of semirigid frames,' *Journal of Structural Engineering, ASCE*, **120**(7), 2156–75.

King, W. S., White, D. W., and Chen, W. F. (1992a). 'A modified plastic hinge method for second-order inelastic analysis of rigid frames,' *Structural Engineering Review*, **4**(1), 31–41.

King, W. S., White, D. W., and Chen, W. F. (1992b). 'Second-order inelastic analysis methods for steel-frame design,' *Journal of Structural Engineering, ASCE*, **118**(2), 408–27.

Kishi, N. and Chen, W. F. (1986). 'Data base of steel beam-to -column connections,' *Structural Engineering Report No.CE-STR-86-26*, Purdue University, USA.

Kishi, N., Chen, W. F., and Goto, Y. (1997). 'Effective length factor of columns in semirigid and unbraced frames,' *Journal of Structural Engineering, ASCE*, **123**(3), 313–20.

Kishi, N., Chen, W. F., Goto, Y., and Komuro, M. (1998). 'Effective length factor of columns in flexibly jointed and braced frames,' *Journal of Constructional Steel Research*, **47**(1–2), 93–118.

Kitjasateanphun, T., Shen, J., Srivanich, W., and Hao, H. (2001). 'Inelastic analysis of steel frames with reduced beam sections,' *The Structural Design of Tall Buildings*, **10**(4), 231–44.

Kiureghian, A. D. and Liu, P. L. (1985). 'Structural reliability under incomplete probability information,' *Report UCB/SESM-85/01*, University of California, Berkeley, California.

Korn, A. and Galambos, T. V. (1968). 'Behavior of elastic-plastic frames,' *Journal of the Structural Division, ASCE*, **94**(5), 1119–42.

Krakovski, M. B. (1995). 'Structural reliability estimation using Monte Carlo simulation and Pearson's curves,' *Structural Engineering and Mechanics*, **3**(3), 201–13.

Krishnamurthy, N., Huang, H.-T., Jeffrey, P. K., and Avery, L. K. (1979). 'Analytical M-θ curves for end-plate connections,' *Journal of the Structural Division, ASCE*, **105**(ST1), 113–45.

Krishnan, S. and Hall, J. F. (2006a). 'Modeling steel frame buildings in three dimensions. I: Panel zone and plastic hinge beam elements,' *Journal of Engineering Mechanics, ASCE*, **132**(4), 345–58.

Krishnan, S. and Hall, J. F. (2006b). 'Modeling steel frame buildings in three dimensions. II: Elastofiber beam element and examples,' *Journal of Engineering Mechanics, ASCE*, **132**(4), 359–74.

Kukreti, A. R., Murray, T. M., and Abolmaali, A. (1987). 'End-plate connection moment-rotation relationship,' *Journal of Constructional Steel Research*, **8**(1), 137–57.

Kumar, S. and Usami, T. (1996). 'An evolutionary-degrading hysteretic model for thin-walled steel structures,' *Engineering Structures*, **18**(7), 504–14.

Kuwamura, H. and Sasaki, M. (1990). 'Control of random yielding-strength for mechanism-based seismic design,' *Journal of Structural Engineering, ASCE*, **116**(1), 98–110.

Landesmann, A. and Batista, E. D. (2005). 'Advanced analysis of steel framed buildings using the Brazilian standard and Eurocode-3,' *Journal of Constructional Steel Research*, **61**(8), 1051–74.

Lau, S. M., Kirby, P. A., and Davison, J. B. (1999). 'Semi-rigid design of partially restrained columns in non-sway steel frames,' *Journal of Constructional Steel Research*, **50**(3), 305–28.

Law, S. S., Chan, T. H. T., and Wu, D. (2001). 'Super-element with semi-rigid joints in model updating,' *Journal of Sound and Vibration*, **239**(1), 19–39.

Law, S. S., Wu, Z. M., and Chan, S. L. (2003). 'Hybrid beam-column element with end frictional joints,' *Journal of Engineering Mechanics, ASCE*, **129**(5), 564–70.

Lee, G. C. and Morrel, M. L. (1974). 'Allowable stress for web-tapered beams with lateral restraints,' *Welding Research Council Bulletin*, **192**, 1–12.

Lee, G. C., Morrel, M. L., and Ketter, R. L. (1972). 'Design of tapered members,' *Welding Research Council Bulletin*, **173**, 1–32.

Lee, S. J. and Lu, L. W. (1989). 'Cyclic test of full-scale composite joint subassemblages,' *Journal of Structural Engineering, ASCE*, **115**(8), 1977–98.

Lee, S. L., Shanmugam, N. E., and Chiew, S. P. (1988). 'Thin-walled box columns under arbitrary end loads,' *Journal of Structural Engineering, ASCE*, **114**(6), 1390–402.

Lee, S. Y., Ke, H. Y., and Kuo, Y. H. (1990). 'Exact static deflection of a non-uniform Bernoulli-Euler beam with general elastic end restraints,' *Computers and Structures*, **36**(1), 91–7.

Leger, P., Paulter, P., and Nuggihalli, R. (1991). 'Elatic analysis of frames considering panel zones deformations,' *Computers and Structrues*, **39**(6), 689–97.

Leu, L. J. and Tsou, C. H. (2001). 'Second-order analysis of planar steel frames considering the effect of spread of plasticity,' *Structural Engineering and Mechanics*, **11**(4), 423–42.

Li, G. Q. (1985). 'New method for calculation of structural reliability,' *Master thesis*, Chongqing Institute of Architecture and Engineering, Chongqing (in Chinese).

Li, G. Q. (1988). 'Elasto-plastic dynamic analysis of multi-story and high-rising steel frames under horizontal bi-direction earthquake and with consideration of couple of in-plane and torsion deformation,' *PhD thesis*, Department of Structural Engineering, Tongji University, Shanghai, China (in Chinese).

Li, G. Q. (1993). 'A simplified model for predicting nonlinear seimic response of high-rise buildings in braced steel frames,' *Proceedings of the 4th Asia-Pacific Conference on Structural Engineering and Construction*, Seoul, Korea, September.

LI, G. Q. (1998). 'Development of tall steel buildings in China,' *Proceedings of the 5th Pacific Structural Steel Conference*, Seoul, Korea, October.

Li, G. Q. (2004). *Structural Design of Multi- Story and High- Rise Steel Buildings*, Press of Chinese Building Industry, Beijing (in Chinese).

Li, G. Q., Huang, J. Y., and Shen, Z. Y. (1993). 'Computer imitaion technique of the spatial hysteretic behavior of steel columns,' *Proceedings of the 2nd Asian-Pacific Conference on Computational Mehcanics*, Sydney, Austrialia, August.

Li, G. Q. and Jiang, S. C. (1999). 'Prediction to nonlinear behaviour of steel frames subjected to fire,' *Fire Safety Journal*, **32**(4), 347–68.

Li, G. Q. and Li, J. J. (2000). 'Effects of shear deformation on the effective length of tapered columns with I-section for steel portal frames,' *Structural Engineering and Mechanics*, **20**(5), 479–89.

Li, G. Q. and Li, J. J. (2002a). 'A semi-analytical simulation method for reliability assessments of structural systems,' *Reliability Engineering and Systems Safety*, **78**(3), 275–81.

Li, G. Q. and Li, J. J. (2002b). 'A tapered Timoshenko-Euler beam element for analysis of steel portal frames,' *Journal of Constructional Steel Research*, **58**, 1531–44.

Li, G. Q. and Li, J. J. (2003). 'A generalized FE formulation for buckling analysis of solid or built-up columns with nonprismatic section,' *Journal of Applied Mechanics and Engineering*, **8**(4), 651–63.

Li, G. Q. and Mativo, J. (2000). 'Approximate estimation of the maximum load of semi-rigid steel frames,' *Journal of Constructional Steel Research*, **54**(2), 213–25.

Li, G. Q. and Shen, Z. Y. (1990). 'Inelastic danamic response of space steel frames under bi-directional earthquake,' *Developments in Structural Engineering*, Chapman and Hall, London.

Li, G. Q. and Shen, Z. Y. (1991a). 'Limit load-bearing capacities of space steel frames,' *Journal of Tongji University*, 36–46, (English issue).

Li, G. Q. and Shen, Z. Y. (1991b). 'Using reliability theory to determine the design rule of eccentrically braced steel frames,' *Industrial Buildings*, **2**, 22–5, (in Chinese).

Li, G. Q. and Shen, Z. Y. (1992). 'Special problems in analysis of steel frames using finite element method,' *Proceedings of the International Conference on Education, Practice Promotion of Computational Methods in Engineering Using Small Computers*, Dalian, China, August.

Li, G. Q. and Shen, Z. Y. (1994). 'Shaking table tests and seismic resistant design of space steel frames,' *Proceedings of International Conference on High-rise Buildings*, Nanjing, China, March.

Li, G. Q. and Shen, Z. Y. (1995). 'An unified matrix approach for nonlinear analysis of steel frames subjected to wind or earthquakes,' *Computers and Structures*, **54**(2), 315–25.

Li, G. Q., Shen, Z. Y., and Huang, J. Y. (1999). 'Spatial hysteretic model and elasto-stiffness of steel columns,' *Journal of Constructional Steel Research*, **50**(3), 283–303.

Li, J. H., Lin, Z. M. et al. (1990). *Probabilistic Limit State Design of Building Structures*, China Architectural Industry Press, Beijing (in Chinese).

Li, J. J. (2001). 'Research on theory of nonlinear analysis and integrated reliability design for steel portal frames with tapered frames,' *PhD thesis*, Department of Structural Engineering, Tongji University (in Chinese).

Li, J. J. and Li, G. Q. (2002). 'Large-Scale testing of steel portal frames comprising tapered beams and columns,' *Advances in Structural Engineering*, **5**(4), 259–69.

Li, J. J. and Li, G. Q. (2004). 'Reliability-based integrated design of steel portal frames with tapered members,' *Structural Safety*, **26**(2), 221–39.

Li, J. J., Li, G. Q., and Chan, S. L. (2003). 'A second-order inelastic model for steel frames of tapered members with slender web,' *Engineering Structures*, **25**(8), 1033–43.

Li, K. S. and Lumb, P. (1985). 'Reliability analysis by numerical integration and curve fitting,' *Structural Safety*, **3**(1), 29–36.

Li, T. Q., S., C. B., and Nethercot, D. A. (1993b). 'Moment curvature relations for steel and composite beams,' *Journal of Singapore Structural Steel Society, Steel Structures*, **4**(1), 35–51.

Liew, J. Y. R. (1992). 'Advanced analysis for frame design,' *Ph. D thesis*, Purdue University, West Lafayette, Ind.

Liew, J. Y. R. (1993a). 'Limit-states design of semi-rigid frames using advanced analysis Part 1: Connection modeling and classification,' *Journal of Constructional Steel Research*, **26**(1), 1–27.

Liew, J. Y. R. (1993b). 'Limit-states design of semi-rigid frames using advanced analysis Part 2: Analysis and design,' *Journal of Constructional Steel Research*, **26**(1), 29–57.

Liew, J. Y. R. (1993c). 'Second-order refined plastic hinge analysis of frame design, Part 1,' *Journal of Structural Engineering, ASCE*, **119**(11), 3196–216.

Liew, J. Y. R. (1993d). 'Second-order refined plastic hinge analysis of frame design, Part 2,' *Journal of Structural Engineering, ASCE*, **119**(11), 3217–37.

Liew, J. Y. R. (1994). 'Notional load plastic hinge method for frame design,' *Journal of Structural Engineering, ASCE*, **120**(5), 1434–54.

Liew, J. Y. R., Chen, H., and Shanmugam, N. E. (2001). 'Inelastic analysis of steel frames with composite beams,' *Journal of Structural Engineering, ASCE*, **127**(2), 194–202.

Liew, J. Y. R., Chen, H., Shanmugam, N. E., and Chen, W. F. (2000a). 'Improved nonlinear plastic hinge analysis of space frame structures,' *Engineering Structures*, **22**(10), 1324–38.

Liew, J. Y. R., Chen, H., Yu, C. H. *et al.* (1997). 'Second-order inelastic analysis of three-dimensional core-braced frames,' *Research Report No. CE024/97*, Department of Civil Engineering. National University of Singapore.

Liew, J. Y. R. and Chen, W. F. (1991). 'Refining the plastic hinge concept for advanced analysis/ design of steel frames,' *Journal of Singapore Structural Steel Society, Steel Structures*, **2**(1), 13–30.

Liew, J. Y. R. and Chen, W. F. (1994). 'Implications of using refined plastic hinge analysis for load and resistance factor design,' *Thin-Walled Structures*, **20**(1–4), 17–47.

Liew, J. Y. R. and Chen, W. F. (1995). 'Analysis and design of steel frames considering panel joint deformations,' *Journal of Structural Engineering, ASCE*, **121**(10), 1531–40.

Liew, J. Y. R., Chen, W. F., and Chen, H. (2000). 'Advanced inelastic analysis of frame structures,' *Journal of Constructional Steel Research*, **55**(1–3), 245–65.

Liew, J. Y. R., Shanmugam, N. E., and Lee, S. L. (1989). 'Behavior of thin-walled steel box columns under biaxial loading,' *Journal of Structural Engineering, ASCE*, **115**(12), 3076–94.

Liew, J. Y. R. and Tang, L. K. (2000). 'Advanced plastic hinge analysis for the design of tubular space frames,' *Engineering Structures*, **22**(7), 769–83.

Liew, J. Y. R., Tang, L. K., and Choo, Y. S. (2002). 'Advanced analysis for performance-based design of steel structures exposed to fires,' *Journal of Structural Engineering, ASCE*, **128**(12), 1584–93.

Liew, J. Y. R., Tang, L. K., Tore, H., and Y. S. Choo. (1998). 'Advanced analysis for the assessment of steel frames in fire,' *Journal for Constructional Steel research*, **47**(1–2), 19–45.

Liew, J. Y. R., White, D. W., and Chen, W. F. (1991). 'Beam-column design in steel frameworks-insight on current methods and trends,' *Journal of Constructional Steel Research*, **18**(4), 269–308.

Lim, J. B. P. and Nethercot, D. A. (2002). 'F. E.-assisted design of the eaves bracket of a cold-formed steel portal frame,' *Steel and Composite Structures*, **2**(6), 411–28.

Lin, T. S. and Corotis, R. B. (1985). 'Reliability of ductile systems with random strength,' *Journal of Structural Engineering, ASCE*, **111**(6), 1306–25.

Lind, N. C. (1988). 'Statistical method for concrete quality control,' *Proceedings of 2nd International Colloquium on Concrete in Developing Countries*, Vol.1, Section 2: 21-26, Bombay, India.

Lindberg, G. M. (1963). 'Vibration of non-uniform beams,' *The Aerospace Quarter*, **14**(2), 387–95.

Liu, E. M. and Ge, M. (2005). 'Analysis and design for stability in the US – An overview,' *Steel and Composite Structures*, **5**(2–3), 103–26.

Liu, H. B., Liu, W. H., and Zhang, Y. L. (2005). 'Calculation analysis of shearing slip for steel-concrete composite beam under concentrated load,' *Applied Mathematics and Mechanics*, **26**(6), 735–40 (English edition).

Liu, L. W., Shen, S. Z., Shen, Z. Y., and Hu, X. R. (1983). *Stability Theory of Structural Steel Members*, Press of China Building Industry, Beijing (in Chinese).

Liu, Y. S. (2004). 'Research on theory of nonlinear analysis and integrated reliability design for planar steel frames,' *PhD Dissertation*, Tongji University, Shanghai, P. R. China (in Chinese).

Long, Y. Q. and Bao, S. H. (1981). *Structural Analysis (Part 2)*, Higher Education Press, Beijin.

López, W. A. (2001). 'Design of unbonded braced frames,' *Proceedings of the 70th Annual Convention*, 23–23, SEAOC, Sacramento, CA.

LRFD. (1997). *Effective Length and Notional Load Approaches for Assessing Frame Stability: Implications for American Steel Design*, Technical Committee on LRFD, ASCE New York.

Luca, A., Faella, C., and Piluso, V. (1991). *Stability of Sway Frames: Different Approaches Around the World*, International Conference on Steel and Aluminium Structures, Singapore, May.

Lui, E. M. and Lopes, A. (1997). 'Dynamic analysis and response of semirigid frames,' *Engineering Structures*, **19**(8), 644–54.

Lundberg, J. E. and Galambos, T. V. (1996). 'Load and resistance factor design of composite columns,' *Structural Safety*, **18**(2–3), 169–78.

Ma, H. F. and Ang, H. S. (1982). 'Reliability analysis of redundant ductile structural systems, Civil engineering studies: System reliability developments in structural engineering,' *Structural Research Series No.494*, University of Illinois, Urbana.

Mabie, H. H. and Rogers, C. B. (1974). 'Transverse vibration of double-tapered cantilever beams with end support and end mass,' *Journal of the Acoustical Society of America*, **55**(9), 986–91.

MacRae, G., Roeder, C. W., Gunderson, C., and Kimura, Y. (2004). 'Brace-beam-volumn vonnections for voncentrically braced grames with voncrete gilled tube columns,' *Journal of Structural Engineering, ASCE*, **130**(2), 233–43.

MacRae, G. A., Kimura, Y., and Roeder, C. (2004). 'Effect of column stiffness on braced frame seismic behavior,' *Journal of Structural Engineering, ASCE*, **130**(3), 381–91.

Mahadevan, S. and Haldar, A. (1991). 'Stochastic FEM-based validation of LRFD,' *Journal of Structural Engineering*, **117**(5), 1393–412.

Mahendran, M. and Moor, C. (1999). 'Three-dimensional modeling of steel portal frame buildings,' *Journal of Structural Engineering, ASCE*, **125**(8), 870–78.

Mahini, M. R. and Seyyedian, H. (2006). 'Effective length factor for columns in braced frames considering axial forces on restraining members,' *Structural Engineering and Mechanics*, **22**(6), 685–700.

Mallett, R. H. and Marcal, P. V. (1968). ' Finite element analysis of nonlinear structure,' *Journal of the Structural Division, ASCE*, **94**(9), 2081–105.

Mamaghani, I. H. P., Usami, T., and Mizuno, E. (1996). 'Inelastic large deflection analysis of structural steel members under cyclic loading,' *Engineering Structures*, **18**(9), 659–68.

Manfredi, G., Fabbrocino, G., and Cosenza, E. (1999). 'Modeling of steel-concrete composite beams under negative bending,' *Journal of Engineering Mechanics, ASCE*, **125**(6), 654–62.

Martin, H. C. (1966). *Introduction to Matrix Methods of Structural Analysis*, McGraw-Hill, New York.

Martinelli, L., Mulas, M. G., and Perotti, F. (1996). 'The seismic response of concentrically braced moment-resisting steel frames,' *Earthquake Engineering and Structural Dynamics*, **25**(11), 1275–99.

Marxhausen, P. D. and Stalnaker, J. J. (2006). 'Buckling of conventionally sheathed stud walls,' *Journal of Structural Engineering, ASCE*, **132**(5), 745–50.

Masarira, A. (2002). 'The effect of joints on the stability behaviour of steel frame beams,' *Journal of Constructional Steel Research*, **58**(10), 1375–90.

Matsui, C., Morino, S., and Tsuda, K. (1980). 'Inelastic behaviour of wide-flange beam-columns under constant vertical and two-dimensional alternating horizontal loads,' *Proceedings of the 7th World Conference on Earthquake Engineering (WCEE)*, 39–46.

Maxwell, S. M., Jenkins, W. M., and Howlett, J. H. (1981). 'A theoretical approach to the analysis of connection behavior,' *Joints in Structural Steelwork*, Pentech Press, London.

MBMA. (1996). *Low Rise Building Systems Manual*, Metal Building Manufacture Association, Cleveland, OH.

McGuire, W., Gallagher, R. H., and Ziemian, R. D. (1999). *Matrix Structural Analysis*, 2nd edn, John Wiley & Sons, Inc, New York.

McNamee, B. M., and Lu L-W (1972). 'Inelastic multistory frame buckling,' *Journal of the Structural Engineering Division, ASCE*, **98**(7), 1613–31.

Medhekar, M. S. and Kennedy, D. J. L. (1997). 'An assessment of the effect of brace overstrength on the seismic response of a single-storey steel building,' *Canadian Journal of Civil Engineering*, **24**(5), 692–704.

Medhekar, M. S. and Kennedy, D. J. L. (1999). 'Seismic response of two-storey buildings with concentrically braced steel frames,' *Canadian Journal of Civil Engineering*, **26**(4), 497–509.

Mehanny, S. S. F. and Deierlein, G. G. (2001). 'Seismic damage and collapse assessment of composite moment frames,' *Journal of Structural Engineering, ASCE*, **127**(9), 1045–53.

Melchers, R. E. (1987). *Structural Reliability Analysis and Prediction*, Ellis Horwood, New York.

Melchers, R. E. (1990a). 'Radial importance sampling for structural reliability,' *Journal of Engineering Mechanics, ASCE*, **116**(1), 189–203.

Melchers, R. E. (1990b). 'Search-based importance sampling,' *Structural Safety*, **9**(2), 117–28.

Mirza, S. A. and Skrabek, B. W. (1991). 'Reliability of short composite beam-column strength interaction,' *Journal of Structural Engineering, ASCE*, **117**(8), 2320–39.

Moan, T. (1997). 'Target levels of reliability-based reassessment of offshore structures, in structural safety and reliability,' *Proceedings of ICSSAR'97*, Shiraishi, N., Shinozuka, M., and Wen, Y. K. (eds), Balkema, Rotterdam, 2049–56.

Moarefzadeh, M. R. and Melchers, R. E. (1999). 'Directional importance sampling for ill-proportional spaces,' *Structural Safety*, **21**(1), 1–23.

Moghaddam, H. and Hajirasouliha, I. (2006). 'An investigation on the accuracy of pushover analysis for estimating the seismic deformation of braced steel frames,' *Journal of Constructional Steel Research*, **64**(4), 343–51.

Moghaddam, H., Hajirasouliha, I., and Doostan, A. (2005). 'Optimum seismic design of concentrically braced steel frames: Concepts and design procedures,' *Journal of Constructional Steel Research*, **61**(2), 151–66.

Moncarz, P. D. and Gerstle, K. H. (1981). 'Steel frames with nonlinear connections,' *Journal of the Structural Division, ASCE*, **107**(ST8), 1427–41.

Morris, G. A. and Fenves, S. J. (1970). 'Elastic-plastic analysis of frameworks,' *Journal of the Structural Division, ASCE*, **96**(5), 931–46.

Morteza, A., Torkamani, M., and Sonmez, M. (2001). 'Inelastic large deflection modeling of beam-columns,' *Journal of Structural Engineering, ASCE*, **127**(8), 876–87.

Moses, F. (1982-1983). 'System reliability developments in structural engineering,' *Structural Safety*, **1**(1), 3–13.

Moses, F. (1990). 'New directions and research needs in system reliability research,' *Structural Safety*, **7**(2–4), 93–100.

Nethercot, D. A. (2000). 'Frame structures: Global performance, static and stability behaviour General Report,' *Journal of Constructional Steel Research*, **55**(1–3), 109–24.

Nethercot, D. A. (2003). *Composite Construction*, Spon Press, New York.

Newmark, N. M., Siess, C. P., and Viest, I. M. (1951). 'Tests and analysis of composite beams with incomplete interaction,' *Proceedings of Society of Experimental Stress Analysis*, **V9**(1), 75–92.

Nie, J., Fan, J., and Cai, C. S. (2004). 'Stiffness and deflection of steel–concrete composite beams under negative bending,' *Journal of Structural Engineering, ASCE*, **130**(11), 1842–51.

Nigam, N. C. (1970). 'Yielding in framed structures under dynamic loads,' *Journal of Engineering Mechanics, ASCE*, **96**(5), 687–709.

Nishiyama, I., Fujimoto, T., Fukumoto, T., and Yoshioka, K. (2004). 'Inelastic force-deformation response of joint shear panels in beam-column moment connections to concrete-filled tubes,' *Journal of Structural Engineering, ASCE*, **130**(2), 244–52.

Norris, C. H., Wilbur, J. B., and Utku, S. (1976). *Elementary Structural Analysis*, 3rd edn, McGraw-Hill, New York.

Nukala, P. K. V. V. and White, D. W. (2004). 'A mixed finite element for three-dimensional nonlinear analysis of steel frames,' *Computer Methods in Applied Mechanics and Engineering*, **193**(23–26), 2507–45.

Oda, H. and Usami, T. (2000). 'Stability design of steel plane frames by second-order elastic analysis,' *Engineering Structures*, **19**(8), 617–27.

Olufsen, A., Leira, B. J., and Moan, T. (1992). 'Uncertainty and reliability analysis of jacket platform,' *Journal of Structural Engineering, ASCE*, **118**(10), 2699–715.

Oral, S. (1995). 'Hybrid-stress finite element for nonuniform filament-wound composite box-beams,' *Computers and Structures*, **56**(4), 667–72.

Oran, C. (1973a). 'Tangent stiffness in plane frames,' *Journal of the Structural Division, ASCE*, **99**(ST6), 973–85.

Oran, C. (1973b). 'Tangent stiffness in space frames,' *Journal of the Structural Division, ASCE*, **99**(ST6), 987–1001.

Orbison, J. G., Mcguire, W., and Abel, F. (1982). 'Yield surface application in nonlinear steel frame analysis,' *Computer Methods in Applied Mechanics and Engineering*, **33**(5), 557–73.

Pan, C. L. and Peng, J. L. (2005). 'Performance of cold-formed steel wall frames under compression,' *Steel and Composite Structures*, **5**(5), 407–20.

Papoulis, A. (1991). *Probability, Random Vibrations, and Stochastic Process*, McGraw-Hill, New York.

Park, H. S. and Kwon, J. H. (2003). 'Optimal drift design model for multi-story buildings subjected to dynamic lateral forces,' *Structural Design of Tall and Special Buildings*, **12**(4), 317–33.

Petrolito, J. and Legge, K. A. (2001). 'Nonlinear analysis of frames with curved members,' *Computers and Structures*, **79**, 727–35.

Phocas, M. C. and Pocanschi, A. (2003). 'Steel frames with bracing mechanism and hysteretic dampers,' *Earthquake Engineering and Structural Dynamics*, **32**(5), 811–25.

Pi, Y. L. and Trahair, N. S. (1994a). 'Nonlinear inelastic analysis of steel beam-columns 1. Theory,' *Journal of Structural Engineering, ASCE*, **120**(7), 2041–61.

Pi, Y. L. and Trahair, N. S. (1994b). 'Nonlinearinelastic analysis of steel beam-columns 2. Applications,' *Journal of Structural Engineering, ASCE*, **120**(7), 2062–85.

Pi, Y. L. and Trahair, N. S. (1999). 'In-plane buckling and design of steel arches,' *Journal of Structural Engineering, ASCE*, **125**(11), 1291–98.

Porco, G., Spadea, G., and Zinno, R. (1994). 'Finite element analysis and parametric study of steel-concrete composite beams,' *Cement and Concrete Composites*, **16**(4), 261–72.

Powell, G. H. (1969). 'Theory of nonlinear elastic structures,' *Journal of the Structural Division, ASCE*, **95**(12), 2687–701.

Pradlwarter, H. J. and Schueller, G. I. (1997). 'On advanced Monte Carlo simulation procedures in stochastic structural dynamic,' *International Journal of Non-linear Mechanics*, **32**(4), 735–44.

Prawel, S. P., Morrel, M. L., and Lee, G. C. (1974). 'Bending and buckling strength of tapered structural members,' *Welding Journal*, **53**(2), 75–84.

Pulido, J. E., Jacobs, T. L., and Lima, E. C. P. (1992). 'Structural reliability using Monte-Carlo simulation with variance reduction techniques on elastic-plastic structures,' *Computers and Structures*, **43**(3), 419–30.

Punniyakotty, N. M., Liew, J. Y. R., and Shanmugam, N. E. (2000). 'Nonlinear analysis of self-erecting framework by cable-tensioning technique,' *Journal of Structural Engineering, ASCE*, **126**(3), 361–70.

Puppo, A. H. and Berterno, R. D. (1992). 'Evaluation of probabilties using orientated simulation,' *Journal of Structural Engineering, ASCE*, **118**(6), 1683–704.

Raftoyiannis, I. G. (2005). 'The effect of semi-rigid joints and an elastic bracing system on the buckling load of simple rectangular steel frames,' *Journal of Constructional Steel Research*, **61**(9), 1205–25.

Rai, D. C. and Goel, S. C. (2003). 'Seismic evaluation and upgrading of chevron braced frames,' *Journal of Constructional Steel Research*, **59**(8), 971–94.

Ramadan, T. and Ghobarah, A. (1995). 'Behavior of bolted link column joints in eccentrically braced frames,' *Canadian Journal of Civiel Engineering*, **22**(4), 745–54.

Ramalingeswara, R. and Ganesan, N. (1995). 'Dynamic response of tapered composite beams using higher order shear deformation theory,' *Journal of Sound and Vibration*, **187**(5), 737–55.

Ranzi, G., Bradford, M. A., and B., U. (2003). 'A general method of analysis of composite beams with partial interaction,' *Steel and Composite Structures*, **3**(3), 169–84.

Ranzi, G., Bradford, M. A., and Uy, B. (2004). 'A direct stiffness analysis of a composite beam with partial interaction,' *International Journal for Numerical Method in Engineering*, **61**(5), 657–72.

Redwood, R. G. and Jain, A. K. (1992). 'Code provisions for seimic design for concentrically braced steel frames,' *Canadian Journal of Civil Engineering*, **19**(6), 1025–31.

Remennikov, A. M. and Walpole, W. R. (1997a). 'Analytical prediction of seismic behaviour for concentrically-braced steel systems,' *Earthquake Engineering and Structural Dynamics*, **26**(8), 859–74.

Remennikov, A. M. and Walpole, W. R. (1997b). 'Modelling the inelastic cyclic behaviour of a bracing member for work-hardening material,' *International Journal of Solids and Structures*, **34**(27), 3491–515.

Ren, W. X. and Zeng, Q. Y. (1997). 'Interactive buckling behavior and ultimate load of I-section steel columns,' *Journal of Structural Engineering, ASCE*, **123**(9), 1210–17.

Reyes-Salazar, A. and Haldar, A. (1999). 'Nonlinear seismic response of steel structures with semi-rigid and composite connections,' *Journal of Constructional Steel Research*, **51**(1), 37–59.

Richards, P. W. and Uang, C. M. (2006). 'Testing protocol for short links in eccentrically braced frames,' *Journal of Structural Engineering, ASCE*, **132**(8), 1183–91.

Roeder, C. W. and Popov, E. P. (1978). 'Eccentrically braced steel frames for earthquakes,' *Journal of the Structural Engineering Division, ASCE*, **104**(ST3), 391–412.

Rosowsky, D. V., Hassan, A. F., and Kumar, N. V. V. P. (1994). 'Calibration of current factors in LRFD for steel,' *Journal of Structural Engineering, ASCE*, **120**(9), 2737–46.

Ruiz, S. E., Urrego, O. E., and Silva, F. L. (1995). 'Influence of the spatial-distribution of energy-dissipating bracing elements on the seismic response of multistory frames,' *Earthquake Engineering and Structural Dynamics*, **24**(11), 1511–25.

Saafan, S. A. (1963). 'Nonlinear behavior of structural plane frame,' *Journal of the Structural Division, ASCE*, **89**(4), 557–79.

Sakumoto, Y. (1998). 'Current research on fire-safe design and fire-protection materials in Japan,' *Proceedings of the 5th Pacific Structural Steel Conference*, Seoul, Korea, October.

Salari, M. R. and Spacone, E. (2001). 'Finite element formulations of one-dimensional elements with bond-slip,' *Engineering Structures*, **23**(7), 815–26.

Salari, M. R., Spacone, E., Shing, P. B., and Frangopol, D. M. (1998). 'Nonlinear analysis of composite beams with deformable shear connectors,' *Journal of Structural Engineering, ASCE*, **124**(10), 1148–58.

Salem, A. H., El Dib, F. F., El Aghoury, M., and Hanna, M. T. (2004). 'Elastic stability of planar steel frames with unsymmetrical beam loading,' *Journal of Structural Engineering, ASCE*, **130**(11), 1852–9.

Salter, J. B., Anderson, D., and May, I. M. (1980). 'Test on tapered steel columns,' *The Structural Engineer*, **58**(6), 189–93.

Sapountzakis, E. J. (2004). 'Dynamic analysis of composite steel-concrete structures with deformable connection,' *Computers and Structures*, **82**(9–10), 717–29.

Sapountzakis, E. J. and Katsikadelis, J. T. (2000). 'Interface forces in composite steel-concrete structure,' *International Journal of Solids and Structures*, **37**(32), 4455–72.

Schafer, B. W. and Bajpai, P. (2005). 'Stability degradation and redundancy in damaged structures,' *Engineering Structures*, **27**(11), 1642–51.

Schneider, S. P. and Amidi, A. (1998). 'Seismic behavior of steel frames with deformable panel zones,' *Journal of Structural Engineering, ASCE*, **124**(1), 35–42.

Sebastian, W. M. and McConnel, R. E. (2000). 'Nonlinear FE analysis of steel-concrete composite structures,' *Journal of Structural Engineering, ASCE*, **126**(6), 662–74.

Sekulovic, M. and Salatic, R. (2001). 'Nonlinear analysis of frames with flexible connections,' *Computers and Structures*, **79**(11), 1097–107.

Sekulovic, M., Salatic, R., and Nefovska, M. (2002). 'Dynamic analysis of steel frames with flexible connections,' *Computers and Structures*, **80**(11), 935–55.

Shanmugam, N. E., Chiew, S. P., and Lee, S. L. (1987). 'Strength of thin-walled square steel box columns,' *Journal of Structural Engineering, ASCE*, **113**(4), 818–31.

Shing, P. B., Bursi, O. S., and Vannan, M. T. (1994). 'Pseudodynamic tests of a concentrically braced frame using substructuring techniques,' *Journal of Constructional Steel Research*, **29**(1–3), 121–48.

Shiomi, H. and Kurata, M. (1983). 'Strength formula for tapered beam-columns,' *Journal of Structural Engineering, ASCE*, **110**(7), 1630–43.

Shu, X. P. and Shen, P. S. (1993). 'Geometrical and material nonlinear analysis of planar steel frames,' *Journal of Hunan University* , **20**(4), 97–103, (in Chinese).

Shugyo, M. (2003). 'Elastoplastic large deflection analysis of three-dimensional steel frames,' *Journal of Structural Engineering, ASCE*, **129**(9), 1259–67.

Singh, H., Paul, D. K., and Sastry, V. V. (1998). 'Inelastic dynamic response of reinforced concrete infilled frames,' *Computers and Structures*, **69**(6), 685–93.

So, A. K. W. and Chan, S. L. (1991). 'Buckling and geometrically nonlinear analysis of frames using one element/ member,' *Journal of Constructional Steel Research*, **20**(4), 271–89.

Soares, R. C., Mohamed, A., Venturini, W. S., and Lemaire, M. (2002). 'Reliability analysis of non-linear reinforced concrete frames using the response surface method,' *Reliability Engineering and System Safety*, **75**(1), 1–16.

Soegiarso, R. and Adeli, H. (1997). 'Optimum load and resistance factor design of steel space-frame structures,' *Journal of Structural Engineering, ASCE*, **123**(2), 184–92.

Soha, I. S. and Chen, W. F. (1987). 'Local buckling and sectional behavior of fabricated tubes,' *Journal of Structural Engineering, ASCE*, **113**(3), 519–33.

Sophianopoulos, D. S. (2003). 'The effect of joint flexibility on the free elastic vibration characteristics of steel plane frames,' *Journal of Constructional Steel Research*, **59**(8), 995–1008.

Sorace, S. (1998). 'Seismic damage assessment of steel frames,' *Journal of Structural Engineering, ASCE*, **124**(5), 531–40.

Sorace, S. and Terenzi, G. (2001). 'Non-linear dynamic design procedure of FV spring-dampers for base isolation – frame building applications,' *Engineering Structures*, **23**(12), 1568–76.

Spacone, E. and El-Tawil, S. (2004). 'Nonlinear analysis of steel-concrete composite structures: State of the art,' *Journal of Structural Engineering, ASCE*, **130**(2), 159–68.

Spyrou, S., Davison, B., Burgess, I., and Plank, R. (2004). 'Experimental and analytical studies of steel joint components at elevated temperatures,' *Fire and Materials*, **28**(2–4), 83–94.

Sridharan, S. and Ali, R. M. A. (1986). 'An improved interactive buckling analysis of thin-walled columns having symmetric sections,' *International Journal of Solids and Structures*, **22**(4), 145–61.

Stelmack, T. W., Marley, M. J., and Gerstle, K. H. (1986). 'Analysis and tests of flexibly connected steel frames,' *Journal of Structural Engineering, ASCE*, **112**(7), 1573–88.

Subedi, N. K. and Coyle, N. R. (2002). 'Improving the strength of fully composite steel-concrete-steel beam elements by increased surface roughness – an experimental study,' *Engineering Structures*, **24**(10), 1349–55.

Sun, F.-F. and Bursi, O. S. (2005). 'Displacement-based and two-field mixed variational formulations for composite beams with shear lag,' *Journal of Engineering Mechanics, ASCE*, **131**(2), 199–210.

Surovek-Maleck, A. E. and White, D. W. (2004). 'Alternative approaches for elastic analysis and design of steel frames. I: Overview,' *Journal of Structural Engineering, ASCE*, **130**(8), 1186–96.

Tabar, A. M. and Deylami, A. (2005). 'Instability of beams with reduced beam section moment connections emphasizing the effect of column panel zone ductility,' *Journal of Constructional Steel Research*, **61**(11), 1475–91.

Takewaki, I. (1997). 'Efficient semi-analytical generator of initial stiffness designs for steel frames under seismic loading. 1. Fundamental frame,' *Structural Design of Tall Buildings*, **6**(2), 151–62.

Taniguchi, H. and Tahanashi, K. (1984). 'Inelastic response behaviour of H-shaped steel columns to bi-directional earthquake motion,' *Proceedings of the 8th World Conference on Earthquake Engineering (WCEE)*, 209–16.

Teh, L. H. (2001). 'Cubic beam elements in practical analysis and design of steel frames,' *Engineering Structures*, **23**(10), 1243–55.

Teh, L. H. (2004). 'Beam element verification for 3D elastic steel frame analysis,' *Computers and Structures*, **82**(15–16), 1167–79.

Tezcan, S. S. (1966). 'Computer analysis of plane and space structures,' *Journal of the Structural Division, ASCE*, **92**(2), 143–73.

Thoft-Christensen, P. (1986). *Application of Structural Systems Reliability Theory*, Springer-Verlag, Berlin.

Thomas, D. L., Wilson, J. M., and Wilson, R. R. (1973). 'Timoshenko beam finite elements,' *Journal of Sound and Vibration*, **31**(3), 315–30.

Tian, Y. S., Wang, J., and Lu, T. J. (2004). 'Racking strength and stiffness of cold-formed steel wall frames,' *Journal of Constructional Steel Research*, **60**(7), 1069–93.

Timoshenko, S. P. and Gere, J. M. (1961). *Theory of Elastic Stability*, McGraw -Hill, New York.

To, C. W. S. (1981). 'A linearly tapered beam finite element incorporating shear deformation and rotary inertia for vibration analysis,' *Journal of Sound and Vibration*, **78**(4), 475–84.

Toma, S. and Chen, W. F. (1992). 'European calibration frames for second-order inelastic analysis,' *Engineering Structures*, **14**(4), 7–14.

Tong, X. (2001). 'Seismic behavior of composite steel frame-reinforced concrete infill wall structural system,' *Ph.D.Dissertation*, University of Minnesota.

Trahair, N. S. and Chan, S. L. (2003). 'Out-of-plane advanced analysis of steel structures,' *Engineering Structures*, **25**(13), 1627–37.

Trahair, N. S. and Hancock, G. J. (2004). 'Steel member strength by inelastic lateral buckling,' *Journal of Structural Engineering, ASCE*, **130**(1), 64–9.

Tranberg, W., Meek, J. L., and Swannell, P. (1976). 'Frame collapse using tangent stiffness,' *Journal of the Structural Engineering Division, ASCE*, **102**(3), 659–75.

Tremblay, R. (2002). 'Inelastic seismic response of steel bracing members,' *Journal of Constructional Steel Research*, **58**(5–8), 665–701.

Tremblay, R., Archambault, M. H., and Filiatrault, A. (2003). 'Seismic response of concentrically braced steel frames made with rectangular hollow bracing members,' *Journal of Structural Engineering, ASCE*, **129**(12), 1626–36.

Tremblay, R., Cote, B., and Leger, P. (1999). 'An evaluation of P-Delta amplification factors in multistorey steel moment resisting frames,' *Canadian Journal of Civil Engineering*, **26**(5), 535–48.

Tremblay, R. and Poncet, L. (2005). 'Seismic performance of concentrically braced steel frames in multistory buildings with mass irregularity,' *Journal of Structural Engineering, ASCE*, **131**(9), 1363–75.

Usami, T. (1993). 'Effective width of locally buckled plates in compression and bending,' *Journal of Structural Engineering, ASCE*, **119**(5), 1358–73.

Usami, T., Gao, S., and Ge, H. (2000). 'Elastoplastic analysis of steel members and frames subjected to cyclic loading,' *Engineering Structures*, **22**(2), 135–45.

Usmani, A. S. and Lamont, S. (2004). 'Key events in the structural response of a composite steel frame structure in fire,' *Fire and Materials*, **28**(2–4), 281–97.

Uy, B. and Bradford, M. A. (1995). 'Local buckling of cold formed steel in composite structural elements at elevated temperatures,' *Journal of Constructional Steel Research*, **34**(1), 53–73.

Val, D., Bljuger, F., and Yankelevsky, D. (1997). 'Reliability evaluation in nonlinear analysis of reinforced concrete structures,' *Structural Safety*, **19**(2), 203–17.

Vellasco, P. C. G. d. S. and Hobbs, R. E. (1996). 'Local web buckling in tapered composite beams,' *Structural Engineering and Mechanics*, **74**(3), 41–6.

Vickery, B. J. and EngSc, B. E. M. (1962). 'The behavior at collapse of simple steel frames with tapered members,' *The Structural Engineers*, **40**(11), 365–76.

Viest, I. M., Colaco, J. P., Furlong, R. W. *et al.* (1997). *Composite Construction Design for Buildings*, ASCE and McGraw Hill, New York.

Vogel, U. (1985). 'Calibrating frames,' *Stahlbau*, **10**, 1–7.

Vu-Quoc, L. and Leger, P. (1992). 'Effect evaluation of the flexibility of tapered I-beams accounting for shear deformation,' *International Journal for Numerical Methods in Engineering*, **33**(3), 553–66.

Wada, A., Saeki, E., Takeuchi, T. and Watanabe, A. (1998). 'Development of unbonded brace,' Nippon Steel Corporation Building Construction and Urban Development Division, Tokyo, Japan.

Wang, C. H. and Foliente, G. C. (2006). 'Seismic reliability of low-rise nonsymmetric woodframe buildings,' *Journal of Structural Engineering, ASCE*, **132**(5), 733–44.

Wang, J. S. H., Chang, K. C., Lee, G. C., and Ketter, R. L. (1989). 'Shaking table tests of pinned-base steel gable frame,' *Journal of Structural Engineering, ASCE*, **115**(12), 3031–43.

Wang, Y. C. (1998). 'Deflection of steel-concrete composite beams with partial shear interaction,' *Journal of Structural Engineering, ASCE*, **124**(10), 1159–65.

Watanabe, A., Hitomi, Y., Yaeki, E. *et al.* (1988). 'Properties of brace encased in buckling-restraining concrete and steel tube,' *Proceedings of the 9th World Conference on Earthquake Engineering*, Tokyo-Kyoto, Japan, IV, 719,724.

Wekezer, J. W. (1989). 'Vibrational analysis of thin-walled bars with open cross sections,' *Journal of Structural Engineering, ASCE*, **115**(12), 2965–78.

Wen, Y. K. (2001). 'Reliability and performance-based design,' *Structural Safety*, **23**(4), 407–28.

White, D. W. (1996). 'Comprehensive performance assessment of building structural systems: Research to practice,' *Engineering Structures*, **18**(10), 778–85.

White, D. W. and Hajjar, J. F. (1997a). 'Accuracy and simplicity of alternative procedures for stability design of steel frames,' *Journal of Constructional Steel Research*, **42**(3), 209–61.

White, D. W. and Hajjar, J. F. (1997b). 'Buckling models and stability design of steel frames: A unified approach,' *Journal of Constructional Steel Research*, **42**(3), 171–208.

White, D. W. and Hajjar, J. F. (1997c). 'Design of steel frames without consideration of effective length,' *Engineering Structures*, **19**(10), 797–810.

Whittaker, A., Gilani, A., and Bertero, V. (1998). 'Evaluation of pre-northbridge steel moment resisting frame joints,' *The Structural Design of Tall Buildings*, **7**(4), 263–83.

Wongkaew, K. and Chen, W. F. (2002). 'Consideration of out-of-plane buckling in advanced analysis for planar steel frame design,' *Journal of Constructional Steel Research*, **58**(5–8), 943–65.

Wood, J. V. and Dawe, J. L. (2006). 'Full-scale test behavior of cold-formed steel roof trusses,' *Journal of Structural Engineering, ASCE*, **132**(4), 616–23.

Wright, E. W. and Gaylord, E. H. (1968). 'Analysis of unbraced multistory steel rigid frames,' *Journal of the Structural Division, ASCE*, **94**(ST5), 1143–63.

Wu, J. R. and Li, Q. S. (2003). 'Structural performance of multi-outrigger-braced tall buildings,' *Structural Design of Tall and Special Buildings*, **12**(2), 155–76.

Wyllie, L. A. and Degenkolb, H. J. (1977). 'Improving the seismic response of braced frames,' *Proceedings of the 6th World Conference of Earthquake Engineering (WCEE)*.

Xiao, Q. and Mahadevan, S. (1994). 'Plasticity effects on frame member reliability,' *Structural Safety*, **16**(3), 201–14.

Xu, L. (2001). 'Second-order analysis for semirigid steel frame design,' *Canadian Journal of Civil Engineering*, **28**(1), 59–76.

Xu, L. and Liu, Y. (2002). 'Story stability of semi-braced steel frames,' *Journal of Constructional Steel Research*, **58**(4), 467–91.

Xue, Q. and Wu, C. W. (2006). 'Preliminary detailing for displacement-based seismic design of buildings,' *Engineering Structures*, **28**(3), 431–40.

Yang, M. S. (1982). 'Seimic behavior of an eccentrically X-braced steel structure,' *Report No. UBC/EERC-82/14*, September.

Yang, M. S. (1984). 'Shaking table studies of an eccentrically X-braced steel structure,' *Proceedings of the 8th World Conference of Earthquake Engineering (WCEE)*.

Yang, T. X. (1979). *Structural Analysis*, People's Education Press, Beijin.

Yao, T. H. J. (1996). 'Response surface method for time-variant reliability analysis,' *Journal of Structural Engineering, ASCE*, **122**(2), 193–201.

Yee, Y. L. and Melchers, R. E. (1986). 'Moment-rotation curves for bolted connections,' *Journal of Structural Engineering, ASCE*, **112**(3), 615–35.

Yuan, Z., Mahendran, M., and Avery, P. (1999). 'A parametric study of non-compact I-section for the pseudo plastic zone analysis method,' *Research Monograph 99-4*, Physical Infrastructure Center, Queensland University of Technology, June.

Yuan, Z., Mahendran, M., and Avery, P. (2000a). 'Finite element modelling of steel I-beams subject to lateral buckling effects under uniform moment and axial force,' *Research Report*, Physical Infrastructure Center, Queensland University of Technology.

Yuan, Z., Mahendran, M., and Avery, P. (2000b). 'Steel frame design using advanced analysis,' *Research Report*, Physical Infrastructure Center, Queensland University of Technology.

Yun, Y. M. and Kim, B. H. (2005). 'Optimum design of plane steel frame structures using second-order inelastic analysis and a genetic algorithm,' *Journal of Structural Engineering, ASCE*, **131**(12), 1820–31.

Zahrai, S. M. and Bruneau, M. (1999). 'Cyclic testing of ductile end diaphragms for slab-on-girder dteel bridges,' *Journal of Structural Engineering, ASCE*, **125**(9), 987–96.

Zhao, Y. G. and Ono, T. (1998). 'System reliability evaluation of ductile frame structures,' *Journal of Structural Engineering, ASCE*, **124**(6), 678–85.

Zhou, S. P., Duan, L., and Chen, W. F. (1991). 'The P-Δ effect on portal steel frames,' *International Conference on Steel and Aluminium Structures*, Singapore, May.

Zhou, W. (2000). 'Reliability evaluations of reinforced concrete columns and steel frames,' *Ph.D Thesis*, University of Western Ontario.

Zhou, W. and Hong, H. P. (2004). 'System and member reliability of steel frames,' *Steel and Composite Structures*, **4**(6), 419–35.

Zhou, Z. H. and Chan, S. L. (1996). 'Refined second-order analysis of frames with members under lateral and axial loads,' *Journal of Structural Engineering, ASCE*, **122**(5), 548–54.

Zhou, Z. H. and Chan, S. L. (1997). 'Second-order analysis of slender steel frames under distributed axial and member loads,' *Journal of Structural Engineering, ASCE*, **123**(9), 1187–93.

Zhou, Z. H. and Chan, S. L. (2004). 'Elastoplastic and large deflection analysis of steel frames by one element per member. I: One hinge along member,' *Journal of Structural Engineering, ASCE*, **130**(4), 538–44.

Ziemian, R. D. (1990). 'Advanced methods of inelastic analysis in the limit states design of steel structures,' *Ph. D dissertation*, School of Civil and Environmental Engineering, Cornell University, Ithaca, NY.

Ziemian, R. D., McGuire, W., and Deierlein, G. G. (1992a). 'Inelastic limit states design, part 1: Planar frame studies,' *Journal of Structural Engineering, ASCE*, **118**(9), 2532–49.

Ziemian, R. D., McGuire, W., and Deierlein, G. G. (1992b). 'Inelastic limit states design, part 2: Three-dimensional frame studies,' *Journal of Structural Engineering, ASCE*, **118**(9), 2550–67.

Ziemian, R. D. and Miller, A. R. (1997). 'Inelastic analysis and design: Frames with members in minor-axis bending,' *Journal of Structural Engineering, ASCE*, **123**(2), 151–56.

Ziemian, R. D., White, D. W., Deierlein, G. G., and McGuire, W. (1990). 'One approach to inelastic analysis and design,' *Proceedings of the 1990 National Steel Conference*, AISC, Chicago.

Zimmerman, J. J., Corotis, R. B., and Ellis, J. H. (1992). 'Structural system reliability considerations with frame instability,' *Engineering Structures*, **14**(6), 371–8.

Zona, A., Barbato, M., and Conte, J. P. (2005). 'Finite element response sensitivity analysis of steel-concrete composite beams with deformable shear connection,' *Journal of Engineering Mechanics, ASCE*, **131**(11), 1126–39.

Zong, Z. and Lam, K. Y. (1998). 'Estimation of complicated distributions using B-spline functions,' *Structural Safety*, **20**(4), 341–56.

Author Index

Subject Index